GAME THEORY

AN INTRODUCTION

Steven Tadelis

PRINCETON UNIVERSITY PRESS

Princeton and Oxford

Copyright © 2013 by Princeton University Press

Published by Princeton University Press, 41 William Street, Princeton, New Jersey 08540
In the United Kingdom: Princeton University Press, 6 Oxford Street, Woodstock,
Oxfordshire OX20 1TW
press.princeton.edu

Library of Congress Cataloging-in-Publication Data

Tadelis, Steve.
    Game theory : an introduction / Steven Tadelis.
        p.   cm.
    Includes bibliographical references and index.
    ISBN 978-0-691-12908-2 (hbk. : alk. paper)
    1. Game theory.   I. Title.
    HB144.T33   2013
    519.3—dc23   2012025166

British Library Cataloging-in-Publication Data is available

This book has been composed in Times Roman and Myriad using ZzTEX by Princeton Editorial
Associates Inc., Scottsdale, Arizona

Printed on acid-free paper.   ∞

Printed in the United States of America

10 9 8 7 6 5 4

To Irit, who always plays fair

Contents

PART II   Static Games of Complete Information

Preface

The study of economics, political science, and the social sciences more generally is an attempt to understand the ways in which people behave and make decisions, as individuals and in group settings. The goal is often to apply our understanding to the analysis of questions pertinent to the functioning of societies and their institutions, such as markets, governments, and legal institutions. Social scientists have developed frameworks and rigorous models that abstract from reality with the intent of focusing attention on the crux of the issues at hand, while ignoring details that seem less relevant and more peripheral. We use these models not only to shed light on what we observe but also to help us predict what we cannot yet see. One of the ultimate goals is to prescribe policy recommendations to the private and public sectors, based on the simplistic yet rigorous models that guide our analysis. In this process we must be mindful of the fact that the strength of our conclusions will depend on the validity of our assumptions, in particular those regarding human behavior and the environment in which people act.

Game theory provides a framework based on the construction of rigorous models that describe situations of conflict and cooperation between *rational* decision makers. Following the tradition of mainstream decision theory and economics, rational behavior is defined as choosing actions that maximize one's payoff (or some form of payoff) subject to the constraints that one faces. This is clearly a caricature of reality, but it is a useful benchmark that in many cases works surprisingly well. Game theory has been successfully applied to many relevant situations, such as business competition, the functioning of markets, political campaigning, jury voting, auctions and procurement contracts, and union negotiations, to name just a few. Game theory has also shed light on other disciplines, such as evolutionary biology and psychology.

This book provides an introduction to game theory. It covers the main ideas of the field and shows how they have been applied to many situations drawn mostly from economics and political science. Concepts are first introduced using simple motivating examples, then developed carefully and precisely to present their general formulation, and finally used in a variety of applications.

The Book's Origins and Intended Audience

As with many textbook authors, it was never my intention to write a textbook. This book grew out of lecture notes that I used for an advanced undergraduate game theory course that I taught at Stanford University from 1997 through 2004. Over the years I was convinced by some colleagues and a persistent executive editor to take those notes and expand them.

Given its origins, the book is aimed at more advanced undergraduates in economics. In writing the book I tried hard to be precise—as one ought to be with a more advanced textbook—while at the same time being reader friendly. Relative to the advanced course that I taught, I added many more examples, both easier and harder than the ones I had used in my class.

I was somewhat frustrated by the books that were available at the time I taught: some were too loose and others were too dense. As a consequence the ideas in this textbook are first presented in a way that most students with minimal mathematical training can follow; they are then further developed to meet the needs and address the curiosity of students who are more rigorously trained. Concepts are presented rigorously but illustrated using examples with varying degrees of complexity from easier to harder. I am therefore quite confident that first-year graduate students in economics and political science will find the book useful as well, especially as a backdrop to more demanding graduate-level textbooks.

This text is meant to be self-contained. Many examples should help the reader absorb the concepts of decision making and strategic analysis. Because precise logical reasoning is at the center of game theory, and of this textbook, a degree of mathematical maturity will be useful to comprehend the material fully. That said, it is not assumed that the reader has a strong mathematical background. A mathematical appendix covers most of what would be needed by someone with a relatively rigorous high school training. Calculus and knowledge of some basic probability theory are required to follow all of the material in this book, but even without this knowledge many of the basic examples and constructions can be appreciated.

The Book's Structure and Suggested Use

The book contains five parts:

- *Part I (Chapters 1–2)—Rational Decision Making:* This part presents the basic ideas of the *rational choice paradigm* used in economics and adopted by many other social science disciplines. Students who have had a basic microeconomics course can easily skip this part of the book.

- *Part II (Chapters 3–6)—Static Games of Complete Information:* The most fundamental aspects of game theory and "normal-form" games are developed in this part of the book. It starts with the notions of dominated and dominant strategies, moves on to consider the consequences of assuming rationality and common knowledge of rationality, and ends with the celebrated concept of Nash equilibrium. All the concepts are first introduced using "pure" (non-stochastic) strategies, and the more demanding concept of mixed strategies is introduced toward the end of this part.

- *Part III (Chapters 7–11)—Dynamic Games of Complete Information:* This part extends the static framework developed earlier to be able to deal with games that unfold over time. The "extensive-form" game is introduced, as well as the concepts of sequential rationality and subgame-perfect equilibrium. These concepts are then used to explore multistage and repeated games, as well as bargaining games. Applications include collusion between price-setting firms, the development of institutions that support anonymous trade, and legislative bargaining.

- *Part IV (Chapters 12–14)—Static Games of Incomplete Information:* This is where the analysis becomes more demanding. This part expands the concepts introduced in Part II to be able to tackle situations in which players are not exactly aware of the characteristics of the other players with whom they interact. The concepts of Bayesian games and Bayesian Nash equilibria are carefully developed and applied to such important contexts as adverse selection, jury voting, and auctions. Some of the more advanced treatments of auctions, and the chapter on mechanism design, are intended for graduate students and very rigorously trained undergraduates.

- *Part V (Chapters 15–18)—Dynamic Games of Incomplete Information:* The last part of the book extends Part IV to deal with games in which information unfolds over time. The idea of sequential rationality is extended to Bayesian games, and equilibrium concepts such as perfect Bayesian and sequential equilibrium are defined and illustrated. Applications include signaling games, the development of reputation, and information-transmission games.

As this outline illustrates, the book contains much more material than can be taught in a quarter or semester undergraduate course. Since it is not suited for an easy "math-free" course, I envision it being used in one or more of the following three courses:

- *Intermediate undergraduate game theory:* Such a course, most likely aimed at undergraduates who either have been exposed to intermediate microeconomics or are comfortable with logical analysis, would include Chapters 3–5 and 7–8. Depending on the instructor's preferences, and the students' level of preparation, some parts of Chapter 6, as well as Chapters 9–12 could be used, and possibly even parts of Chapters 15–16. Students who have not been exposed to rational choice theory will benefit from covering Chapters 1–2.

- *Advanced undergraduate game theory:* This is the course I taught at Stanford. It was aimed at undergraduates who had had intermediate microeconomics with calculus, and were familiar with probabilities and random variables. It included all of the material described for the intermediate course, as well as Chapter 6, most parts of Chapters 9–12, and parts of Chapters 13 and 15–16.

- *Graduate game theory:* This book would not be suited for an advanced graduate course on game theory, for which there are several excellent texts, such as Myerson (1991), Osborne and Rubinstein (1994), and most notably Fudenberg and Tirole (1991). It would be quite useful, however, for first-year Ph.D. students in economics and political science who are covering a first course on game theory. The topics would cover everything described in the advanced undergraduate course, as well as all of Chapters 12 and 15–16 and parts of Chapters 13–14 and 17–18.

Regardless of the way in which the book is used, all the topics are motivated with simple examples that illustrate the main concepts, many of which are used to slowly and carefully explain the main ideas. The easiest examples are tailored to appeal to students in an intermediate undergraduate class, while some examples will be suited only for the most advanced undergraduates as well as for graduate students. The book contains over 150 exercises, which should be more than enough to drive the various points home. About half the exercises have solutions that are freely accessible online, while solutions to the rest are available only to instructors. For more information, visit this book's page on the Princeton University Press web site at http://press.princeton .edu/titles/10001.html.

The Book's Style

The book is casual yet precise in tone, and sometimes a bit demanding. Where mathematical concepts and notation are introduced, they are explained, and where applicable the reader is referred to the mathematical appendix. Given that I trust instructors to assign the book to its intended audience, I expect most students not to need the appendix. However, because my intention is for a curious reader to be able to use the book without an instructor, the mathematical appendix may come in handy.

Many authors struggle with the fact that the English language does not have a singular pronoun that is sex neutral. I have noticed that in recent years many authors have been careful to use a mix of male and female players, and some have decided to tip the balance toward the female pronoun. Alas, I don't trust myself to be careful enough to catch all the uses of "he" and "she" were I to attempt to use both, and I am sure that I would sometimes make an error. Hence I decided to be old fashioned and use the singular male pronoun. I hope that readers will not find this practice insensitive.

Acknowledgments

One person is responsible for me choosing an academic career: Yossi Greenberg. Yossi introduced me to game theory and planted the seed of academic curiosity deep in the soil of my mind. I continued learning game theory from Binyamin Shitovitz and Dov Monderer as an undergraduate, and in graduate school I had the pleasure and privilege of learning so much more game theory from Eric Maskin and Drew Fudenberg. I owe my passion for the subject to the wonderful professors I have had during the years of my training.

Teaching curious students at Stanford only fueled that passion further. Many wonderful students passed through the game theory course that I taught, some of whom have continued on to their own academic careers and from whom I am now learning new things. One student has had a particularly important impact on this book. In 2003 Wendy Sheu took her excellent written notes from my class, with all my examples, definitions, and tangent comments, and typed them up for me. These notes formed the skeleton of the lecture notes that I used over the next three years, and from those notes came this textbook—after many more hours than I am willing to admit to myself.

While I was developing the notes at Stanford, I had the good fortune of employing two excellent Ph.D. students as my teaching assistants, David Miller and Dan Quint, who have both offered many valuable suggestions that improved the notes. In addition, Victor Bennett, Peter Hammond, Igal Hendel, Matt Jackson, and Steve Matthews have kindly agreed to use parts of this book in its earlier stages and have provided valuable feedback. Several anonymous reviewers selected by Princeton University Press offered excellent direction and comments. The final manuscript of this textbook was read by Orie Shelef, whose careful reading is second to none and whose thoughtful comments proved to be invaluable.

I appreciate the time and effort made by the following people who pointed out typos in the first printing of the book: Orie Shelef, Jingyi Mao, Joseph Hall, and Jozsef Sakovics.

Of course, the editors at Princeton University Press played an important role in making this book a reality. Tim Sullivan's relentless pursuits convinced me to embark on the journey of transforming class notes into a textbook. After Tim left the Press, Seth Ditchik provided the necessary encouragement to make me follow through on my commitment. There were many days, and even more nights, when I regretted that decision, although I am now pleased with the outcome. I was once told that there are three things one should do in the course of one's life: have a child, plant a tree, and write a book.[1] I must thank Tim and Seth for helping me scratch the third and last item off that list! I am also thankful to Peter Strupp and his team at Princeton Editorial Associates for providing outstanding copyediting and production services that improved the book tremendously.

Last but not least, I am grateful to my wife, Irit, whose help and encouragement touch my work in so many ways. Her sharp mind, exceptional drive, and superb organizational skills are an inspiration. Without her constant support I surely would accomplish only a fraction of what I manage to do. And when I think of game theory, I can't deny that my two sons, Nadav and Noam, constantly teach me that when it comes to strategy, there is still so much more that I have to learn!

---

1. Some attribute this saying to the Cuban national hero, poet, and writer José Marti.

PART I

RATIONAL DECISION MAKING

1

The Single-Person
Decision Problem

Imagine yourself in the morning, all dressed up and ready to have breakfast. You might be lucky enough to live in a nice undergraduate dormitory with access to an impressive cafeteria, in which case you have a large variety of foods from which to choose. Or you might be a less-fortunate graduate student, whose studio cupboard offers the dull options of two half-empty cereal boxes. Either way you face the same problem: what should you have for breakfast?

This trivial yet ubiquitous situation is an example of a *decision problem*. Decision problems confront us daily, as individuals and as groups (such as firms and other organizations). Examples include a division manager in a firm choosing whether or not to embark on a new research and development project; a congressional representative deciding whether or not to vote for a bill; an undergraduate student deciding on a major; a baseball pitcher contemplating what kind of pitch to deliver; or a lost group of hikers confused about which direction to take. The list is endless.

Some decision problems are trivial, such as choosing your breakfast. For example, if Apple Jacks and Bran Flakes are the only cereals in your cupboard, and if you hate Bran Flakes (they belong to your roommate), then your decision is obvious: eat the Apple Jacks. In contrast, a manager's choice of whether or not to embark on a risky research and development project or a lawmaker's decision on a bill are more complex decision problems.

This chapter develops a *language* that will be useful in laying out rigorous foundations to support many of the ideas underlying strategic interaction in games. The language will be formal, having the benefit of being able to represent a host of different problems and provide a set of tools that will lend structure to the way in which we think about decision problems. The formalities are a vehicle that will help make ideas precise and clear, yet in no way will they overwhelm our ability and intent to keep the more practical aspect of our problems at the forefront of the analysis.

In developing this formal language, we will be forced to specify a set of assumptions about the behavior of decision makers or players. These assumptions will, at times, seem both acceptable and innocuous. At other times, however, the assumptions will be almost offensive in that they will require a significant leap of faith. Still, as the analysis unfolds, we will see the conclusions that derive from the assumptions that we make, and we will come to appreciate how sensitive the conclusions are to these assumptions.

As with any theoretical framework, the value of our conclusions will be only as good as the sensibility of our assumptions. There is a famous saying in computer science—"garbage in, garbage out"—meaning that if invalid data are entered into a system, the resulting output will also be invalid. Although originally applied to computer software, this statement holds true more generally, being applicable, for example, to decision-making theories like the one developed herein. Hence we will at times challenge our assumptions with facts and question the validity of our analysis. Nevertheless we will argue in favor of the framework developed here as a useful benchmark.

1.1 Actions, Outcomes, and Preferences

Consider the examples described earlier: choosing a breakfast, deciding about a research project, or voting on a bill. These problems all share a similar structure: an individual, or player, faces a situation in which he has to choose one of several alternatives. Each choice will result in some outcome, and the consequences of that outcome will be borne by the player himself (and sometimes other players too).

For the player to approach this problem in an intelligent way, he must be aware of three fundamental features of the problem: What are his possible choices? What is the result of each of those choices? How will each result affect his well-being? Understanding these three aspects of a problem will help the player choose his best action. This simple observation offers us a first working definition that will apply to *any decision problem:*

The Decision Problem A **decision problem** consists of three features:

1. **Actions** are all the alternatives from which the player can choose.

2. **Outcomes** are the possible consequences that can result from any of the actions.

3. **Preferences** describe how the player ranks the set of possible outcomes, from most desired to least desired. The **preference relation** $\succsim$ describes the player's preferences, and the notation $x \succsim y$ means "x is at least as good as y."

To make things simple, let's begin with our rather trivial decision problem of choosing between Apple Jacks and Bran Flakes. We can define the set of actions as $A = \{a, b\}$, where a denotes the choice of Apple Jacks and b denotes the choice of Bran Flakes.[1] In this simple example our actions are practically synonymous with the outcomes, yet to make the distinction clear we will denote the set of outcomes by $X = \{x, y\}$, where x denotes eating Apple Jacks (the consequence of *choosing* Apple Jacks) and y denotes eating Bran Flakes.

---

1. More on the concept of a set and the appropriate notation can be found in Section 19.1 of the mathematical appendix.

1.1.1 Preference Relations

Turning to the less familiar notion of a **preference relation,** imagine that you prefer eating Apple Jacks to Bran Flakes. Then we will write $x \succsim y$, which should be read as "x is at least as good as y." If instead you prefer Bran Flakes, then we will write $y \succsim x$, which should be read as "y is at least as good as x." Thus our preference relation is just a shorthand way to express the player's ranking of the possible outcomes.

We follow the common tradition in economics and decision theory by expressing preferences as a "weak" ranking. That is, the statement "x is at least as good as y" is consistent with x being *better* than y or *equally as good as y*. To distinguish between these two scenarios we will use the **strict preference relation,** $x \succ y$, for "x is strictly better than y," and the **indifference relation,** $x \sim y$, for "x and y are equally good."

It need not be the case that actions are synonymous with outcome, as in the case of choosing your breakfast cereal. For example, imagine that you are in a bar with a drunken friend. Your actions can be to let him drive home or to order him a cab. The outcome of letting him drive is a certain accident (he's *really* drunk), and the outcome of ordering him a cab is arriving safely at home. Hence for this decision problem your actions are physically different from the outcomes.

In these examples the action set is *finite,* but in some cases one might have infinitely many actions from which to choose. Furthermore there may be infinitely many outcomes that can result from the actions chosen. A simple example can be illustrated by me offering you a two-gallon bottle of water to quench your thirst. You can choose how much to drink and return the remainder to me. In this case your action set can be described as the interval $A = [0, 2]$: you can choose any action a as long as it belongs to the interval $[0, 2]$, which we can write in two ways: $0 \leq a \leq 2$ or $a \in [0, 2]$.[2] If we equate outcomes with actions in this example then $X = [0, 2]$ as well. Finally it need not be the case that more is better. If you are thirsty then drinking a pint may be better than drinking nothing. However, drinking a gallon may cause you to have a stomachache, and you may therefore prefer a pint to a gallon.

Before proceeding with a useful way to represent a player's preferences over various outcomes, it is important to stress that we will make two important assumptions about the player's ability to think through the decision problem.[3] First, we require the player to be able to rank *any two outcomes* from the set of outcomes. To put this more formally:

The Completeness Axiom   The preference relation $\succsim$ is **complete:** any two outcomes $x, y \in X$ can be ranked by the preference relation, so that either $x \succsim y$ or $y \succsim x$.

At some level the completeness axiom is quite innocuous. If I show you two foods, you should be able to rank them according to how much you like them (including being indifferent if they are equally tasty and nutritious). If I offer you two cars, you should be able to rank them according to how much you enjoy driving them, their safety

---

2. The notation symbol $\in$ means "belongs to." Hence "$x, y \in X$" means "elements x and y belong to the set X." If you are unfamiliar with sets and these kinds of descriptions please refer to Section 19.1 of the mathematical appendix.

3. These assumptions are referred to as "axioms," following the language used in the seminal book by von Neumann and Morgenstern (1944) that laid many of the foundations for both decision theory and game theory.

specifications, and so forth. If I offer you two investment portfolios, you should be able to rank them according to the extent to which you are willing to balance risk and return. In other words, the completeness axiom *does not let you be indecisive between any two outcomes.*[4]

The second assumption we make guarantees that a player can rank *all* of the outcomes. To do this we introduce a rather mild consistency condition called *transitivity:*

The Transitivity Axiom The preference relation $\succsim$ is **transitive:** for any three outcomes x, y, $z \in X$, if $x \succsim y$ and $y \succsim z$ then $x \succsim z$.

Faced with several outcomes, completeness guarantees that any two can be ranked, and transitivity guarantees that there will be no contradictions in the ranking, which could create an indecisive cycle. To observe a violation of the transitivity axiom, consider a player who strictly prefers Apple Jacks to Bran Flakes, $a \succ b$, Bran Flakes to Cheerios, $b \succ c$, and Cheerios to Apple Jacks, $c \succ a$. When faced with any two boxes of cereal, say $A = \{a, b\}$, he has no problem choosing his preferred cereal a. What happens, however, when he is presented with all three alternatives, $A = \{a, b, c\}$? The poor guy will be unable to decide which of the three to choose, because for any given box of cereal, there is another box that he prefers. Therefore, by requiring that the player have complete and transitive preferences, we basically guarantee that among *any set of outcomes,* he will always have at least one *best outcome* that is as good as or better than any other outcome in that set.

To foreshadow what will be our premise for decision making, a preference relation that is complete and transitive is called a **rational preference relation.** We will be concerned only with players who have such rational preferences, for without such preferences we can offer neither predictive nor prescriptive insights.

Remark As noted by the Marquis de Condorcet in 1785, it is possible to have a group of rational individual players who, when put together to make decisions as a group, will become an "irrational" group. For example, imagine three roommates, called players 1, 2, and 3, who have to choose one box of cereal for their apartment kitchen. Player 1's preferences are given by $a \succ_1 c \succ_1 b$, player 2's are given by $c \succ_2 b \succ_2 a$, and player 3's are given by $b \succ_3 a \succ_3 c$. Imagine that our three players make choices in a democratic way and use majority voting to reach a decision. What will be the resulting preferences of the group, $\succ_G$? When faced with the pair a and c, players 1 and 3 will vote for Apple Jacks, hence $a \succ_G c$. When faced with the pair c and b, players 1 and 2 will vote for Cheerios, hence $c \succ_G b$. When faced with the pair a and b, players 2 and 3 will vote for Bran Flakes, hence $b \succ_G a$. As a result, our three rational players will not be able to reach a conclusive decision using the group preferences that result from majority voting! This type of group indecisiveness resulting from majority voting is often referred to as the *Condorcet Paradox.* Because we will not be analyzing group decisions, it is not something we will confront, but it is useful to be mindful of such phenomena, in which imposing individual rationality does not imply "group rationality."

---

4. In other words, this axiom prohibits the kind of problem referred to as "Buridan's ass." One version describes a situation in which an ass is placed between two identical stacks of hay, assuming that the ass will always go to whichever stack is closer. However, since the stacks are both the same distance from the ass, it will not be able to choose between them and will die of hunger.

1.1.2 Payoff Functions

When we restrict attention to players with rational preferences, not only do we get players who behave in a consistent and appealing way, but as an added bonus we can replace the preference relation with a much friendlier, and more operational, apparatus. Consider the following simple example. Imagine that you open a lemonade stand on your neighborhood corner. You have three possible actions: choose low-quality lemons (l), which imply a cost of $10 and a revenue from sales of $15; choose medium-quality lemons (m), which imply a cost of $15 and a revenue from sales of $25; or choose high-quality lemons (h), which imply a cost of $28 and a revenue from sales of $35. Thus the action set is $A = \{l, m, h\}$, and the outcome set is given by net profits and is $X = \{5, 10, 7\}$, where the action l yields a profit of $5, the action m yields a profit of $10, and the action h yields a profit of $7. Assuming that obtaining higher profits is strictly better, we have $10 \succ 7 \succ 5$. Hence you should choose alternative m and make a profit of $10.

Notice that we took a rather obvious profit-maximizing problem and fit it into our framework for a decision problem. We derived the preference relation that is consistent with maximizing profit, the objective of any for-profit business. Arguably it would be more natural and probably easier to comprehend the problem if we looked at the actions and their associated profits. In particular we can define the **profit function** in the obvious way: every action $a \in A$ yields a profit $\pi(a)$. Then, instead of considering a preference relation over profit outcomes, we can just look at the profit from each action directly and choose an action that maximizes profits. In other words, we can *use the profit function to evaluate actions and outcomes.*

As this simple example demonstrates, a profit function is a more direct way for a player to rank his actions. The question then is, can we find similar ways to approach decision problems that are not about profits? It turns out that we can do exactly that if we have players with rational preferences, and to do that we define a payoff function.[5]

Definition 1.1 A **payoff function** $u : X \to \mathbb{R}$ represents the preference relation $\succsim$ if for any pair $x, y \in X$, $u(x) \geq u(y)$ if and only if $x \succsim y$.

To put the definition into words, we say that the preference relation $\succsim$ is represented by the payoff function $u : X \to \mathbb{R}$ that assigns to each outcome in X a real number, if and only if the function assigns a higher value to higher-ranked outcomes.

It is important to notice that representing preferences with payoff functions is convenient, but that payoff values by themselves have no meaning whatsoever. Payoff is an *ordinal* construct: it is used to order the alternatives from most to least desirable. For example, if I like Apple Jacks more than Bran Flakes, then I can construct the payoff function $u(\cdot)$ so that $u(a) = 5$ and $u(b) = 3$. I can also use a different payoff function $\tilde{u}(\cdot)$ that represents the same preferences as follows: $\tilde{u}(a) = 100$ and $\tilde{u}(b) = -237$. Just as Fahrenheit and Celsius are two different ways to describe hotter and colder temperatures, there are many ways to represent preferences with payoff functions.

Using payoff functions instead of preferences will allow us to operationalize a theory of how decision makers with rational preferences ought to behave, and how they often will behave. They will choose actions that maximize a payoff function that

---

5. Recall that a function relates each of its inputs to exactly one output. For more on this see Section 19.2 of the mathematical appendix.

represents their preferences. One last question we need to ask is whether we know for sure that this method will work: is it true that players will surely have a payoff function representing their preferences? One case is easy and worth going through briefly. In what follows, we provide a formal proposition and a formal, yet fairly easy to follow, proof.

Proposition 1.1 *If the set of outcomes X is finite then any rational preference relation over X can be represented by a payoff function.*

Proof The proof is by construction. Because the preference relation is complete and transitive, we can find a least-preferred outcome $\underline{x} \in X$ such that all other outcomes $y \in X$ are at least as good as $\underline{x}$, that is, $y \succsim \underline{x}$ for all other $y \in X$. Now define the "worst outcome equivalence set," denoted X_1, to include $\underline{x}$ and any other outcome for which the player is indifferent between it and $\underline{x}$. Then, from the remaining elements of $X \backslash X_1$,[6] define the "second worst outcome equivalence set," X_2, and continue in this fashion until the "best outcome equivalence set," X_n, is created. Because X is finite and $\succsim$ is rational, such a finite collection of n equivalence sets exists. Now consider n arbitrary values $u_n > u_{n-1} > \cdots > u_2 > u_1$, and assign payoffs according to the function defined by: for any $x \in X_k$, $u(x) = u_k$. This payoff function represents $\succsim$. Hence we have proved that such a function exists. ∎

This proposition is useful: for many realistic situations, we can create payoff functions that work in a similar way as profit functions, giving the player a useful tool to see which actions are best and which ought to be avoided. We will not explore this issue further, but payoff representations exist in many other cases that include infinitely many outcomes. The treatment of such cases is beyond the scope of this textbook, but you are welcome to explore one of the many texts that offer a more complete treatment of the topic, which is referred to under the title "representation theorems." (See, e.g., Kreps [1990a, pp. 18–37, and 1988] for an in-depth treatment of this topic.)

As we have seen so far, the formal structure of a decision problem offers a coherent framework for analysis. For decades, however, teachers, students, and practitioners have instead used the intuitive and graphically simple tool of *decision trees*.

Imagine that, in addition to Apple Jacks (a) and Bran Flakes (b), your breakfast options include a muffin (m) and a scone (s). Your preferences are given as $s \succ a \succ m \succ b$. (Recall that we now consider preferences over outcomes as directly over actions.) Consider the following payoff representation: $v(s) = 4$, $v(a) = 3$, $v(m) = 2$, and $v(b) = 1$. We can write down the corresponding decision tree, which is depicted in Figure 1.1.

To read this simple decision tree, notice that the player resides at the "root" of the tree on the left, and that the tree then branches off, each branch representing a possible action. In the example of choosing breakfast, each action results in a final payoff, and these payoffs are written to correspond to each of the action branches. Our rational decision maker will look down the tree, consider the payoff from each branch, and choose the branch with the highest payoff.

The node at which the player has to make a choice is called a **decision node.** The nodes at the end of the tree where payoffs are attached are called **terminal nodes.** As

---

6. The notation $A \backslash B$ means "the elements that are in A but are not in B," or sometimes "the set A less the set B."

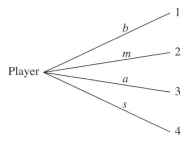

FIGURE 1.1 A simple breakfast decision tree.

the next chapter demonstrates, the structure of a decision tree will become slightly more involved and useful to capture more complex decision problems. We will return to similar trees in Chapter 7, where we consider the strategic interaction between many possible players, which is the main focus of this book.

1.2 The Rational Choice Paradigm

We now introduce *Homo economicus* or "economic man." *Homo economicus* is "rational" in that he chooses actions that maximize his well-being as defined by his payoff function over the resulting outcomes.[7] The assumption that the player is rational lies at the foundation of what is known as the **rational choice paradigm.** Rational choice theory asserts that when a decision maker is choosing between potential actions he will be guided by rationality to choose his best action. This can be assumed to be true for individual human behavior, as well as for the behavior of other entities, such as corporations, committees, or nation-states.

It is important to note, however, that by adopting the paradigm of rational choice theory we are imposing some implicit assumptions, which we now make explicit.

Rational Choice Assumptions The player fully understands the decision problem by knowing:

1. all possible actions, A;
2. all possible outcomes, X;
3. exactly how each action affects which outcome will materialize; and
4. his rational preferences (payoffs) over outcomes.

Perhaps at a first glance this set of assumptions may seem a bit demanding, and further contemplation may make you feel that it is impossible to satisfy for most decision problems. Still, it is a benchmark for a world in which decision problems are completely understood by the player, in which case he can approach the problems in a systematic and structured way. If we let go of any of these four knowledge

---

7. A naive application of the *Homo economicus* model assumes that our player knows what is best for his long-term well-being and can be relied upon to always make the right decision for himself. We take this naive approach throughout the book, though we will sometimes question how appropriate this approach is.

requirements then we cannot impose the notion of rational choice. If (1) is unknown then the player may be unaware of his best course of action. If (2) or (3) are unknown then he may not correctly foresee the actual consequences of his actions. Finally if (4) is unknown then he may incorrectly perceive the effect of his choice's consequence on his well-being.

To operationalize this paradigm of rationality we must choose among *actions,* yet we have defined preferences—and payoffs—over *outcomes* and not actions. It would be useful, therefore, if we could define preferences—and payoffs—over actions instead of outcomes. In the simple examples of choosing a cereal or how much water to drink, actions and outcomes were synonymous, yet this need not always be the case. Consider the situation of letting your friend drive drunk, in which the actions and outcomes are not the same. Still each action led to one and only one outcome: letting him drive leads to an accident, and getting him a cab leads to safe arrival. Hence, even though preferences and payoff were defined over outcomes, this *one-to-one correspondence,* or function, between actions and outcomes means that we can consider the preferences and payoffs to be over actions, and we can use this correspondence between actions and outcomes to define the payoff over actions as follows: if $x(a)$ is the outcome resulting from action a, then the payoff from action a is given by $v(a) = u(x(a))$, the payoff from $x(a)$. We will therefore use the notation $v(a)$ to represent the payoff from action a.[8] Now we can precisely define a rational player as follows:

Definition 1.2   A player facing a decision problem with a payoff function $v(\cdot)$ over actions is rational if he chooses an action $a \in A$ that maximizes his payoff. That is, $a^* \in A$ is chosen if and only if $v(a^*) \geq v(a)$ for all $a \in A$.

We now have a formal definition of *Homo economicus:* a player who has rational preferences and is rational in that he understands all the aspects of his decision problem and always chooses an option that yields him the highest payoff from the set of possible actions.

So far we have seen some simple examples with finite action sets. Consider instead an example with a continuous action space, which requires some calculus. Imagine that you're at a party and are considering engaging in social drinking. Given your physique, you'd prefer some wine, both for taste and for the relaxed feeling it gives you, but too much will make you sick. There is a one-liter bottle of wine, so your action set is $A = [0, 1]$, where $a \in A$ is how much you choose to drink. Your preferences are represented by the following payoff function over actions: $v(a) = 2a - 4a^2$, which is depicted in Figure 1.2. As you can see, some wine is better than no wine (0.1 liter gives you some positive payoff, while drinking nothing gives you zero), but drinking a whole bottle will be worse than not drinking at all ($v(1) = -2$). How much should you drink? Your maximization problem is

$$\max_{a \in [0,1]} \quad 2a - 4a^2.$$

Taking the derivative of this function and equating it to zero to find the solution, we obtain that $2 - 8a = 0$, or $a = 0.25$, which is a bit more than two normal glasses of

---

8. To be precise, let $x : A \to X$ be the function that maps actions into outcomes, and let the payoff function over outcomes be $u : X \to \mathbb{R}$. Define the payoff over actions as the composite function $v = u \circ x : A \to \mathbb{R}$, where $v(a) = u(x(a))$.

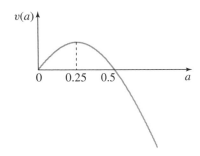

FIGURE 1.2 The payoff from drinking wine.

wine.[9] Thus, by considering how much wine to drink as a decision problem, you were able to find your optimal action.

1.3 Summary

- A simple decision problem has three components: actions, outcomes, and preferences over outcomes.

- A rational player has complete and transitive preferences over outcomes and hence can always identify a best alternative from among his possible actions. These preferences can be represented by a payoff (or profit) function over outcomes and the corresponding payoffs over actions.

- A rational player chooses the action that gives him the highest possible payoff from the possible set of actions at his disposal. Hence by maximizing his payoff function over his set of alternative actions, a rational player will choose his optimal decision.

- A decision tree is a simple graphic representation for decision problems.

1.4 Exercises

1.1 **Your Decision:** Think of a simple decision you face regularly and formalize it as a decision problem, carefully listing the actions and outcomes without the preference relation. Then assign payoffs to the outcomes and draw the decision tree.

1.2 **Going to the Movies:** There are two movie theaters in your neighborhood: Cineclass, which is located one mile from your home, and Cineblast, located three miles from your home. Each is showing three films. Cineclass is showing *Casablanca, Gone with the Wind,* and *Dr. Strangelove,* while Cineblast is showing *The Matrix, Blade Runner,* and *Aliens.* Your problem is to decide which movie to go to.

---

9. To be precise, we must also make sure that first, the second derivative is negative for the solution $a = 0.25$ to be a local maximum, and second, the value of $v(a)$ is not greater at the two boundaries $a = 0$ and $a = 1$. For more on maximizing the value of a function, see Section 19.3 of the mathematical appendix.

    a. Draw a decision tree that represents this problem without assigning payoff values.

    b. Imagine that you don't care about distance and that your preferences for movies are alphabetic (i.e., you like *Aliens* the most and *The Matrix* the least). Using payoff values 1 through 6 complete the decision tree you drew in part (1). Which option would you choose?

    c. Now imagine that your car is in the shop and that the cost of walking each mile is equal to one unit of payoff. Update the payoffs in the decision tree. Would your choice change?

1.3 **Fruit or Candy:** A banana costs $0.50 and a piece of candy costs $0.25 at the local cafeteria. You have $1.25 in your pocket and you value money. The money-equivalent value (payoff) you get from eating your first banana is $1.20, and that of each additional banana is half the previous one (the second banana gives you a value of $0.60, the third $0.30, and so on). Similarly the payoff you get from eating your first piece of candy is $0.40, and that of each additional piece is half the previous one ($0.20, $0.10, and so on). Your value from eating bananas is not affected by how many pieces of candy you eat and vice versa.

    a. What is the set of possible actions you can take given your budget of $1.25?

    b. Draw the decision tree that is associated with this decision problem.

    c. Should you spend all your money at the cafeteria? Justify your answer with a rational choice argument.

    d. Now imagine that the price of a piece of candy increases to $0.30. How many possible actions do you have? Does your answer to (c) change?

1.4 **Alcohol Consumption:** Recall the example in which you needed to choose how much to drink. Imagine that your payoff function is given by $\theta a - 4a^2$, where θ is a parameter that depends on your physique. Every person may have a different value of θ, and it is known that in the population (1) the smallest θ is 0.2; (2) the largest θ is 6; and (3) larger people have higher θs than smaller people.

    a. Can you find an amount that no person should drink?

    b. How much should you drink if your $\theta = 1$? If $\theta = 4$?

    c. Show that in general smaller people should drink less than larger people.

    d. Should any person drink more than one 1-liter bottle of wine?

1.5 **Buying a Car:** You plan on buying a used car. You have $12,000, and you are not eligible for any loans. The prices of available cars on the lot are given as follows:

Make, model, and year	Price
Toyota Corolla 2002	$9,350
Toyota Camry 2001	10,500
Buick LeSabre 2001	8,825
Honda Civic 2000	9,215
Subaru Impreza 2000	9,690

For *any given year,* you prefer a Camry to an Impreza, an Impreza to a Corolla, a Corolla to a Civic, and a Civic to a LeSabre. For *any given year,* you are willing to pay up to $999 to move from any given car to the next preferred one. For example, if the price of a Corolla is z, then you are willing to buy it rather than a Civic if the Civic costs more than $(z - 999)$, but you would prefer to buy the Civic if it costs less than this amount. Similarly you prefer the Civic at z to a Corolla that costs more than $(z + 1000)$, but you prefer the Corolla if it costs less. For *any given car,* you are willing to move to a model a year older if it is cheaper by at least $500. For example, if the price of a 2003 Civic is z, then you are willing to buy it rather than a 2002 Civic if the 2002 Civic costs more than $(z - 500)$, but you would prefer to buy the 2002 Civic if it costs less than this amount.

    a.   What is your set of possible alternatives?
    b.   What is your preference relation between the alternatives in (a) above?
    c.   Draw a decision tree and assign payoffs to the terminal nodes associated with the possible alternatives. What would you choose?
    d.   Can you draw a decision tree with different payoffs that represents the same problem?

1.6   **Fruit Trees:** You have room for up to two fruit-bearing trees in your garden. The fruit trees that can grow in your garden are either apple, orange, or pear. The cost of maintenance is $100 for an apple tree, $70 for an orange tree, and $120 for a pear tree. Your food bill will be reduced by $130 for each apple tree you plant, by $145 for each pear tree you plant, and by $90 for each orange tree you plant. You care only about your total expenditure in making any planting decisions.

    a.   What is the set of possible actions and related outcomes?
    b.   What is the payoff of each action/outcome?
    c.   Draw the associated decision tree. What will a rational player choose?
    d.   Now imagine that the reduction in your food bill is half for the second tree of the same kind. (You like variety.) That is, the first apple tree still reduces your food bill by $130, but if you plant two apple trees your food bill will be reduced by $130 + $65 = $195, and similarly for pear and orange trees. What will a rational player choose now?

1.7   **City Parks:** A city's mayor has to decide how much money to spend on parks and recreation. City codes restrict this spending to no more than 5% of the budget, and the yearly budget of the city is $20,000,000. The mayor wants to please his constituents, who have diminishing returns from parks. The money-equivalent benefit from spending c on parks is $v(c) = \sqrt{400c} - \frac{1}{80}c$.

    a.   What is the action set for the city's mayor?
    b.   How much should the mayor spend?
    c.   The movie *An Inconvenient Truth* has shifted public opinion, and now people are more willing to pay for parks. The new preferences of the people are given by $v(c) = \sqrt{1600c} - \frac{1}{80}c$. What now is the action set for the mayor, and how much spending should he choose to cater to his constituents?

2

Introducing Uncertainty and Time

Now that we have a coherent and precise language to describe decision problems, we move on to be more realistic about the complexity of many such problems. The cereal example was fine to illustrate a simple decision problem and to get used to our formal language, but it is certainly not very interesting.

Consider a division manager who has to decide on whether a research and development (R&D) project is worthwhile. What will happen if he does not go ahead with it? Maybe over time his main product will become obsolete and outdated, and the profitability of his division will no longer be sustainable. Then again, maybe profits will still continue to flow in. What happens of he does go ahead with the project? It may lead to vast improvements in the product line and offer the prospect of sustained growth. Or perhaps the research will fail and no new products will emerge, leaving behind only a hefty bill for the expensive R&D endeavor. In other words, both actions have uncertainty over what outcomes will materialize, implying that the choice of a best action is not as obvious as in the cereal example.

How should the player approach this more complex problem? As you can imagine, using language like "maybe this will happen, or maybe that will happen" is not very useful for a rational player who is trying to put some structure on his decision problem. We must introduce a method through which the player can compare uncertain consequences in a meaningful way. For this approach, we will use the concept of stochastic (random) outcomes and probabilities, and we will describe a framework within which payoffs are defined over random outcomes.

2.1   Risk, Nature, and Random Outcomes

Put yourself in the shoes of our division manager who is deciding whether or not to embark on the R&D project. Denote his actions as g for going ahead or s for keeping the status quo, so that $A = \{g, s\}$. To make the problem as simple as possible, imagine that there are only two final outcomes: his product line is successful, which is equivalent to a profit of 10 (choose your denomination), or his product line is obsolete, which is equivalent to a profit of 0, so that $X = \{0, 10\}$. However, as already explained, there is no one-to-one correspondence here between actions and outcomes. Instead there is uncertainty about which outcome will prevail, and the uncertainty is tied to the choice made by the player, the division manager.

In order to capture this uncertainty in a precise way, we will use the well-understood notion of randomness, or risk, as described by a random variable. Use of random variables is the common way to precisely and consistently describe random prospects in mathematics and statistics. We will not use the most formal mathematical representation of a random variable but instead present it in its most useful depiction for the problems we will address. Section 19.4 of the mathematical appendix has a short introduction to random variables that you can refer to if this notion is completely new to you. Be sure to make yourself familiar with the concept: it will accompany us closely throughout this book.

2.1.1 Finite Outcomes and Simple Lotteries

Continuing with the R&D example, imagine that a successful product line is more likely to be created if the player chooses to go ahead with the R&D project, while it is less likely to be created if he does not. More precisely, the odds are 3 to 1 that success happens if g is chosen, while the odds are only 50-50 if s is chosen. Using the language of probabilities, we have the following description of outcomes following actions: If the player chooses g then the probability of a payoff of 10 is 0.75 and the probability of a payoff of 0 is 0.25. If, however, the player chooses s then the probability of a payoff of 10 is 0.5, as is the probability of a payoff of 0.

We can therefore think of the player as if he is choosing between two **lotteries.** A lottery is exactly described by a random payoff. For example, the state lottery offers each player either several million dollars or zero, and the likelihood of getting zero is extremely high. In our example, the choice of g is like choosing a lottery that pays zero with probability 0.25 and pays 10 with probability 0.75. The choice of s is like choosing a lottery that pays either zero or 10, each with an equal probability of 0.5.

It is useful to think of these lotteries as choices of another player that we will call "Nature." The probabilities of outcomes that Nature chooses depend on the actions chosen by our decision-making player. In other words, Nature chooses a probability distribution over the outcomes, and the probability distribution is conditional on the action chosen by our decision-making player.

We can utilize a decision tree to describe the player's decision problem that includes uncertainty. The R&D example is described in Figure 2.1. First the player takes an action, either g or s. Then, conditional on the action chosen by the player, Nature (denoted by N) will choose a probability distribution over the outcomes 10 and 0. The branches of the player are denoted by his actions, and the branches of Nature's

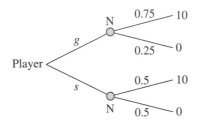

FIGURE 2.1 The R&D decision problem.

choices are denoted by their corresponding probabilities, which are conditional on the choice made by the player.

We now introduce a definition that generalizes the kind of randomness that was demonstrated by the R&D example. Consider a decision problem with n possible outcomes, $X = \{x_1, x_2, \ldots, x_n\}$.

Definition 2.1 A simple lottery over outcomes $X = \{x_1, x_2, \ldots, x_n\}$ is defined as a probability distribution $p = (p(x_1), p(x_2), \ldots, p(x_n))$, where $p(x_k) \geq 0$ is the probability that x_k occurs and $\sum_{k=1}^{n} p(x_k) = 1$.

By the definition of a probability distribution over elements in X, the probability of each outcome cannot be a negative number, and the sum of all probabilities over all outcomes must add up to 1. In our R&D example, following a choice of g, the lottery that Nature chooses is $p(10) = 0.75$ and $p(0) = 0.25$. Similarly, following a choice of s, the lottery that Nature chooses is $p(10) = p(0) = 0.5$.

Remark To be precise, the lottery that Nature chooses is conditional on the action taken by the player. Hence, given an action $a \in A$, the conditional probability that $x_k \in X$ occurs is given by $p(x_k|a)$, where $p(x_k|a) \geq 0$, and $\sum_{k=1}^{n} p(x_k|a) = 1$ for all $a \in A$.

Note that our trivial decision problem of choosing a cereal can be considered as a decision problem in which the probability over outcomes after any choice is equal to 1 for some outcome and 0 for all other outcomes. We call such a lottery a **degenerate lottery.** You can now see that decision problems with no randomness are just a very special case of those with randomness. Thus we have enriched our language to include more complex decision problems while encompassing everything we have developed earlier.

2.1.2   Simple versus Compound Lotteries

Arguably a player should care only about the probabilities of the various final outcomes that are a consequence of his actions. It seems that the exact way in which randomness unfolds over time should not be consequential to a player's well-being, but that only distributions over final outcomes should matter.

To understand this concept better, imagine that we make the R&D decision problem a bit more complicated. As before, if the player chooses not to embark on the R&D project (s) then the product line is successful with probability 0.5. If he chooses to go ahead with R&D (g) then two further stages will unfold. First, it will be determined whether the R&D effort was successful or not. Second, the outcome of the R&D phase will determine the likelihood of the product line's success. If the R&D effort is a failure then the success of the product is as likely as if no R&D had been performed; that is, the product line succeeds with probability 0.5. If the R&D effort is a success, however, then the probability of a successful product line jumps to 0.9. To complete the data for this example, we assume that R&D succeeds with probability 0.625 and fails with probability 0.375.

In this modified version of our R&D problem we have Nature moving once after the choice s and twice in a row after the choice g: once through the outcome of the R&D phase and then through the determination of the product line's success. This new decision problem is depicted in Figure 2.2.

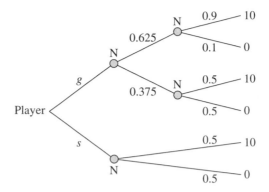

FIGURE 2.2 The modified R&D decision problem.

It seems like the two decision problems in Figures 2.1 and 2.2 are of different natures (no pun intended). Then again, let's consider what a decision problem ought to be about: actions, distributions over outcomes, and preferences. It is apparent that the player's choice of s in both Figure 2.1 and Figure 2.2 leads to the *same distribution* over outcomes. What about the choice of g? In Figure 2.2 this is followed by two random stages. However, the outcomes are still either 10 or 0. What are the probabilities of each outcome?

There are two ways that 10 can be obtained after the choice of g: First, with probability 0.625 the R&D project succeeds, and then with probability 0.9 the payoff 10 will be obtained. Hence the probability of "R&D success followed by 10" is equal to $0.625 \times 0.9 = 0.5625$. Second, with probability 0.375 the R&D project fails, and then with probability 0.5 the payoff 10 will be obtained. Hence the probability of "R&D failure followed by 10" is equal to $0.375 \times 0.5 = 0.1875$. Thus if the player chooses g then the probability of obtaining 10 is just the sum of the probabilities of these two *exclusive events*, which equals $0.5625 + 0.1875 = 0.75$. It follows that if the player chooses g then the probability of obtaining a payoff of 0 is 0.25, the complement of the probability of obtaining 10 (you should check this).

What then is the difference between the two decision problems? The first, simpler, R&D problem has a simple lottery following the choice of g. The second, more complex, problem has a *simple lottery over simple lotteries* following the choice of g. We call such lotteries over lotteries **compound lotteries.** Despite this difference, we impose on the player a rather natural sense of rationality. In his eyes the two decision problems are the same: he has the same set of actions, each one resulting in the same probability distributions over final outcomes. This innocuous assumption will make it easier for the player to evaluate and compare the benefits from different lotteries over outcomes.

2.1.3 Lotteries over Continuous Outcomes

Before moving on to describe how the player will evaluate lotteries over outcomes, we will go a step further to describe random variables, or lotteries, over continuous-outcome sets. To start, consider the following example. You are growing 10 tomato vines in your backyard, and your crop, measured in pounds, will depend on two inputs. The first is how much you water your garden per day and the second is the weather. Your action set can be any amount of water up to 50 gallons (50 gallons will completely flood your backyard), so that $A \in [0, 50]$, and your outcome set can be any amount of crop that 10 vines can yield, which is surely no more than 100 pounds, hence $X = [0, 100]$. Temperatures vary daily, and they vary continuously. This implies that your final yield, given any amount of water, will also vary continuously.

In this case we will describe the uncertainty not with a discrete probability, as we did for the R&D case, but instead with a **cumulative distribution function** (CDF) defined as follows:[1]

Definition 2.2 A simple lottery over an interval $X = [\underline{x}, \overline{x}]$ is given by a cumulative distribution function $F : X \to [0, 1]$, where $F(\widehat{x}) = \Pr\{x \leq \widehat{x}\}$ is the probability that the outcome is less than or equal to $\widehat{x}$.

For those of you who have seen continuous random variables, this is not new. If you have not, Section 19.4 of the mathematical appendix may fill in some of the gaps.[2] The basic idea is simple. Because we have infinitely many possible outcomes, it is somewhat meaningless to talk about the probability of growing a certain exact weight of tomatoes. In fact it is correct to say that the probability of producing any particular predefined weight is zero. However, it is meaningful to talk about the probability of being below a certain weight x, which is given by the CDF $F(x)$, or similarly the probability of being above a certain weight x, which is given by the complement $1 - F(x)$.

Remark Just as in the case of finite outcomes, we wish to consider the case in which the distribution over outcomes is conditional on the action taken. Hence, to be precise, we need to use the notation $F(x|a)$.

Now that we have concluded with a description of what randomness is, we can move along to see how our decision-making player evaluates random outcomes.

2.2 Evaluating Random Outcomes

From now on we will consider the choice of an action $a \in A$ as the *choice of a lottery* over the outcomes in X. If the decision problem does not involve any randomness, then these lotteries are degenerate. This implies that we can stick to our notation of defining a decision problem by the three components of actions, outcomes, and preferences. The novelty is that each action is a lottery over outcomes.

The next natural question is: how will a player faced with the R&D problem in Figure 2.1 choose between his options of going forward or staying the course? Upon reflection, you may have already reached a conclusion. Despite the fact that his different choices lead to different lotteries, it seems that the two lotteries that follow g and s are easy to compare. Both have the same set of outcomes, a profit of 10 or a profit of 0. The choice g has a higher chance at getting the profit of 10, and hence we would expect anyone in their right mind to choose g. This implicitly assumes, however, that there are no costs to launching the R&D project.

---

1. The definition considers the outcome set to be a finite interval $X = [\underline{x}, \overline{x}]$. We can use the same definition for any subset of the real numbers, including the real line $(-\infty, \infty)$. An example of a lottery over the real line is the normal "bell-shape" distribution.
2. You are encouraged to learn this material since it will be useful, but one can continue through most of Parts I–III of this book without this knowledge.

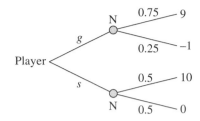

FIGURE 2.3 The R&D problem with costs.

Let's consider a less obvious revision of the R&D problem, and imagine that there is a real cost of pursuing the R&D project equivalent to 1. Hence the outcome of success yields a profit of 9 instead of 10, and the outcome of failure yields a profit of -1 instead of 0. This new problem is depicted in Figure 2.3. Now the comparison is not as obvious: is it better to have a coin toss between 10 and 0, or to have a good shot at 9, with some risk of losing 1?

2.2.1 Expected Payoff: The Finite Case

To our advantage, there is a well-developed methodology for evaluating how much a lottery is worth for a player, how different lotteries compare to each other, and how lotteries compare to "sure" payoffs (degenerate lotteries). This methodology, called "expected utility theory," was first developed by John von Neumann and Oskar Morgenstern (1944), two of the founding fathers of game theory, and explored further by Leonard Savage (1951). It turns out that there are some important assumptions that make this method of evaluation valid. (The foundations that validate expected payoff theory are beyond the scope of this text, and are rather technical in nature.)[3]

The intuitive idea is about averages. It is common for us to think of our actions as sometimes putting us ahead and sometimes dealing us a blow. But if *on average* things turn out on the positive side, then we view our actions as pretty good because the gains will more than make up for the losses. We want to take this idea, with its intuitive appeal, and use it in a precise way to tackle a single decision problem. To do this we introduce the following definition:

Definition 2.3 Let $u(x)$ be the player's payoff function over outcomes in $X = \{x_1, x_2, \ldots, x_n\}$, and let $p = (p_1, p_2, \ldots, p_n)$ be a lottery over X such that $p_k = \Pr\{x = x_k\}$. Then we define the player's **expected payoff from the lottery** p as

$$E[u(x)|p] = \sum_{k=1}^{n} p_k u(x_k) = p_1 u(x_1) + p_2 u(x_2) + \cdots + p_n u(x_n).$$

The idea of an expected payoff is naturally related to the intuitive idea of averages: if we interpret a lottery as a list of "weights" on payoff values, so that numbers that appear with higher probability have more weight, then the expected payoff of a lottery is nothing other than the weighted average of payoffs for each realization of the lottery.

---

3. The key idea was introduced by von Neumann and Morgenstern (1944) and is based on the "Independence Axiom." A nice treatment of the subject appears in Kreps (1990a, Chapter 3).

That is, payoffs that are more likely to occur receive higher weight while payoffs that are less likely to occur receive lower weight.

Using the definition of expected payoff we can revisit the R&D problem in Figure 2.3. First assume that the payoff to the player is equal to his profit, so that $u(x) = x$. By choosing g, the expected payoff to the player is

$$v(g) = E[u(x)|g] = 0.75 \times 9 + 0.25 \times (-1) = 6.5.$$

In contrast, by choosing s his expected payoff is

$$v(s) = E[u(x)|s] = 0.5 \times 10 + 0.5 \times 0 = 5.$$

Hence his best action using expected profits as a measure of preferences over actions is to choose g. You should be able to see easily that in the original R&D game in Figure 2.1 the expected payoff from s is still 5, while the expected payoff from g is 7.5, so that g was also his best choice, as we intuitively argued earlier.

Notice that we continue to use our notation $v(a)$ to define the expected payoff of an action given the distribution over outcomes that the action causes. This is a convention that we will use throughout this book, because the object of our analysis is what a player should do, and this notation implies that his ranking should be over his actions.

2.2.2 Expected Payoff: The Continuous Case

Consider the case in which the outcomes can be any one of a continuum of values distributed on some interval X. The definition of expected utility will be analogous, as follows:

Definition 2.4 Let $u(x)$ be the player's payoff function over outcomes in the interval $X = [\underline{x}, \overline{x}]$ with a lottery given by the cumulative distribution $F(x)$, with density $f(x)$. Then we define the player's expected payoff as[4]

$$E[u(x)] = \int_{\underline{x}}^{\overline{x}} u(x) f(x) dx.$$

To see an example with continuous actions and outcomes, recall the tomato growing problem in Section 2.1.3, in which your choice is how much water to use in the set $A = [0, 50]$ and the outcome is the weight of your crop that will result in the set $X = [0, 100]$. Imagine that given a choice of water $a \in A$, the distribution over outcomes is uniform over the quantity support $[0, 2a]$. (Alternatively the distribution of x conditional on a is given by $x|a \sim U[0, 2a]$.) For example, if you use 10 gallons of water, the output will be uniformly distributed over the weight interval $[0, 20]$, with the cumulative distribution function given by $F(x|a = 10) = \frac{x}{20}$ for $0 \leq x \leq 20$,

---

4. More generally, if there are continuous distributions that do not have a density because $F(\cdot)$ is not differentiable, then the expected utility is given by

$$E[u(x)] = \int_{x \in X} u(x) dF(x).$$

This topic is covered further in Section 19.4 of the mathematical appendix.

and $F(x|a=10) = 1$ for all $x > 20$. More generally the cumulative distribution function is given by $F(x|a) = \frac{x}{2a}$ for $0 \leq x \leq 2a$, and $F(x|a) = 1$ for all $x > 2a$. The density is given by $f(x|a) = \frac{1}{2a}$ for $0 \leq x \leq 2a$, and $f(x|a) = 0$ for all $x > 2a$. Thus if your payoff from quantity x is given by $u(x)$ then your expected payoff from any choice $a \in A$ is given by

$$v(a) = E[u(x)|a] = \int_0^{2a} u(x)f(x|a)dx = \frac{1}{2a}\int_0^{2a} u(x)dx.$$

Given a specific function to replace $u(\cdot)$ we can compute $v(a)$ for any $a \in [0, 50]$. As a concrete example, let $u(x) = 18\sqrt{x}$. Then we have

$$v(a) = \frac{1}{2a}\int_0^{2a} 18x^{\frac{1}{2}}dx = \frac{9}{a}\left[\frac{2}{3}x^{\frac{3}{2}}\right]_0^{2a} = \frac{6}{a}(2a)^{\frac{3}{2}} = 12\sqrt{2a}.$$

2.2.3 Caveat: It's Not Just the Order Anymore

Recall that when we introduced the idea of payoffs in Section 1.1.2, we argued that any payoff function that preserves the order of outcomes as ranked by the preference relation $\succsim$ will be a valid representation for the preference relation $\succsim$. It turns out that this statement is no longer true when we step into the realm of expected payoff theory as a paradigm for evaluating random outcomes.

Looking back at the R&D problem in Figure 2.3, we took a leap when we equated the player's payoff with profit. This step may seem innocuous: it is pretty reasonable to assume that, other things being equal, a rational player will prefer more money to less. Hence for the player in the R&D problem we have $10 \succ 9 \succ 0 \succ -1$, a preference relation that is indeed captured by our imposed payoff function where $u(x) = x$.

What would happen if payoffs were not equated with profits? Consider a different payoff function to represent these preferences. In fact, consider only a slight modification as follows: $u(10) = 10$, $u(9) = 9$, $u(0) = 0$, and $u(-1) = -8$. The order of outcomes is unchanged, but what happens to the expected payoffs? $E[u(s)] = 5$ is unchanged, but now

$$v(g) = E[u(x)|g] = 0.75 \times 9 + 0.25 \times (-8) = 4.75.$$

Thus even though the order of preferences has not changed, the player would now prefer to choose s instead of g, just because of the different payoff number we assigned to the profit outcome of -1.

The reason behind this reversal of choice has important consequences. When we choose to use expected payoff then the *intensity of preferences* matters—something that is beyond the notion of simple order. We can see this from our intuitive description of expected payoff. Recall that we used the intuitive notion of "weights": payoffs that appear with higher probability have more weight in the expected payoff function. But then, if we change the number value of the payoff of some outcome without changing its order in the payoff representation, we are effectively changing its weight in the expected payoff representation.

This argument shows that, unlike payoff over *certain* outcomes, which is meant to represent *ordinal preferences* $\succsim$, the expected payoff representation involves a *cardinal ranking,* in which values matter just as much as order. At some level this implies that we are making assumptions that are not as innocuous about decision making when we extend our rational choice model to include preferences over lotteries and choices among lotteries. Nevertheless we will follow this prescription as a benchmark for putting structure on decision problems with uncertainty. We now briefly explore some implications of the intensity of preferences in evaluating random outcomes.

2.2.4 Risk Attitudes

Any discussion of the evaluation of uncertain outcomes would be incomplete without addressing a player's attitudes toward risk. By treating the value of outcomes as "payoffs" and by invoking the expected payoff criterion to evaluate lotteries, we have effectively circumvented the need to discuss risk, because by assumption all that people care about is their expected payoff.

To illustrate the role of risk attitudes, it will be useful to distinguish between monetary rewards and their associated payoff values. Imagine that a player faces a lottery with three monetary outcomes: $x_1 = \$4$, $x_2 = \$9$, and $x_3 = \$16$ with the associated probabilities p_1, p_2, and p_3. If the player's payoff function over money x is given by some function $u(x)$ then his expected payoff is

$$E[u(x)|p] = \sum_{k=1}^{3} p_k u(x_k) = p_1 u(x_1) + p_2 u(x_2) + p_3 u(x_3).$$

Now consider two different lotteries: $p' = (p_1', p_2', p_3') = \left(\frac{7}{12}, 0, \frac{5}{12}\right)$ and $p'' = (p_1'', p_2'', p_3'') = (0, 1, 0)$. That is, the lottery p' randomizes between \$4 and \$16 with probabilities $\frac{7}{12}$ and $\frac{5}{12}$, respectively, while the lottery p'' picks \$9 for sure. Which lottery should the player prefer? The obvious answer will depend on the expected payoff of each lottery. If $\frac{7}{12}u(4) + \frac{5}{12}u(16) > u(9)$, then p' will be preferred to p'', and vice versa. This answer, by itself, tells us nothing about risk, but taken together with the special way in which p' and p'' relate to each other, it tells us a lot about the player's risk attitudes.

The lotteries p' and p'' were purposely constructed so that the average payoff of p' is equal to the sure payoff from p'': $\frac{7}{12} \times 4 + \frac{5}{12} \times 16 = 9$. Hence, *on average*, both lotteries offer the player the same amount of money, but one is a sure thing while the other is uncertain. If the player chooses p' instead of p'', he faces the risk of getting \$5 less, but he also has the chance of getting \$7 more. How then do his choices imply something about his attitude toward risk?

Imagine that the player is indifferent between the two lotteries, implying that $\frac{7}{12}u(4) + \frac{5}{12}u(16) = u(9)$. In this case we say that the player is **risk neutral,** because replacing a sure thing with an uncertain lottery that has the same expected monetary payout has no effect on his well-being. More precisely we say that a player is risk neutral if he is willing to exchange any sure payout with any lottery that promises *the same expected* monetary payout.

Alternatively the player may prefer not to be exposed to risk for the same expected payout, so that $\frac{7}{12}u(4) + \frac{5}{12}u(16) < u(9)$. In this case we say that the player is **risk averse.** More precisely a player is risk averse if he is *not* willing to exchange a sure payout with any (nondegenerate) lottery that promises *the same expected* monetary payout. Finally a player is **risk loving** if the opposite is true: he *strictly prefers* any lottery that promises the same expected monetary payout.

Remark Interestingly risk attitudes are related to the fact that the payoff representation of preferences matters above and beyond the rank order of outcomes, as discussed in Section 2.2.3. To see this imagine that $u(x) = x$. This immediately implies that the player is risk neutral: $\frac{7}{12}u(4) + \frac{5}{12}u(16) = 9 = u(9)$. In addition it is obvious from $u(\cdot)$ that the preference ranking is $\$16 \succ \$9 \succ \$4$. Now imagine that we use a different payoff representation for the same preference ranking: $u(x) = \sqrt{x}$. Despite the fact that the ordinal ranking is preserved, we now have $\frac{7}{12}u(4) + \frac{5}{12}u(16) = \frac{17}{6} < 3 = u(9)$. Hence a player with this modified payoff function, which *preserves the ranking* among the outcomes, will exhibit different risk attitudes.

2.2.5 The St. Petersburg Paradox

Some trace the first discussion of risk aversion to the St. Petersburg Paradox, so named in Daniel Bernoulli's original presentation of the problem and his solution, published in 1738 in the *Commentaries of the Imperial Academy of Science of Saint Petersburg*. The decision problem goes as follows.

You pay a fixed fee to participate in a game of chance. A "fair" coin (each side has an equal chance of landing up) will be tossed repeatedly until a "tails" first appears, ending the game. The "pot" starts at $1 and is doubled every time a "head" appears. You win whatever is in the pot after the game ends. Thus you win $1 if a tail appears on the first toss, $2 if it appears on the second, $4 if it appears on the third, and so on. In short, you win 2^{k-1} dollars if the coin is tossed k times until the first tail appears. (In the original introduction, this game was set in a hypothetical casino in St. Petersburg, hence the name of the paradox.)

The probability that the first "tail" occurs on the kth toss is equal to the probability of the "head" appearing $k-1$ times in a row and the "tail" appearing once. The probability of this event is $\left(\frac{1}{2}\right)^k$, because at any given toss the probability of any side coming up is $\frac{1}{2}$. We now calculate the expected *monetary value* of this lottery, which takes expectations over the possible events as follows: You win $1 with probability $\frac{1}{2}$; $2 with probability $\frac{1}{4}$; $4 with probability $\frac{1}{8}$, and so on. The expected value of this lottery is

$$\sum_{k=1}^{\infty} \frac{1}{2^k} \times 2^{k-1} = \sum_{k=1}^{\infty} \frac{1}{2} = \infty.$$

Thus the expected monetary value of this lottery is infinity! The reason is that even though large sums are very unlikely, when these events happen they are huge. For example, the probability that you will win more than $1 million is less than one in 500,000!

When Bernoulli presented this example, it was very clear that no reasonable person would pay more than a few dollars to play this lottery. So the question is: where is the paradox? Bernoulli suggested a few answers, one being that of decreasing marginal payoff for money, or a concave payoff function over money, which is basically risk aversion. He correctly anticipated that the value of this lottery should not be measured in its expected monetary value, but instead in the monetary value of its *expected payoff*.

Throughout the rest of this book we will make no more references to risk preferences but instead assume that every player's preferences can be represented using expected payoffs. For a more in-depth exposition of attitudes toward risk, see Chapter 3 in Kreps (1990a) and Chapter 6 in Mas-Colell et al. (1995).

2.3   Rational Decision Making with Uncertainty

2.3.1   Rationality Revisited

We defined a rational player as one who chooses an action that maximizes his payoff among the set of all possible actions. Recall that the four rational choice assumptions in Section 1.2 included a requirement that the player know "exactly how each action affects which outcome will materialize."

For this knowledge to be meaningful and to guarantee that the player is correctly perceiving the decision problem when outcomes can be stochastic, it must be the case that he fully understands how each action translates into a lottery over the set of possible outcomes. In other words, the player knows that by choosing actions he is choosing lotteries, and he knows exactly what the probability of each outcome is, conditional on his choice of an action.

Understanding the requirements for rational decision making under uncertainty, together with the adoption of expected payoff as a means of evaluating random outcomes, offers a natural way to define rationality for decision problems with random outcomes:

Definition 2.5   A player facing a decision problem with a payoff function $u(\cdot)$ over outcomes is rational if he chooses an action $a \in A$ that maximizes his expected payoff. That is, $a^* \in A$ is chosen if and only if $v(a^*) = E[u(x)|a^*] \geq E[u(x)|a] = v(a)$ for all $a \in A$.

That is, the player, who understands the stochastic consequences of each of his actions, will choose an action that offers him the highest expected payoff. In the R&D problems described in Figures 2.1 and 2.3 the choice that maximizes expected payoff was to go ahead with the project and choose g.

2.3.2   Maximizing Expected Payoffs

As another illustration of maximizing expected payoff with a finite set of actions and outcomes, consider the following example. Imagine that you have been working after college and now face the decision of whether or not to get an MBA at a prestigious institution. The cost of getting the MBA is 10. (Again, you can decide on the denomination, but rest assured that this sum includes the income lost over the course of the two years you will be studying!) Your future value is your stream of income, which depends on the strength of the labor market for the next decade. If the labor market is strong then your income value from having an MBA is 32, while your income value from your current status is 12. If the labor market is average then your income value from having an MBA is 16, while your income value from your current status is 8. If the labor market is weak then your income value from having an MBA is 12, while your income value from your current status is 4. After spending some time researching the topic, you learn that the labor market will be strong with probability 0.25, average with probability 0.5, and weak with probability 0.25. Should you get the MBA?

This decision problem is depicted in Figure 2.4. Notice that following the decision of whether or not to get an MBA, we subtract the cost of the degree from the income benefit in each of the three states of nature. To solve this decision problem we first evaluate the expected payoff from each action. We have

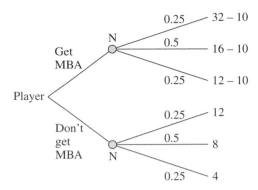

FIGURE 2.4 The MBA degree decision problem.

$$v(\text{Get MBA}) = 0.25 \times 22 + 0.5 \times 6 + 0.25 \times 2 = 9$$

$$v(\text{Don't get MBA}) = 0.25 \times 12 + 0.5 \times 8 + 0.25 \times 4 = 8.$$

Thus we conclude that, given the parameters of the problem, it is worth getting the MBA.

To illustrate the maximization of expected payoffs when there is a continuous set of actions and outcomes, consider the following example, which builds on the tomato growing problem from Section 2.2.2, with $A = [0, 50]$, $X = [0, 100]$ and the distribution of x conditional on a is uniform given by $x|a \sim U[0, 2a]$. We showed in the example in Section 2.2.2 that if the player's payoff from quantity x is given by $u(x)$ then his expected payoff from any choice $a \in A$ is given by

$$v(a) = E[u(x)|a] = \frac{1}{2a} \int_0^{2a} u(x)dx.$$

To account for the cost of water, assume that choosing $a \in A$ imposes a payoff cost of $2a$ (you can think of 2 being the cost of a gallon of water). Also assume that the payoff value from quantity x is given by the function $u(x) = 18\sqrt{x}$. (The square root function implies that the added value of every extra unit is less than the added value of the previous unit because this function is concave.) Then the player wishes to maximize his *expected net payoff*. This will be obtained by choosing the amount of water a that maximizes the difference between the expected benefit from choosing some $a \in A$ (given by $E[18\sqrt{x}|a]$) and the actual cost of $2a$. Thus the player's mathematical representation of the decision problem is given by

$$\max_{a \in [0,50]} \frac{1}{2a} \int_0^{2a} 18\sqrt{x}dx - 2a.$$

Solving for the integral, this is equivalent to maximizing the objective function $12\sqrt{2a} - 2a$. Differentiating this gives us the first-order condition for finding an optimum, which is

$$\frac{12}{\sqrt{2a}} - 2 = 0,$$

resulting in the solution $a = 18$, the quantity of water that maximizes the player's net expected payoff.[5] Plugging this back into the objective function yields the expected net payoff of $12\sqrt{2 \times 18} - 2 \times 18 = 36$.

2.4 Decisions over Time

The setup we have used up to now fits practically any decision problem in which you need to choose an action that has consequences for your well-being. Notice, however, that the approach we have used has a timeless, static structure. Our decision problems have shared the feature that you have to make a choice, after which some outcome will materialize and the payoff will be obtained. Many decision problems, however, are more involved; they require some decisions to follow others, and the sequencing of the decision making is imposed by the problem at hand.

2.4.1 Backward Induction

Once again, let's consider a modification of the R&D problem described in Figure 2.2, with a slight twist. Imagine that after the stage at which Nature first randomizes over whether or not the R&D project succeeds, the player faces another decision of whether to engage in a marketing campaign (m) or do nothing (d). The campaign can be launched only if the R&D project was executed. The marketing campaign will double the profits if the new product line is a success but not if it is a failure, and the campaign costs 6. The resulting modified decision problem is presented in Figure 2.5. For example, if the R&D effort succeeds and the marketing campaign is launched then profits from the product line will be 20 (double the original 10 in Figure 2.2), but the cost of the campaign, equal to 6, must be accounted for. This explains the payoff of $20 - 6 = 14$.

Now the question of whether the player should choose g or s is not as simple as before, because it may depend on what he will do later, after the fate of the R&D project is determined by Nature. Our assumption that the player is rational has some strong implications: he will be rational *at every stage* at which he faces a decision. At the beginning of the problem the player knows that he will act optimally to maximize his expected payoff at later stages, hence he can predict what he will do there.

This logic is the simple idea behind the optimization procedure known as **dynamic programming** or **backward induction.** To explain this procedure it is useful to separate the player's decision nodes into separate groups as follows: Group 1 will include all the nodes after which no more decision nodes exist, so that only Nature or final payoffs follow such nodes. For example, in Figure 2.5 Group 1 would include both decision nodes at which the player must decide between m and d. Then define Group 2 nodes as follows: a node k will belong to Group 2 if and only if the only decision nodes that follow any action at node k are decision nodes of Group 1. Define higher-order groups similarly.

Now consider all the nodes in Group 1 and figure out the optimal action of the player at each of these nodes. Once this is done we can compute the expected payoff from optimizing at that node, and that will be the value of the decision node.

---

5. We also need to check the second-order condition, that the second derivative of the objective function is negative at the proposed candidate. This is indeed the case, since the second derivative of the objective function is $-12(2a)^{\frac{3}{2}}$, which is always negative (the objective function is concave).

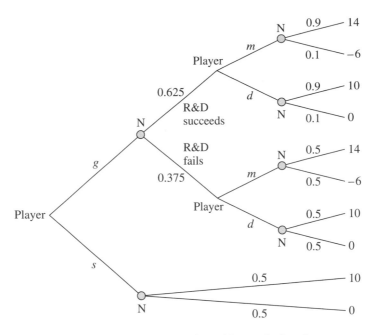

FIGURE 2.5   The R&D problem with a marketing phase.

Effectively we can throw away all the branches that follow the decision nodes of
Group 1 and, assuming rationality, associate these nodes with the expected payoff
from acting optimally. We can continue this process "backward" through the tree
until we cover all the decision nodes of the player.

We can use the decision problem in Figure 2.5 to illustrate this procedure. First,
we can compute the expected payoff of the player from choices m and d at the node
after it has been determined by Nature that the R&D project was a success. We have

$$E[u(x)|\text{R\&D succeeds and } m] = 0.9 \times (20 - 6) + 0.1 \times (-6) = 12$$

$$E[u(x)|\text{R\&D succeeds and } d] = 0.9 \times 10 + 0.1 \times 0 = 9,$$

which implies that at this node the player will choose m in anticipation of an expected
payoff of 12. Now consider his same choice problem at the node after it has been
determined by Nature that the R&D project was a failure. We have

$$E[u(x)|\text{R\&D fails and } m] = 0.5 \times (20 - 6) + 0.5 \times (-6) = 4$$

$$E[u(x)|\text{R\&D fails and } d] = 0.5 \times 10 + 0.5 \times 0 = 5,$$

which implies that at this node the player will choose D in anticipation of an expected
payoff of 5.

Imposing these rational decisions in the Group 1 nodes allows us to rewrite the
decision tree as a simpler tree that already folds in the optimal decisions of the player
at the Group 1 nodes. This "reduced" decision tree is depicted in Figure 2.6. The
player's choice at the beginning of the tree is now easy to analyze. Taking into account
his optimal actions after the R&D project's fate is determined, his expected payoff
from choosing g is

$$v(g) = 0.625 \times 12 + 0.375 \times 5 = 9.375.$$

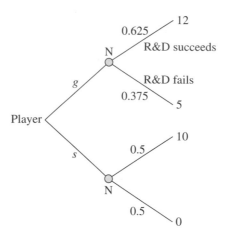

FIGURE 2.6 The marketing staged reduction of the R&D problem.

Because his expected payoff from choosing s is 5, it is clear that, anticipating his future decisions, his first decision should be to choose g.

2.4.2 Discounting Future Payoffs

In the R&D example as we have analyzed it so far, a player treated his costs and benefits equally. That is, even though the costs of the R&D project were incurred at the beginning, and the benefits came some (unspecified) time later, a dollar "today" was worth a dollar "tomorrow." However, this is often not the way current and future payments are evaluated. The convention in decision analysis is to *discount future payoffs* so that a dollar tomorrow is worth less than a dollar today.

For those who have had some experience with finance, the motivation for discounting future financial payoffs is simple. Imagine that you can invest money today in an interest-bearing savings account that yields 2% interest a year. If you invest $100 today then you can receive $100 \times 1.02 = \$102$ in a year, $100 \times (1.02)^2 = \$104.04$ in two years, and similarly $100 \times (1.02)^t$ in t years. As we can see, any amount today will be worth more and more in *nominal* terms as we move further into the future. As a consequence, the opposite should be true: any amount x that is expected in t years will be worth $v = \frac{x}{(1.02)^t}$ today precisely because we need only to invest v today in order to get x in t years. More generally, if the interest rate is $r\%$ per period, then any amount x that is received in t periods is discounted and is worth only $\frac{x}{(1+r)^t}$ today.

Another motivation for discounting future payoffs is uncertainty over the future coupled with expected future values. Most people are quite certain that they will be alive and well a year from today. That said, there is always that small chance that one's future may be cut short due to illness or accident. (This is the reason that life insurance companies use actuarial tables.) Imagine that a player assesses that with probability $\delta \in (0, 1)$ he will be alive and well in one period (a year, a month, and so forth), while with probability $1 - \delta$ he will not. This implies that if he is offered a payoff of x in one period then his expected utility is $v = \delta x + (1 - \delta)0 = \delta x < x$. Similarly, if he is promised a payoff of x in t periods, then he would be willing to trade that promise for a payoff of $v = \delta^t x$ today.

More generally, if a player expects to receive a stream of payments $x_1, x_2, \ldots, x_T$ over the periods $t = 1, 2, \ldots, T$, and he evaluates payments with the utility function

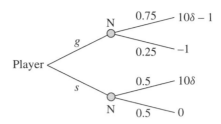

FIGURE 2.7 The R&D decision problem with discounting.

$u(x)$ in every period, then in the first period $t = 1$ his *discounted sum of future payoffs* is defined by

$$v(x_1, x_2, \ldots, x_T) = u(x_1) + \delta u(x_2) + \cdots + \delta^{T-1} u(x_T) = \sum_{t=1}^{T} \delta^{t-1} u(x_t).$$

The reason that discounting future payoffs is important is that changes in the discount rate will change the way decisions today are traded off against future payoffs. Consider the simple R&D problem. A successful product line, worth 10 in one period, will occur with probability 0.75 if the player chooses g, while only with probability 0.5 if he chooses s. The cost today of the R&D project is 1. If future payoffs are discounted at a rate of δ, then the problem is described by the decision tree depicted in Figure 2.7.

The expected payoffs for the player from choosing g or s are given by

$$v(g) = E[u(x)|g] = 0.75 \times (10\delta - 1) + 0.25 \times (-1) = 7.5\delta - 1$$

and

$$v(s) = E[u(x)|s] = 0.5 \times 10\delta + 0.5 \times 0 = 5\delta.$$

It is easy to see that the optimal decision will depend on the discount factor δ. In particular the player will choose to go ahead with the R&D development if and only if $7.5\delta - 1 > 5\delta$, or $\delta > 0.4$. The intuition is quite straightforward: the cost of the investment equal to 1 is borne today, and hence evaluated at a value of 1 regardless of the discount factor δ. Future payoffs, however, are discounted at the rate of δ. Hence as δ gets smaller the value of the future benefits from R&D decreases, while the value of today's costs remains the same, and once δ drops below the critical value of 0.4 the costs are no longer covered by the added benefits.

2.5 Applications

2.5.1 The Value of Information

When decisions lead to stochastic outcomes, our rational player chooses his action to maximize his expected payoff so that *on average* he is making the right choice. If the player actually knew the choice of Nature ahead of time then he might have chosen something else. Recall the example in Section 2.3.2, the decision of whether or not to get an MBA, which is depicted in Figure 2.4. As our previous analysis indicated, the expected payoff from getting an MBA is 9, while the expected payoff from not getting an MBA is only 8. Hence our rational decision maker will leave his job and go back to school, making what is, *on average,* the right choice.

Now imagine that an all-knowing oracle appears before the player the moment before he is about to resign from his job and says, "I know what the state of the labor market will be and I can tell you for a price." If you are the player, the questions that you need to answer are, first, is this information valuable, and second, how much is it worth?

It is quite easy to answer the first question: the information is valuable if it may cause you to change the decision you would have made without it. Looking at the decision problem, the answer is clear: If you learn that the labor market is strong then you will not change your decision. In contrast, if you learn that the labor market is weak then you would rather forgo the MBA program because it gives you a payoff of 2 while staying at your current job gives you a payoff of 4. The same decision would be made if you learn that the labor market is average in its strength. Hence it is quite clear that the oracle has what may be considered valuable information.

Now to our second question: how much is this information worth? This question is also not too hard to answer by considering the decision problem the player *would face* after the oracle's announcement. In particular we can calculate the expected payoff that the player anticipated before making a decision *without the advice of the oracle* and compare it to the expected payoff the player anticipated before making a decision *with the oracle's advice.* The difference is clearly due to the oracle's information, and this will be our measure of the value of the information.

We concluded that the choice the player would have made without the oracle's advice was to get an MBA, and that the expected payoff of this choice was 9, which is the expected value of the decision problem depicted in Figure 2.4. With the oracle's advice, though, the player can make a decision *after* learning the state of the labor market. How does this affect his expected payoff? The correct way to calculate this is to maintain the probability distribution over the three states of Nature, but to take into account that the player will make different choices that *depend on the state of Nature,* unlike the case in which he has to make a choice that applies to *all states of Nature.*

This new decision problem is shown in Figure 2.8, and as we argued, the player will condition his behavior on the oracle's advice. In particular the labor market will be strong with probability 0.25, and in this case the player will get an MBA and will have a payoff of 22. The labor market will be average with probability 0.5, and in this case the player will not get an MBA and will have a payoff of 8. Finally the labor market will be weak with probability 0.25, and in this case the player will not get an MBA and will have a payoff of 4. This implies that, *before* hearing the oracle's advice, with the anticipation of using the oracle's advice, the player will have an expected payoff of

$$E[u] = 0.25 \times 22 + 0.5 \times 8 + 0.25 \times 4 = 10.5.$$

Thus we can conclude that with the oracle's advice the expected payoff to the player is 10.5, which is 1.5 more than his expected payoff without the oracle's advice. This answers our second question: the oracle's information is worth 1.5 to the player.

In general when a decision maker is faced with the option of whether or not to acquire information and how much to pay for it, then by comparing the decision problem with the added information to the decision problem without the additional information, the player will be able to calculate the value of the information. Of

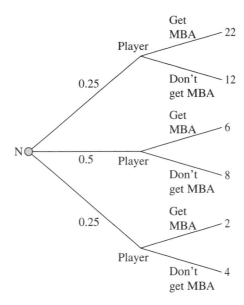

FIGURE 2.8 The MBA decision problem with the oracle's information.

course, this approach assumes that the player knows exactly what kind of information he can receive, how each piece of information will affect his payoffs, and what are the probabilities that each event will happen. Despite relying on a very demanding set of assumptions, this approach is valuable in offering a framework for decision making and valuing information.

2.5.2 Discounted Future Consumption

Usually we receive income or monetary gifts every so often, but we need to consume over several periods of time in between these income events. For example, you may receive a paycheck every month, but after paying your monthly costs, like rent and utilities, you need to buy groceries every week. If you spend too much during the first week, you may go hungry toward the end of the month. This kind of problem is known as *choosing consumption over time*.

Imagine a player who has $\$K$ today that need to be consumed over the next two periods, $t = 1, 2$. The utility over consuming $\$x$ in any period is given by the concave utility function $u(x)$, with $u'(x) > 0$ and $u''(x) < 0$. At period $t = 1$, the player values his utility from consuming x_2 in period $t - 2$ at the discounted value of $\delta u(x_2)$, so that at period $t = 1$ the player maximizes his present value of utility given by

$$\max_{x_1} \ u(x_1) + \delta u(K - x_1),$$

and the player's first-order condition is therefore[6]

$$u'(x_1) = \delta u'(K - x_1). \tag{2.1}$$

---

6. Because $u(\cdot)$ is concave, the second-order condition (that the second derivative be zero) is satisfied, implying that the solution to the first-order condition is the maximum.

For example, if $u(x) = \ln(x)$, so that $u'(x) = \frac{1}{x}$, the solution to the player's problem will be

$$\frac{1}{x_1} = \frac{\delta}{K - x_1}$$

or

$$x_1 = \frac{K}{1 + \delta}. \tag{2.2}$$

Turning back to the player's first-order condition in equation (2.1), the solution has a nice intuitive flavor that will be familiar to students of economics. Because the player needs to allocate a scarce resource, K, across the two periods, the optimal solution equates the *marginal values* across the two periods, taking into account that the second period is discounted by the factor δ. Turning to the specific form using the natural log function in (2.2), we can see that when period $t = 2$ is not discounted ($\delta = 1$) then the player splits K equally between the two periods. This is the case in which the player perfectly *smooths* his first-period resource across the two periods. As discounting becomes stronger, more of the scarce resource is consumed in the first period, and when the second period is completely discounted (δ goes to zero) then the player consumes all of the resource in the first period.

2.6 Theory versus Practice

We required the player to be rational in that he fully understands the decision problem he is facing. Hence if we present the same decision problem in different ways to the player, then as long as our presentation is loyal to the facts and includes all the relevant information, he should be able to decipher the true problem regardless of the way we present it to him. Yet Amos Tversky and Daniel Kahneman (1981) have shown that "framing," or the way in which a problem is presented, can affect the choices made by a decision maker. This would imply that our fundamental assumptions of rational choice do not hold.

Tversky and Kahneman demonstrated systematic "preference reversals" when the same problem was presented in different ways, in particular for the "Asian disease" decision problem. Physician participants were asked to "imagine that the U.S. is preparing for the outbreak of an unusual Asian disease, which is expected to kill 600 people. Two alternative programs to combat the disease have been proposed. Assume the exact scientific estimates of the consequences of the programs are as follows." The first group of participants was presented with a decision problem between two programs:

Program A 200 people will be saved.
Program B There is a $\frac{1}{3}$ probability that 600 people will be saved and a $\frac{2}{3}$ probability that no people will be saved.

When presented with these alternatives, 72% of participants preferred program A and 28% preferred program B. The second group of participants was presented with the following choice:

Program C 400 people will die.
Program D There is a $\frac{1}{3}$ probability that nobody will die, and a $\frac{2}{3}$ probability that 600 people will die.

For this decision problem, 78% preferred program D, while the remaining 22% preferred program C. Notice, however, that programs A and C *are identical,* as are programs B and D. A change in the framing of the decision problem between the two groups of participants produced a "preference reversal," with the first group preferring program A/C and the second group preferring program B/D.

Many argue today that framing biases affect a host of decisions made by people in their daily lives. Preference reversals and other associated phenomena have been key subjects for research in the flourishing field of behavioral economics, which is primarily involved in the study of behavior that contradicts the predictions of rational choice theory.

Nevertheless the rather naive and simple framework of rational choice theory goes a long way toward helping us understand the decisions of many individuals. Furthermore many argue that the "rational" decision makers will drive out the irrational ones when the stakes are high enough, and hence when we look at behavior that persists in situations for which some people have the opportunity to learn the environment, that behavior will often be consistent with rational choice theory. I will use the rational choice framework throughout this book, and I will leave its defense to others.

2.7 Summary

- When prospects are uncertain, a rational decision maker needs to put structure on the problem in the form of probability distributions over outcomes that we call *lotteries.*

- Whether the acts of Nature evolve over time or whether they are chosen once and for all, a rational player cares only about the distribution over final outcomes. Hence any series of *compound lotteries* can be compressed to its equivalent *simple lottery.*

- When evaluating lotteries, we use the expected payoff criterion. Hence every lottery is evaluated by the expected payoff it offers the player.

- Unlike certain outcomes in which only the order of preferences matters, when random outcomes are evaluated with expected payoffs, the magnitude of payoffs matters as well. The difference in payoff values between outcomes will also be related to a player's risk preferences.

- Rational players will always choose the action that offers them the highest expected payoff.

- When decisions need to be made over time, a rational player will solve his problem "backwards" so that early decisions take into account later decisions.

- Payoffs that are received in future periods will often be discounted in earlier periods to reflect impatience, costs of capital, or uncertainty over whether future periods will be relevant.

2.8 Exercises

2.1 **Getting an MBA:** Recall the decision problem in Section 2.3.2, and now assume that the probability of a strong labor market is p, that of an average labor market is 0.5, and that of a weak labor market is $0.5 - p$. All the other values are the same.

    a.   For which values of p will you decide not to get an MBA?

    b.   If $p = 0.4$, what is the highest price the university can charge for you to be willing to go ahead and get an MBA?

2.2   **Recreation Choices:** A player has three possible activities from which to choose: going to a football game, going to a boxing match, or going for a hike. The payoff from each of these alternatives will depend on the weather. The following table gives the agent's payoff in each of the two relevant weather events:

Alternative	Payoff if rain	Payoff if shine
Football game	1	2
Boxing match	3	0
Hike	0	1

Let p denote the probability of rain.

    a.   Is there an alternative that a rational player will never take regardless of p?

    b.   What is the optimal decision as a function of p?

2.3   **At the Dog Races:** You're in Las Vegas, and you must decide what to do at the dog-racing betting window. You may choose not to participate or you may bet on one of two dogs as follows. Betting on Snoopy costs $1, and you will be paid $2 if he wins. Betting on Lassie costs $1, and you will be paid $11 if she wins. You believe that Snoopy has probability 0.7 of winning and that Lassie has probability 0.1 of winning (there are other dogs on which you are not considering betting). Your goal is to maximize the expected monetary return of your action.

    a.   Draw the decision tree for this problem.

    b.   What is your best course of action, and what is your expected value?

    c.   Someone offers you gambler's "anti-insurance," which you may accept or decline. If you accept it, you get paid $2 up front and you agree to pay back 50% of any winnings you receive. Draw the new decision tree and find the optimal action.

2.4   **Drilling for Oil:** An oil drilling company must decide whether or not to engage in a new drilling venture before regulators pass a law that bans drilling on that site. The cost of drilling is $1 million. The company will learn whether or not there is oil on the site only after drilling has been completed and all drilling costs have been incurred. If there is oil, operating profits are estimated at $4 million. If there is no oil, there will be no future profits.

    a.   Using p to denote the likelihood that drilling results in oil, draw the decision tree for this problem.

    b.   The company estimates that $p = 0.6$. What is the expected value of drilling? Should the company go ahead and drill?

    c.   To be on the safe side, the company hires a specialist to come up with a more accurate estimate of p. What is the minimum value of p for which it would be the company's best response to go ahead and drill?

2.5   **Discount Prices:** A local department store sells products at a given initial price, and every week a product goes unsold, its price is discounted by 25% of the original price. If it is not sold after four weeks, it is sent back to the regional warehouse. A set of kitchen knives was just put out for $200. Your willingness to pay for the knives (your dollar value) is $180, so if you buy them at a price P, your payoff is $u = 180 - P$. If you don't buy the knives, the chances that they will be sold to someone else, conditional on not having been sold the week before, are as follows:

Week 1	0.2
Week 2	0.4
Week 3	0.6
Week 4	0.8

For example, if you do not buy the knives during the first two weeks, the likelihood that they will be available at the beginning of the third week is the likelihood that they do not sell in either week 1 or week 2, which is $0.8 \times 0.6 = 0.48$.

a. Draw your decision tree for the four weeks after the knives are put out for sale.

b. At the beginning of which week, if any, should you run to buy the knives?

c. Find a willingness to pay for the knives that would make it optimal to buy at the beginning of the first week.

d. Find a willingness to pay that would make it optimal to buy at the beginning of the fourth week.

2.6 **Real Estate Development:** A real estate developer wishes to build a new development. Regulations require an environmental impact study that will yield an "impact score," which is an index number based on the impact the development will likely have on such factors as traffic, air quality, and sewer and water usage. The developer, who has lots of experience, knows that the score will be no less than 40 and no more than 70. Furthermore he knows that any score between 40 and 70 is as likely as any other score between 40 and 70 (use continuous values). The local government's past behavior implies that there is a 35% chance that it will approve the development if the impact score is less than 50 and a 5% chance that it will approve the development if the score is between 50 and 55; if the score is greater than 55 then the project will surely be halted. The value of the development to the developer is $20 million. Assuming that the developer is risk neutral, what is the maximum cost of the impact study such that it is still worthwhile for the developer to have it conducted?

2.7 **Toys:** WakTek is a manufacturer of electronic toys, with a specialty in remote-controlled miniature vehicles. WakTek is considering the introduction of a new product, a remote-controlled hovercraft called WakAtak. Preliminary designs have already been produced at a cost of $2 million. To introduce a marketable product requires the building of a dedicated production line at a cost of $12 million. In addition, before the product can be launched a prototype must be built and tested for safety. The prototype can be crafted in the absence of a production line, at a cost of $0.5 million, but if the prototype is created after the production line is built then its cost is negligible (you can treat it as zero). There is uncertainty over what safety rating WakAtak will get. This could have a significant impact on demand, as a lower safety rating will increase the minimum age required of users. Safety testing costs $1 million.

The outcome of the safety test is estimated to have a 65% chance of resulting in a minimum user age of 8 years, a 30% chance of a minimum age of 15 years, and a 5% chance that the product will be declared unsafe—in which case it could not be sold at all. (The cost of improving the safety rating of a finished design is deemed prohibitive.) After successful safety testing the product could be launched at a cost of $1.5 million.

There is also uncertainty over demand, which will have a crucial impact on the eventual profits. Currently the best estimate is that the finished product, if available to the 8- to 14-year demographic, has a 50-50 chance of resulting in profits of either $10 million or $5 million from that demographic. Similarly there is a 50-50 chance of either $14 million or $6 million in profits from the 15-year-or-above demographic. These demand outcomes are independent across the demographics. The profits do not take into account the costs previously defined; they are measured in expected present-value terms, so they are directly comparable with the costs.

a. What is the optimal plan of action for WakTek? What is the current expected economic value of the WakAtak project?
b. Suddenly it turns out that the original estimate of the cost of safety testing was incorrect. Analyze the sensitivity of WakTek's optimal plan of action to the cost of safety testing.
c. Suppose WakTek also has the possibility of conducting a market survey, which would tell it exactly which demand scenario is true. This market research costs $1.5 million if done simultaneously for both demographics and $1 million if done for only one demographic. How, if at all, is your answer to part (a) affected?
d. Suppose that demand is not independent across demographics after all, but instead is perfectly correlated (i.e., if demand is high in one demographic then it is for sure high in the other one as well). How, if at all, would that change your answer to part (c)?

2.8 **Juice:** Bozoni is a Swiss maker of fruit and vegetable juice, whose products are sold at specialty stores throughout Western Europe. Bozoni is considering whether to add cherimoya juice to its line of products. "It would be one of our more difficult varieties to produce and distribute," observes Johann Ziffenboeffel, Bozoni's CEO. "The cherimoya would be flown in from New Zealand in firm, unripe form, and it would need its own dedicated ripening facility here in Europe." Three successful steps are absolutely necessary for the new cherimoya variety to be worth producing. The industrial ripening process must be shown to allow the delicate flavors of the cherimoya to be preserved; the testing of the ripening process requires the building of a small-scale ripening facility. Market research in selected limited regions around Europe must show that there is sufficient demand among consumers for cherimoya juice. And cherimoya juice must be shown able to withstand the existing tiny gaps in the cold chain (the temperature-controlled supply chain) between the Bozoni plant and the end consumers (these gaps would be prohibitively expensive to fix). Once these three steps have been completed, there would be about €2,500,000 worth of expenses in launching the new variety of juice. A successful new variety will then yield profits, in expected present-value terms, of €42.5 million.

The three necessary steps can be done in parallel or sequentially, and in any order. Data about these three steps are given in the following table:

Step	Probability of success	Cost
Ripening process	0.7	€1,000,000
Market research	0.3	5,000,000
Cold chain	0.6	500,000

"Probability of success" refers to how likely it is that the step will be successful. If it is not successful, then that means that cherimoya juice cannot be sold at a profit. All probabilities are independent of each other (i.e., whether a given step is successful or not does not affect the probabilities that the other steps will be successful). "Cost" refers to the cost of undertaking the step (regardless of whether it is successful or not).

a. Suppose Mr. Ziffenboeffel calls you and asks your advice about the project. In particular he wants to know (i) should he undertake the three necessary steps in parallel (i.e., all at once) or should he undertake them sequentially, and (ii) if sequentially, what's the correct order in which the steps should be done? What answers do you give him?

b. Mr. Ziffenboeffel calls you back. Since the table was produced, Bozoni has found a small research firm that can perform the necessary tests for the ripening process at a lower cost than Bozoni's in-house research department. At the same time, the European Union (EU) has tightened the criteria for getting approval for new food-producing facilities, which raises the costs of these tests. Mr. Ziffenboeffel would like to know how your answer to (a) changes as a function of the cost of the ripening test. What do you tell him?

c. Mr. Ziffenboeffel calls you back yet again. The good news is that the cost of adhering to the EU regulations and the savings from outsourcing the ripening tests balance each other out, so the cost of the test remains €1,000,000. Now the problem is that his marketing department is suggesting that the probability that the market research will result in good news about demand for the juice could be different in light of recent data on the sales of other subtropical fruit products. He would like to know how your answer to (a) changes as a function of the probability of a positive result from the market research. What do you tell him?

2.9 **Steel:** AK Steel Holding Corporation is a producer of flat-rolled carbon, stainless, and electrical steels and tubular products through its wholly owned subsidiary, AK Steel Corporation. The recent surge in demand for steel significantly increased AK's profits,[7] and it is now engaged in a research project to improve its production of rolled steel. The research involves three distinct

---

7. See "Demand Sends AK Steel Profit Up 32%," *New York Times,* July 23, 2008, http://www .nytimes.com/2008/07/23/business/23steel.html?partner=rssnyt&emc=rss.

steps, each of which must be successfully completed before the firm can implement the cost-saving new production process. If the research is completed successfully, it will save the firm $4 million. Unfortunately there is a chance that one or more of the research steps might fail, in which case the entire project would be worthless. The three steps are done sequentially, so that the firm knows whether one step was successful before it has to invest in the next step. Each step has a 0.8 probability of success and costs $500,000. The risks of failure in the three steps are uncorrelated with one another. AK Steel is a risk-neutral company. (In case you are worried about such things, the interest rate is zero.)

 a. Draw the decision tree for the firm.
 b. If the firm proceeds with this project, what is the probability that it will succeed in implementing the new production process?
 c. If the research were costless, what would be the firm's expected gain from it before the project began?
 d. Should the firm begin the research, given that each step costs $500,000?
 e. Once the research has begun, should the firm quit at any point even if it has had no failures? Should it ever continue the research even if it has had a failure?

After the firm has successfully completed the first two steps, it discovers an alternate production process that would cost $150,000 and would lower production costs by $1 million with certainty. This process, however, is a substitute for the three-step cost-saving process; they cannot be used simultaneously. Furthermore, to have this process available, the firm must spend the $150,000 before it knows if it will successfully complete the third step of the three-step research project.

 f. Draw the augmented decision tree that includes the possibility of pursuing this alternate production process.
 g. If the firm continues the three-step project, what is the chance it would get *any* value from also developing the alternate production process?
 h. If developing the alternate production process were costless and if the firm continues the three-step project, what is the expected value that it would get from having the alternate production process available (at the beginning of the third research step)? (This is known as the *option value* of having this process available.)
 i. Should the firm

  i. Pursue only the third step of the three-step project?
  ii. Pursue only the alternate production process?
  iii. Pursue both the third step of the three-step project and the alternate production process?

 j. If the firm had known of the alternate production process before it began the three-step research project, what should it have done?

2.10 **Surgery:** A patient is very sick and will die in 3 months if he goes untreated. The only available treatment is risky surgery. The patient is expected to live for 12 months if the surgery is successful, but the probability that the surgery will fail and the patient will die immediately is 0.3.

 a. Draw a decision tree for this decision problem.

    b.   Let $v(x)$ be the patient's payoff function, where x is the number of months until death. Assuming that $v(12) = 1$ and $v(0) = 0$, what is the lowest payoff the patient can have for living 3 months so that having surgery is a best response?

For the rest of the problem, assume that $v(3) = 0.8$.

    c.   A test is available that will provide some information that predicts whether or not surgery will be successful. A positive test implies an increased likelihood that the patient will survive the surgery as follows:

> *True-positive rate:* The probability that the results of this test will be positive if surgery is to be successful is 0.90.
>
> *False-positive rate:* The probability that the results of this test will be positive if the patient will not survive the operation is 0.10.

       What is the probability of a successful surgery if the test is positive?

    d.   Assuming that the patient has the test done, at no cost, and the result is positive, should surgery be performed?

    e.   It turns out that the test may have some fatal complications; that is, the patient may die during the test. Draw a decision tree for this revised problem.

    f.   If the probability of death during the test is 0.005, should the patient opt to have the test prior to deciding on the operation?

2.11 **To Run or Not to Run:** You're a sprinter, and in practice today you fell and hurt your leg. An x-ray suggests that it's broken with probability 0.2. Your problem is deciding whether you should participate in next week's tournament. If you run, you think you'll win with probability 0.1. If your leg is broken and you run, then it will be further damaged and your payoffs are as follows:

    $+100$ if you win the race and your leg isn't broken
    $+50$ if you win and your leg is broken
    0 if you lose and your leg isn't broken
    -50 if you lose and your leg is broken
    -10 if you don't run and if your leg is broken
    0 if you don't run and your leg isn't broken

    a.   Draw the decision tree for this problem.

    b.   What is your best choice of action and its expected payoff?

You can gather some more information by having more tests, and you can gather more information about whether you'll win the race by talking to your coach.

    c.   What is the value of perfect information about the state of your leg?

    d.   What is the value of perfect information about whether you'll win the tournament?

    e.   As stated previously, the probability that your leg is broken and the probability that you will win the tournament are independent. Can you use a decision tree in the case that the probability that you will win the race depends on whether your leg is broken?

2.12 **More Oil:** Chevron, the number 2 U.S. oil company, was facing a tough decision. The new oil project dubbed "Tahiti" was scheduled to produce its first commercial oil in mid-2008, yet it was still unclear how productive it would be. "Tahiti is one of Chevron's five big projects," said Peter Robertson, vice chairman of the company's board, to the *Wall Street Journal*.[8] Nevertheless it was unclear whether the project would result in the blockbuster success Chevron was hoping for. As of June 2007 $4 billion had been invested in the high-tech deep sea platform, which sufficed to perform early well tests. Aside from offering information on the type of reservoir, the tests would produce enough oil to just cover the incremental costs of the testing (beyond the $4 billion investment).

Following the test wells, Chevron predicted one of three possible scenarios. The optimistic one was that Tahiti sits on one giant, easily accessible oil reservoir, in which case the company expected to extract 200,000 barrels a day after expending another $5 billion in platform setup costs, with a cost of extraction of about $10 a barrel. This would continue for 10 years, after which the field would have no more economically recoverable oil. Chevron believed this scenario had a 1-in-6 chance of occurring. A less rosy scenario, twice as likely as the optimistic one, was that Chevron would have to drill two more wells at an additional cost of $0.5 billion each (above and beyond the $5 billion setup costs), in which case production would be around 100,000 barrels a day with a cost of extraction of about $30 a barrel, and the field would still be depleted after 10 years. The worst-case scenario involves the oil being tucked away in numerous pockets, requiring expensive water injection techniques, which would involve upfront costs of another $4 billion (above and beyond the $5 billion setup costs) and extraction costs of $50 a barrel; production would be estimated to be at about 60,000 barrels a day for 10 years. Bill Varnado, Tahiti's project manager, was quoted as giving this least desirable outcome odds of 50-50.

The current price of oil is $70 a barrel. For simplicity assume that the price of oil and all costs will remain constant (adjusted for inflation) and that Chevron faces a 0% cost of capital (also adjusted for inflation).

    a.   If the test wells would not produce information about which one of three possible scenarios would result, should Chevron invest the setup costs of $5 billion to be prepared to produce under whichever scenario is realized?

    b.   If the test wells do produce accurate information about which of three possible scenarios is true, what is the added value of performing these tests?

2.13   **Today, Tomorrow, or the Day After:** A player has $100 today that needs to be consumed over the next three periods, $t = 1, 2, 3$. The utility over consuming $\$x_t$ in period t is given by the utility function $u(x) = \ln(x)$, and at period $t = 1$ the player values his net present value from all consumption as $u(x_1) + \delta u(x_2) + \delta^2 u(x_3)$, where $\delta = 0.9$.

    a.   How will the player plan to spend the $100 over the three periods of consumption?

    b.   Imagine that the player knows that in period $t = 2$ he will receive an additional gift of $20. How will he choose to allocate his original $100 initially, and how will he spend the extra $20?

8. "Chevron Bets Big on Gulf Output," *Wall Street Journal,* June 27, 2007, http://online.wsj.com/article/SB118291402301349620.html.

PART II

STATIC GAMES OF COMPLETE INFORMATION

3

Preliminaries

The language and tools of analysis that we have developed so far seem to be ideal to depict and analyze a wide variety of decision problems that a rational individual, or an entity with well-defined objectives, could face. The essence of our framework argues that any decision problem is best understood when we set it up in terms of the three elements of which it is made up: the possible actions, the deterministic or probabilistic relationship between actions and outcomes, and the decision maker's preferences over the possible outcomes. We proceeded to argue that a decision maker will choose those actions that are in his best interest.

This framework offers many attractive features: it is precise, well structured, and generally applicable, and most importantly it lends itself to systematic and consistent analysis. It does, however, suffer from one drawback: the world of a decision problem was described as a world in which the outcomes that determine our well-being are consequences of our own actions and some randomness that is beyond our control.

Let's consider for a moment a decision problem that you may be facing now if you are using this text as part of a university course, which you are taking for a grade. It is, I believe, safe to assume that your objective is some combination of learning the material and obtaining a good grade in the course, with higher grades being preferred over lower ones. This objective determines your preferences over outcomes, which are the set of all possible combinations of how much you learned and what grade you obtained. Your set of possible actions is deciding how hard to study, which includes such elements as deciding how many lectures to attend, how carefully to read the text, how hard to work on your problem sets, and how much time to spend preparing for the exams. Hence you are now facing a well-defined decision problem.

To complete the description of your decision problem, I have yet to explain how the outcome of your success is affected by the amount of work you choose to put into your course work. Clearly as an experienced student you know that the harder you study the more you learn, and you are also more likely to succeed on the exams. There is some uncertainty over how hard a given exam will be; that may depend on many random events, such as how you feel on the day of the exam and what mood the professor was in when the exam was written.

Still something seems to be missing. Indeed you must surely know that grades are often set on a curve, so that your grade relies on your success on the exam as an absolute measure of not only how much you got right but also how much the *other students in the class* got right. In other words, if you're having a bad day on an exam, your only hope is that everyone else in your class is having a worse day!

The purpose of this example is to point out that our framework for a decision problem will be inadequate if your outcomes, and as a consequence your well-being, will depend on the choices made by other decision makers. Perhaps we can just treat the other players in this decision problem as part of the randomness of nature: maybe they'll work hard, maybe not, maybe they'll have a bad day, maybe not, and so on. This, however, would not be part of a rational framework, for it would not be sensible for you to treat your fellow players as mere random "noise." Just as you are trying to optimize your decisions, so are they. Each player is trying to guess what others are doing, and how to act accordingly. In essence, you and your peers are engaged in a *strategic environment* in which you have to think hard about what other players are doing in order to decide what is best for you—knowing that the other players are going through the same difficulties.

We therefore need to modify our decision problem framework to help us describe and analyze strategic situations in which players who interact understand their environment, how their actions affect the outcomes that they and their counterparts will face, and how these outcomes are assessed by the other players. It is useful, therefore, to start with the simplest set of situations possible, and the simplest language that will capture these strategic situations, which we refer to as **games.** We will start with **static games of complete information,** which are the most fundamental games, or environments, in which such strategic considerations can be analyzed.

A static game is similar to the very simple decision problems in which a player makes a once-and-for-all decision, after which outcomes are realized. In a static game, a *set of players* independently choose once-and-for-all actions, which in turn cause the realization of an outcome. Thus a static game can be thought of as having two distinct steps:

Step 1: Each player *simultaneously and independently* chooses an action.

By *simultaneously and independently,* we mean a condition broader and more accommodating than players all choosing their actions at the *exact same* moment. We mean that players must take their actions without observing what actions their counterparts take and without interacting with other players to coordinate their actions. For example, imagine that you have to study for your midterm exam *two days before the midterm* because of an athletic event in which you have to participate on the day before the exam. Assume further that I plan on studying *the day before the midterm,* which will be after your studying effort has ended. If I don't know how much you studied, then by choosing my action after you I have no informational advantage over you; it is *as if* we are making our choices simultaneously and independently of each other. This idea will receive considerable attention as we proceed.

Step 2: Conditional on the players' choices of actions, payoffs are distributed to each player.

That is, once the players have all made their choices, these choices will result in a particular outcome, or probabilistic distribution over outcomes. The players have preferences over the outcomes of the game given by some payoff function over outcomes. For example, if we are playing rock-paper-scissors and I draw paper while you simultaneously draw scissors, then the outcome is that you win and I lose, and the payoffs are what winning and losing mean in our context—something tangible, like $0.10, or just the intrinsic joy of winning versus the suffering of losing.

Steps 1 and 2 settle what we mean by *static*. What do we mean by *complete information*? The loose meaning is that *all players understand the environment they are in*—that is, the game they are playing—in every way. This definition is very much related to our assumptions about rational choice in Section 1.2. Recall that when we had a single-person decision problem we argued that the player must know four things: (1) all his possible actions, A; (2) all the possible outcomes, X; (3) exactly how each action affects which outcome will materialize; and (4) what his preferences are over outcomes. How should this be adjusted to fit a game in which many such players interact?

Games of Complete Information   A **game of complete information** requires that the following four components be common knowledge among all the players of the game:

1. all the possible actions of all the players,
2. all the possible outcomes,
3. how each combination of actions of all players affects which outcome will materialize, and
4. the preferences of each and every player over outcomes.

This is by no means an innocuous set of assumptions. In fact, as we will discuss later, they are quite demanding and perhaps almost impossible to justify completely for many real-world "games." However, as with rational choice theory, we use these assumptions because they provide structure and, perhaps surprisingly, describe and predict many phenomena quite well.

You may notice that a new term snuck into the description of games of complete information: *common knowledge.* This is a term that we often use loosely: "it's common knowledge that he gives hard exams" or "it's common knowledge that green vegetables are good for your health." It turns out that *what exactly common knowledge means* is by no means common knowledge. To make it clear,

Definition 3.1   An event E is **common knowledge** if (1) everyone knows E, (2) everyone knows that everyone knows E, and so on *ad infinitum.*

On the face of it, this may seem like an innocuous assumption, and indeed it may be in some cases. For example, if you and I are both walking in the rain together, then it is safe to assume that the event "it is raining" is common knowledge between us. However, if we are both sitting in class and the professor says "tomorrow there is an exam," then the event "there is an exam tomorrow" may not be common knowledge. Despite me knowing that I heard him say it, perhaps you were daydreaming at the time, implying that I *cannot be sure* that you heard the statement as well.

Thus requiring common knowledge is not as innocuous as it may seem, but without this assumption it is quite impossible to analyze games within a structured framework. This difficulty arises because we are seeking to depict a situation in which players can engage in *strategic reasoning*. That is, I want to predict how you will make your choice, given *my belief* that you understand the game. Your understanding incorporates *your belief* about my understanding, and so on. Hence common knowledge will assist us dramatically in our ability to perform this kind of reasoning.

3.1 Normal-Form Games with Pure Strategies

Now that we understand the basic ingredients of a static game of complete information, we develop a formal framework to represent it in a parsimonious and general way, which captures the strategic essence of a game. As with the simple decision problem, the players will have actions from which to choose, and the combination of their choices will result in outcomes over which the players have preferences. For now we will restrict attention to players choosing certain (deterministic) actions that together cause certain (deterministic) outcomes. That is, players will not choose actions stochastically, and there will be no "Nature" player who will randomly select outcomes given a combination of actions that the players will choose.

One of the most common ways of representing a game is described in the following definition of the normal-form game:

A **normal-form game** consists of three features:

1. A set of **players.**
2. A set of **actions** for each player.
3. A **set of payoff functions** for each player that give a payoff value to each combination of the players' chosen actions.

This definition is similar to that of the single-person decision problem that we introduced in Chapter 1, but here we incorporate the fact that many players are interacting. Each has a set of possible actions, the combination (profile) of actions that the players choose will result in an outcome, and each has a payoff from the resulting outcome.

We now introduce the commonly used concept of a **strategy.** A strategy is often defined as *a plan of action intended to accomplish a specific goal.* Imagine a candidate in a local election going to meet a group of potential voters at the home of a neighborhood supporter. Before the meeting, our aspiring politician should have a plan of action to deal with the possible questions he will face. We can think of this plan as a list of the form "if they ask me question q_1 then I will respond with answer a_1; if they ask me question q_2 then I will respond with answer a_2; . . . " and so on. A different candidate may, and often will, have a different strategy of this kind.

The concept of a strategy will escort us throughout this book, and for this reason we now give it both formal notation and a definition:

Definition 3.2 A **pure strategy** for player i is a deterministic plan of action. The set of all **pure strategies** for player i is denoted S_i. A **profile of pure strategies** $s = (s_1, s_2, \ldots, s_n)$, $s_i \in S_i$ for all $i = 1, 2, \ldots, n$, describes a particular combination of pure strategies chosen by all n players in the game.

A brief pause to consider the term "pure" is in order. As mentioned earlier, for the time being and until we reach Chapter 6, we restrict our attention to the case in which players choose deterministic actions. This is what we mean by "pure" strategies: you choose a *certain* plan of action. To illustrate this idea, imagine that you have an exam in three hours, and you must decide how long to study for the exam and how long to just relax, knowing that your classmates are facing the same choice. If, say, you measure time in intervals of 15 minutes, then there are a total of 12 time units in the three-hour window. Your set of pure strategies is then $S_i = \{1, 2, \ldots, 12\}$, where each $s_i \in S_i$ determines how many 15-minute units you will spend studying for the exam. For example, if you choose $s_i = 7$ then you will spend 1 hour and 45 minutes studying and 1 hour and 15 minutes relaxing. An alternative to choosing one of your pure strategies would be for you to choose actions *stochastically*. For example, you can take a die and say "I will roll the die and study for as many 15-minute units as the number on the die indicates." This means that you are stochastically (or randomly) choosing between *any one* of the six pure strategies of studying for 15 minutes, 30 minutes, and so on for up to 1 hour and 30 minutes.

You may wonder why anyone would choose randomly among plans of action. As an example, dwell on the following situation. You meet a friend to go to lunch. Your strategy can be to offer the names of two restaurants that you like and then have your friend decide. But what should you do if he says, "You go ahead and choose"? One option is for you to be prepared with a choice. Another is for you to take out a coin and flip it, so that it is *not you* who is choosing; instead you are randomizing between the two choices.[1] For now, we will restrict attention to pure strategies in which such stochastic play is not possible. That said, stochastic choices play a critical role in game theory. We will introduce *stochastic* or *mixed strategies* in Chapter 6 and continue to use them throughout the rest of the book.

To some extent applying the concept of a strategy or a plan of action to a static game of complete information is overkill, because the players choose actions *once and for all* and *simultaneously*. Thus the only set of relevant plans for player i is the set of his possible actions. This change of focus from actions to strategies may therefore seem redundant. That said, focusing on strategies instead of actions will set the stage for games in which there will be relevance to conditioning one's actions on events that unfold over time, as we will see in Chapter 7. Hence what now seems merely semantic will later be quite useful and important. We now formally define a normal-form game as follows.[2]

Definition 3.3 A **normal-form game** includes three components as follows:

1. A finite **set of players,** $N = \{1, 2, \ldots, n\}$.
2. A **collection of sets of pure strategies,** $\{S_1, S_2, \ldots, S_n\}$.
3. A **set of payoff functions,** $\{v_1, v_2, \ldots, v_n\}$, each assigning a payoff value to each combination of chosen strategies, that is, a set of functions $v_i : S_1 \times S_2 \times \cdots \times S_n \to \mathbb{R}$ for each $i \in N$.

This representation is very general, and it will capture many situations in which each of the players $i \in N$ must simultaneously choose a possible strategy $s_i \in S_i$. Recall again that by *simultaneous* we mean the more general construct in which

---

1. From my experience, once you offer to take out the coin then your friend is very likely to say, "Oh never mind, let's go to x." By taking out the coin you are effectively telling your friend, "If you have a preference for one of the places, now is your last chance to reveal it." This takes away your friend's option of "being nice" by letting you choose since it is not you who is choosing. I always find this strategy amusing since it works so well.

2. Recall that a finite set of elements will be written as $A = \{a, b, c, d\}$, where A is the set and a, b, c, and d are the elements it includes. Writing $a \in A$ means "a is an element of the set A." If we have two sets, A and B, we define the *Cartesian product* of these sets as $A \times B$. If $a \in A$ and $h \in B$ then we can write $(a, h) \in A \times B$. For more on this subject, refer to Section 19.1 of the mathematical appendix.

each player is choosing a strategy without knowing the choices of the other players. After strategies are selected, each player will realize his payoff, given by $v_i(s_1, s_2, \ldots, s_n) \in \mathbb{R}$, where $(s_1, s_2, \ldots, s_n)$ is the strategy profile that was selected by the agents. Thus from now on the normal-form game will be a triple of sets: $\langle N, \{S_i\}_{i=1}^{n}, \{v_i(\cdot)\}_{i=1}^{n} \rangle$, where N is the set of players, $\{S_i\}_{i=1}^{n}$ is the set of all players' strategy sets, and $\{v_i(\cdot)\}_{i=1}^{n}$ is the set of all players' payoff functions over the strategy profiles of all the players.[3]

3.1.1 Example: The Prisoner's Dilemma

The Prisoner's Dilemma is perhaps the best-known example in game theory, and it often serves as a parable for many different applications in economics and political science. It is a static game of complete information that represents a situation consisting of two individuals (the players) who are suspects in a serious crime, say, armed robbery. The police have evidence of only petty theft, and to nail the suspects for the armed robbery they need testimony from at least one of the suspects.

The police decide to be clever, separating the two suspects at the police station and questioning each in a different room. Each suspect is offered a deal that reduces the sentence he will get if he confesses, or "finks" (F), on his partner in crime. The alternative is for the suspect to say nothing to the investigators, or remain "mum" (M), so that they do not get the incriminating testimony from him. (As the Mafia would put it, the suspect follows the "omertà"—the code of silence.)

The payoff of each suspect is determined as follows: If both choose mum, then both get 2 years in prison because the evidence can support only the charge of petty theft. If, say, player 1 mums while player 2 finks, then player 1 gets 5 years in prison while player 2 gets only 1 year in prison for being the sole cooperator. The reverse outcome occurs if player 1 finks while player 2 mums. Finally, if both fink then both get only 4 years in prison. (There is some reduction of the 5-year sentence because each would blame the other for being the mastermind behind the robbery.)

Because it is reasonable to assume that more time in prison is worse, we use the payoff representation that equates each year in prison with a value of -1. We can now represent this game in its normal form as follows:

Players: $N = \{1, 2\}$.

Strategy sets: $S_i = \{M, F\}$ for $i \in \{1, 2\}$.

Payoffs: Let $v_i(s_1, s_2)$ be the payoff to player i if player 1 chooses s_1 and player 2 chooses s_2. We can then write payoffs as

$$v_1(M, M) = v_2(M, M) = -2$$
$$v_1(F, F) = v_2(F, F) = -4$$
$$v_1(M, F) = v_2(F, M) = -5$$
$$v_1(F, M) = v_2(M, F) = -1.$$

This completes the normal-form representation of the Prisoner's Dilemma. We will soon analyze how rational players would behave if they were faced with this game.

---

3. $\{S_i\}_{i=1}^{n}$ is another way of writing $\{S_1, S_2, \ldots, S_n\}$, and similarly for $\{v_i(\cdot)\}_{i=1}^{n}$.

3.1.2   Example: Cournot Duopoly

A variant of this example was first introduced by Augustin Cournot (1838). Two identical firms, players 1 and 2, produce some good. Assume that there are no fixed costs of production, and let the variable cost to each firm i of producing quantity $q_i \geq 0$ be given by the cost function, $c_i(q_i) = q_i^2$ for $i \in \{1, 2\}$. Demand is given by the function $q = 100 - p$, where $q = q_1 + q_2$. Cournot starts with the benchmark of firms that operate in a competitive environment in which each firm takes the market price, p, as given, and believes that its behavior cannot influence the market price. Under this assumption, as every economist knows, the solution will be the competitive equilibrium in which each firm produces at a point at which price equals marginal costs, so that the profits on the marginally produced unit are zero. In this particular case, each firm would produce $q_i = 25$, the price would be $p = 50$, and each firm would make 625 in profits.[4]

Cournot then argues that this competitive equilibrium is naive because rational firms should understand that the price is not given, but rather determined by their actions. For example, if firm 1 realizes its effect on the market price, and produces $q_1 = 24$ instead of $q_1 = 25$, then the price will have to increase to $p(49) = 51$ for demand to equal supply because total supply will drop from 50 to 49. The profits of firm 1 will now be $v_1 = 51 \times 24 - 24^2 = 648 > 625$. Of course, if firm 1 realizes that it has such an effect on price, it should not just set $q_1 = 24$ but instead look for the best choice it can make. However, its best choice depends on the quantity that firm 2 will produce—what will that be? Clearly firm 2 should be as sophisticated, and thus we will have to find a solution that considers both *the actions and the counteractions* of these rational and sophisticated firms.

For now, however, let's focus on the representation of the normal form of the game proposed by Cournot. The actions are choices of quantity, and the payoffs are the profits. Hence the following represents the normal form:

Players: $N = \{1, 2\}$.

Strategy sets: $S_i = [0, \infty]$ for $i \in \{1, 2\}$ and firms choose quantities $s_i \in S_i$.

Payoffs: For $i, j \in \{1, 2\}$, $i \neq j$,

$$v_i(s_i, s_j) = \begin{cases} (100 - s_i - s_j)s_i - s_i^2 & \text{if } s_i + s_j < 100 \\ -s_i^2 & \text{if } s_i + s_j \geq 100. \end{cases}$$

Notice that the payoff function is a little tricky because it has to be well defined for *any* pair of strategies (quantities) that the players choose. We are implicitly assuming that prices cannot fall below zero, so that if the firms together produce a quantity that is greater than 100, the price will be zero (because $p = 100 - s_1 - s_2$) and each firm's payoffs are its costs.

3.1.3   Example: Voting on a New Agenda

Consider three players on a committee who have to vote on whether to remain at the status quo (whatever it is) or adopt a new policy. For example, they could be three

---

4. Those who have taken a course in microeconomics know that the marginal cost is the derivative of the cost function and hence is equal to $2q_i$. Equating this to the price gives us each firm's supply function, $2q_i = p$ or $q_i = \frac{p}{2}$, and adding up the two supply functions yields the market supply, $q = p$. Equating this to demand yields $p = 100 - p$, resulting in the competitive price of $p = 50$, and plugging this into the supply function yields $q_i = 25$ for $i = 1, 2$.

housemates who currently have an agreement under which they clean the house once every two weeks (the status quo) and they are considering cleaning it every week (the new policy). They could also be the members of the board of a firm who have to vote on changing the CEO's compensation, or they could be a committee of legislators who must vote on whether to adopt new regulations.

Each can vote "yes" (Y), "no" (N), or "abstain" (A). We can set the payoff from the status quo to be 0 for each player. Players 1 and 2 prefer the new policy, so let their payoff value for it be 1, while player 3 dislikes the new policy, so let his payoff from it be -1. Assume that the choice is made by majority voting as follows: if there is no majority of Y over N then the status quo prevails; otherwise the majority is decisive.

We can represent this game in normal form as follows:

Players: $N = \{1, 2, 3\}$.

Strategy sets: $S_i = \{Y, N, A\}$ for $i \in \{1, 2, 3\}$.

Payoffs: Let P denote the set of strategy profiles for which the new agenda is chosen (at least two "yes" votes), and let Q denote the set of strategy profiles for which the status quo remains (no more than one "yes" vote). Formally,

$$P = \left\{ \begin{array}{ll} (Y, Y, N), & (Y, N, Y), \\ (Y, Y, A), & (Y, A, Y), \\ (Y, A, A), & (A, Y, A), \\ (Y, Y, Y), & (N, Y, Y), \\ (A, Y, Y), & (A, A, Y) \end{array} \right\} \quad \text{and}$$

$$Q = \left\{ \begin{array}{llll} (N, N, N), & (N, N, Y), & (N, Y, N), & (Y, N, N), \\ (A, A, A), & (A, A, N), & (A, N, A), & (N, A, A), \\ (A, Y, N), & (A, N, Y), & (N, A, Y), & (Y, A, N), \\ (N, Y, A), & (Y, N, A), & (N, N, A), & (N, A, N), \\ (A, N, N) \end{array} \right\}.$$

Then payoffs can be written as

$$v_i(s_1, s_2, s_3) = \left\{ \begin{array}{ll} 1 & \text{if } (s_1, s_2, s_3) \in P \\ 0 & \text{if } (s_1, s_2, s_3) \in Q \end{array} \right. \quad \text{for } i \in \{1, 2\},$$

$$v_3(s_1, s_2, s_3) = \left\{ \begin{array}{ll} -1 & \text{if } (s_1, s_2, s_3) \in P \\ 0 & \text{if } (s_1, s_2, s_3) \in Q. \end{array} \right.$$

This completes the normal-form representation of the voting game.

3.2 Matrix Representation: Two-Player Finite Game

As the voting game demonstrates, games that are easy to describe verbally can sometimes be tedious to describe formally. The value of a formal representation is clarity, because it forces us to specify who the players are, what they can do, and how their actions affect each and every player. We could take some shortcuts to make our life easier, and sometimes we will, but such convenience can come at the cost of misspecifying the game. It turns out that for two-person games in which each player has a finite number of strategies, there is a convenient representation that is easy to read.

In many cases, players may be constrained to choose one of a finite number of actions. This is the case for the Prisoner's Dilemma, rock-paper-scissors, the voting game described previously, and many more strategic situations. In fact, even when players have infinitely many actions to choose from, we may be able to provide a good approximation by restricting attention to a finite number of actions. If we think of the Cournot duopoly example, then for any product that comes in well-defined units (a car, a computer, or a shirt), we can safely assume that we are limited to integer units (an assumption that reduces the strategy set to the natural numbers—after all, fractional shirts will not sell very well). Furthermore, the demand function $p = 100 - q$ suggests that flooding the market with more than 100 units will cause the price of the product to drop to zero. This means that we have effectively restricted the strategy set to a finite number of strategies (101, to be accurate, for the quantities $0, 1, \ldots, 100$).

Being able to distinguish games with finite action sets is useful, so we define a finite game as follows:

Definition 3.4 A **finite game** is a game with a finite number of players, in which the number of strategies in S_i is finite for all players $i \in N$.

As it turns out, any two-player finite game can be represented by a matrix that will capture all the relevant information of the normal-form game. This is done as follows:

Rows Each row represents one of player 1's strategies. If there are k strategies in S_1 then the matrix will have k rows.

Columns Each column represents one of player 2's strategies. If there are m strategies in S_2 then the matrix will have m columns.

Matrix entries Each entry in this matrix contains a two-element vector (v_1, v_2), where v_i is player i's payoff when the actions of both players correspond to the row and column of that entry.

As the following examples show, this is a much simpler way of representing a two-player finite game because all the information will appear in a concise and clear way. Note, however, that neither the Cournot duopoly nor the voting example described earlier can be represented by a matrix. The Cournot duopoly is not a finite game (there are an infinite number of actions for each player), and the voting game has more than two players.[5]

It will be useful to illustrate this with two familiar examples.

3.2.1 Example: The Prisoner's Dilemma

Recall that in the Prisoner's Dilemma each player had two actions, M (mum) and F (fink). Therefore, our matrix will have two rows (for player 1) and two columns (for player 2). Using the payoffs for the prisoner's dilemma given in the example above, the matrix representation of the Prisoner's Dilemma is

---

5. We can represent the voting game using three 3×3 matrices: the rows of each matrix represent the actions of player 1, the columns those of player 2, and each matrix corresponds to an action of player 3. However, the convenient features of two-player matrix games are harder to use for three-player, multiple-matrix representations—not to mention the rather cumbersome structure of multiple matrices.

Player 2

	M	F
M	−2, −2	−5, −1
F	−1, −5	−4, −4

Player 1

Notice that all the relevant information appears in this matrix.

3.2.2 Example: Rock-Paper-Scissors

Consider the famous child's game rock-paper-scissors. Recall that rock (R) beats scissors (S), scissors beats paper (P), and paper beats rock. Let the winner's payoff be 1 and the loser's be -1, and let the payoff for each player from a tie (i.e., they both choose the same action) be 0. This is a game with two players, $N = \{1, 2\}$, and three strategies for each player, $S_i = \{R, P, S\}$. Given the payoffs already described, we can write the matrix representation of this game as follows:

Player 2

	R	P	S
R	0, 0	−1, 1	1, −1
P	1, −1	0, 0	−1, 1
S	−1, 1	1, −1	0, 0

Player 1

Remark Such a matrix is sometimes referred to as a *bi-matrix*. In a traditional matrix, by definition, each entry corresponding to a row-column combination must be a single number, or element, while here each entry has a vector of two elements—the payoffs for each of the two players. Thus we formally have *two matrices*, one for each player. We will nonetheless adopt the common abuse of terminology and call this a matrix.

3.3 Solution Concepts

We have focused our attention on how to describe a game formally and fit it into a well-defined structure. This approach, of course, adds value only if we can use the structure to provide some analysis of what will or should happen in the game. Ideally we would like to be able to either advise players on how to play or try to predict how players will play. To accomplish this, we need some method to *solve* the game, and in this section we outline some criteria that will be helpful in evaluating potential methods to analyze and solve games.

As an example, consider again the Prisoner's Dilemma and imagine that you are player 1's lawyer, and that you wish to advise him about how to behave. The game may be represented as follows:

Player 2

	M	F
M	−2, −2	−5, −1
F	−1, −5	−4, −4

Player 1

Being a thoughtful and rational adviser, you make the following observation for player 1: "If player 2 chooses F, then playing F gives you -4, while playing M gives you -5, so F is better." Player 1 will then bark at you, "My buddy will never squeal on me!" You, however, being a loyal adviser, must coolly reply as follows: "If you're right, and player 2 chooses M, then playing F gives you -1, while playing M gives you -2, so F is still better. In fact, it seems like F is always better!"

Indeed if I were player 2's lawyer, then the same analysis would work for him, and this is the "dilemma": each player is better off playing F regardless of his opponent's actions, but this leads the players to receive payoffs of -4 each, while if they could only agree to both choose M, then they would obtain -2 each. Left to their own devices, and to the advocacy of their lawyers, the players should not be able to resist the temptation to choose F. Even if player 1 believes that player 2 will play M, he is better off choosing F (and vice versa).

Perhaps your intuition steers you to a different conclusion. You might want to say that they are friends, having stolen together for some time now, and therefore that they care for one another. In this case one of our assumptions is incorrect: the payoffs in the matrix may not represent their true payoffs, and if taken into consideration, altruism would lead both players to choose M instead of F. For example, to capture the idea of altruism and mutual caring, we can assume that a year in prison for each player is worth -1 to himself and imposes $-\frac{1}{2}$ on the other player's payoff. (You care about your friend, but not as much as you care about yourself.) In this case, if player 1 chooses F and player 2 chooses M then player 1 gets $-3\frac{1}{2}$ $\left(-\frac{1}{2}\right.$ for each of the 5 years player 2 goes to jail, and -1 for player 1's year in jail) and player 2 gets $-5\frac{1}{2}$ $\left(-\frac{1}{2}\right.$ for the year player 1 is in jail and -5 for the 5 years he spends in jail). The matrix representing the "altruistic" Prisoner's Dilemma is given by the following:

Player 2

		M	F
	M	$-3, -3$	$-5\frac{1}{2}, -3\frac{1}{2}$
Player 1 F		$-3\frac{1}{2}, -5\frac{1}{2}$	$-6, -6$

The altruistic game will predict cooperative behavior: regardless of what player 2 does, it is always better for player 1 to play M, and the same holds true for player 2. This shows us that our results will, as they always do, depend crucially on our assumptions.[6] This is another manifestation of the "garbage in, garbage out" caveat—we have to get the game parameters right if we want to learn something from our analysis.

Another classic game is the *Battle of the Sexes,* introduced by R. Duncan Luce and Howard Raiffa (1957) in their seminal book *Games and Decisions.* The story goes as follows. Alex and Chris are a couple, and they need to choose where to meet this evening. The catch is that the choice needs to be made while each is at work, and they have no means of communicating. (There were no cell phones or email in 1957, and even landline phones were not in abundance.) Both players prefer being together over not being together, but Alex prefers opera (O) to football (F), while Chris prefers the opposite. This implies that for each player being together at the venue of choice is

---

6. Another change in assumptions might be that player 2's brother is a psychopath. If player 1 finks, then player 2's brother will kill him, giving player 1 a utility of, say, $-\infty$ from choosing to fink.

better than being together at the other place, and this in turn is better than being alone. Using the payoffs of 2, 1 and 0 to represent this order, the game is summarized in the following matrix:

		Chris	
		O	F
Alex	O	2, 1	0, 0
	F	0, 0	1, 2

What can you recommend to each player now? Unlike the situation in the Prisoner's Dilemma, the best action for Alex depends on what Chris will do and vice versa. If we want to predict or prescribe actions for this game, we need to make assumptions about the *behavior* and the *beliefs* of the players. We therefore need a **solution concept** that will result in predictions or prescriptions.

A solution concept is a method of analyzing games with the objective of restricting the set of *all possible outcomes* to those that are *more reasonable than others*. That is, we will consider some reasonable and consistent assumptions about the behavior and beliefs of players that will divide the space of outcomes into "more likely" and "less likely." Furthermore, we would like our solution concept to apply to a large set of games so that it is widely applicable.

Consider, for example, the solution concept that prescribes that each player choose the action that is always best, regardless of what his opponents will choose. As we saw earlier in the Prisoner's Dilemma, playing F is always better than playing M. Hence this solution concept will predict that in this game both players will choose F. For the Battle of the Sexes, however, there is *no strategy* that is always best: playing F is best if your opponent plays F, and playing O is best if your opponent plays O. Hence for the Battle of the Sexes, this simple solution concept is not useful and offers no guidance.

We will use the term **equilibrium** for any one of the strategy profiles that emerges as one of the solution concept's predictions. We will often think of equilibria as the *actual predictions* of our theory. A more forgiving meaning would be that equilibria are the *likely predictions,* because our theory will often not account for all that is going on. Furthermore, in some cases we will see that more than one equilibrium prediction is possible for the same game. In fact, this will sometimes be a strength, and not a weakness, of the theory.

3.3.1 Assumptions and Setup

To set up the background for equilibrium analysis, it is useful to revisit the assumptions that we will be making throughout:

1. **Players are "rational":** A *rational* player is one who chooses his action, $s_i \in S_i$, to maximize his payoff consistent with his beliefs about what is going on in the game.

2. **Players are "intelligent":** An *intelligent* player knows everything about the game: the actions, the outcomes, and the preferences of all the players.

3. **Common knowledge:** The fact that players are rational and intelligent is common knowledge among the players of the game.

To these three assumptions, which we discussed briefly at the beginning of this chapter, we add a fourth, which constrains the set of outcomes that are reasonable:

4. **Self-enforcement:** Any prediction (or equilibrium) of a solution concept must be *self-enforcing.*

The requirement that any equilibrium must be self-enforcing is at the core of our analysis and at the heart of **noncooperative game theory.** We will assume throughout this book that the players engage in noncooperative behavior in the following sense: each player is in control of his own actions, and he will stick to an action only if he finds it to be in his best interest. That is, if a profile of strategies is to be an equilibrium, we will require each player to be happy with his own choice given how the others make their own choices. As you can probably figure out, the profile (F, F) is self-enforcing in the Prisoner's Dilemma game: each player is happy playing F. Indeed, we will see that this is a very robust outcome in terms of equilibrium analysis.

The requirement of self-enforcing equilibria is a natural one if we take the game to be the complete description of the environment. If there are outside parties that can, through the use of force or sanctions, enforce profiles of strategies, then the game we are using is likely to be an inadequate depiction of the actual environment. In this case we ought to model the third party as a player who has actions (strategies) that describe the enforcement.

3.3.2 Evaluating Solution Concepts

In developing a theory that predicts the behavior of players in games, we must evaluate our theory by how well it does as a *methodological* tool. That is, for our theory to be widely useful, it must describe a method of analysis that applies to a rich set of games, which describe the strategic situations in which we are interested. We will introduce three criteria that will help us evaluate a variety of solution concepts: existence, uniqueness, and invariance.

3.3.2.1 Existence: How Often Does It Apply? A solution concept is valuable insofar as it applies to a wide variety of games, and not just to a small and select family of games. A solution concept should apply generally and should not be developed in an ad hoc way that is specific to a certain situation or game. That is, when we apply our solution concept to different games we require it to result in the *existence* of an equilibrium solution.

For example, consider an ad hoc solution concept that offers the following prediction: "Players always choose the action that they think their opponent will choose." If this is our "theory" of behavior, then it will fail to apply to many—maybe most—strategic situations. In particular when players have different sets of actions (e.g., one chooses a software package and the other a hardware package) then this theory would be unable to predict which outcomes are more likely to emerge as equilibrium outcomes.

Any proposed theory for a solution concept that relies on very specific elements of a game will not be general and will be hard to adapt to a wide variety of strategic situations, making the proposed theory useless beyond the very special situations it was tailored to address. Thus one goal is to have a method that will be general enough to apply to many strategic situations; that is, it will prescribe a solution that will *exist* for most games we can think of.

3.3.2.2 Uniqueness: How Much Does It Restrict Behavior? Just as we require our solution concept to apply broadly, we require that it be meaningful in that it restricts the set of possible outcomes to a smaller set of reasonable outcomes. In fact one might argue that being able to pinpoint a single *unique* outcome as a prediction would be ideal. *Uniqueness* is then an important counterpart to *existence.*

For example, if the proposed solution concept says "anything can happen," then it always exists: regardless of the game we apply this concept to, "anything can happen" will always say that the solution is one of the (sometimes infinite) possible outcomes. Clearly this solution concept is useless. A good solution concept is one that balances existence (so that it works for many games) with uniqueness (so that we can add some intelligent insight into what can possibly happen).

It turns out that the nature of games makes the uniqueness requirement quite hard to meet. The reason, as we will learn to appreciate, lies in the nature of strategic interaction in a noncooperative environment. To foreshadow the reasons behind this observation, notice that a player's best action will often depend on what other players are doing. A consequence is that there will often be several combinations of strategies that will support each other in this way.

3.3.2.3 *Invariance: How Sensitive Is It to Small Changes?*

Aside from existence and uniqueness, a third more subtle criterion is important in qualifying a solution concept as a reasonable one, namely that the solution concept be *invariant* to small changes in the game's structure. However, the term "small changes" needs to be qualified more precisely.

Adding a player to a game, for instance, may not be a small change if that player has actions that can wildly change the outcomes of the game. Thus adding or removing a player cannot innocuously be considered a small change. Similarly if we add or delete strategies from the set of actions that are available to a player, we may hinder his ability to guarantee himself some outcomes, and therefore this too should not be considered a small change to the game. We are left with only one component to fiddle with: the payoff functions of the players. It is reasonable to argue that if the payoffs of a game are modified only slightly, then this is a small change to the game that should not affect the predictions of a "robust" solution concept.

For example, consider the Prisoner's Dilemma. If instead of 5 years in prison, imposing a pain of -5 for the players, it imposed a pain of -5.01 for player 1 and -4.99 for player 2, we should be somewhat discouraged if our solution concept suddenly changed the prediction of what players will or ought to do. Thus *invariance* is a robustness property with which we require a solution concept to comply. In other words, if two games are "close," so that the action sets and players are the same yet the payoffs are only slightly different, then our solution concept should offer predictions that are not wildly different for the two games. Put formally, if for a small enough value $\varepsilon > 0$ we alter the payoffs of every outcome for every player by no more than ε, then the solution concept's prediction should not change.

3.3.3 Evaluating Outcomes

Once we subscribe to any particular solution concept, as social scientists we would like to evaluate the properties of the *solutions,* or predictions, that the solution concept will prescribe. This process will offer insights into what we expect the players of a game to achieve when they are left to their own devices. In turn, these insights can guide us toward possibly changing the environment of the game so as to improve the social outcomes of the players.

We have to be precise about the meaning of "to improve the social outcomes." For example, many people may agree that it would be socially better for the government to take $10 away from the very rich Bill Gates and give that $10 to an orphan in Latin America. In fact even Gates himself might have approved of this transfer, especially if the money would have saved the child's life. However, Gates may or may not have *liked* the idea, especially if such government intervention would imply that over time most of his wealth would be dissipated through such transfers.

Economists use a particular criterion for evaluating whether an outcome is *socially undesirable*. An outcome is considered to be socially undesirable if there is a different outcome that would make some people better off *without harming anyone else*. As social scientists we wish to avoid outcomes that are socially undesirable, and we therefore turn to the criterion of *Pareto optimality,* which is in tune with the idea of efficiency or "no waste." That is, we would like all the possible value deriving from a given interaction to be distributed among the players. To put this formally:[7]

Definition 3.5 A strategy profile $s \in S$ **Pareto dominates** strategy profile $s' \in S$ if $v_i(s) \geq v_i(s') \forall i \in N$ and $v_i(s) > v_i(s')$ for at least one $i \in N$ (in which case, we will also say that s' is **Pareto dominated** by s). A strategy profile is **Pareto optimal** if it is not Pareto dominated by any other strategy profile.

As social scientists, strategic advisers, or policy makers, we hope that players will act in accordance with the *Pareto criterion* and find ways to coordinate on Pareto-optimal outcomes, or avoid those that are Pareto dominated.[8] However, as we will see time and time again, this result will not be achievable in many games. For example, in the Prisoner's Dilemma we made the case that (F, F) should be considered as a very likely outcome. In fact, as we will argue several times, it is the *only* likely outcome. One can see, however, that it is Pareto dominated by (M, M). (Notice that (M, M) is not the only Pareto-optimal outcome. (M, F) and (F, M) are also Pareto-optimal outcomes because no other profile dominates any of them. Don't confuse Pareto optimality with the best "symmetric" outcome that leaves all players "equally" happy.)

3.4 Summary

- A normal-form game includes a finite set of players, a set of pure strategies for each player, and a payoff function for each player that assigns a payoff value to each combination of chosen strategies.

- Any two-player finite game can be represented by a matrix. Each row represents one of player 1's strategies, each column represents one of player 2's strategies, and each cell in the matrix contains the payoffs for both players.

- A solution concept that proposes predictions of how games will be played should be widely applicable, should restrict the set of possible outcomes to a small set of reasonable outcomes, and should not be too sensitive to small changes in the game.

- Outcomes should be evaluated using the Pareto criterion, yet self-enforcing behavior will dictate the set of reasonable outcomes.

---

7. The symbol $\forall$ denotes "for all."

8. The criterion is named after the Italian economist Vilfredo Pareto. In general economists and other advocates of rational choice theory view this criterion as noncontroversial. However, this view is not necessarily held by everyone. For example, consider two outcomes: In the first, two players get $5 each. In the second, player 1 gets $6 while player 2 gets $60. The Pareto criterion clearly prefers the second outcome, but some other social criterion with equity considerations may disagree with this ranking.

3.5   Exercises

3.1   **eBay:** Hundreds of millions of people bid on eBay auctions to purchase goods from all over the world. Despite being carried out on line, in spirit these auctions are similar to those that have been conducted for centuries. Is an auction a game? Why or why not?

3.2   **Penalty Kicks:** Imagine a kicker and a goalie who confront each other in a penalty kick that will determine the outcome of a soccer game. The kicker can kick the ball left or right, while the goalie can choose to jump left or right. Because of the speed of the kick, the decisions need to be made simultaneously. If the goalie jumps in the same direction as the kick, then the goalie wins and the kicker loses. If the goalie jumps in the opposite direction of the kick, then the kicker wins and the goalie loses. Model this as a normal-form game and write down the matrix that represents the game you modeled.

3.3   **Meeting Up:** Two old friends plan to meet at a conference in San Francisco, and they agree to meet by "the tower." After arriving in town, each realizes that there are two natural choices: Sutro Tower or Coit Tower. Not having cell phones, each must choose independently which tower to go to. Each player prefers meeting up to not meeting up, and neither cares where this would happen. Model this as a normal-form game and write down the matrix form of the game.

3.4   **Hunting:** Two hunters, players 1 and 2, can each choose to hunt a stag, which provides a rather large and tasty meal, or hunt a hare—also tasty, but much less filling. Hunting stags is challenging and requires mutual cooperation. If either hunts a stag alone, then the stag will get away, while hunting the stag together guarantees that the stag will be caught. Hunting hares is an individualistic enterprise that is not done in pairs, and whoever chooses to hunt a hare will catch one. The payoff from hunting a hare is 1, while the payoff to each from hunting a stag together is 3. The payoff from an unsuccessful stag hunt is 0. Represent this game as a matrix.

3.5   **Matching Pennies:** Players 1 and 2 each put a penny on a table simultaneously. If the two pennies come up the same side (heads or tails) then player 1 gets both pennies; otherwise player 2 gets both pennies. Represent this game as a matrix.

3.6   **Price Competition:** Imagine a market with demand $p(q) = 100 - q$. There are two firms, 1 and 2, and each firm i has to simultaneously choose its price p_i. If $p_i < p_j$, then firm i gets all of the market while no one demands the good of firm j. If the prices are the same then both firms split the market demand equally. Imagine that there are no costs to produce any quantity of the good. (These are two large dairy farms, and the product is manure.) Write down the normal form of this game.

3.7   **Public Good Contribution:** Three players live in a town, and each can choose to contribute to fund a streetlamp. The value of having the streetlamp is 3 for each player, and the value of not having it is 0. The mayor asks each player to contribute either 1 or nothing. If at least two players contribute then the lamp will be erected. If one player or no players contribute then the lamp will not be erected, in which case any person who contributed will not get his money back. Write down the normal form of this game.

Rationality and
Common Knowledge

In this chapter we study the implications of imposing the assumptions of rationality as well as common knowledge of rationality. We derive and explore some solution concepts that result from these two assumptions and seek to understand the restrictions that each of the two assumptions imposes on the way in which players will play games.

4.1 Dominance in Pure Strategies

It will be useful to begin by introducing some new notation. We denoted the payoff of a player i from a profile of strategies $s = (s_1, s_2, \ldots, s_{i-1}, s_i, s_{i+1}, \ldots, s_n)$ as $v_i(s)$. It will soon be very useful to refer specifically to the strategies of a player's opponents in a game. For example, the actions chosen by the players who are *not* player i are denoted by the profile

$$(s_1, s_2, \ldots, s_{i-1}, s_{i+1}, \ldots, s_n) \in S_1 \times S_2 \times \cdots \times S_{i-1} \times S_{i+1} \times \cdots \times S_n.$$

To simplify we will hereafter use a common shorthand notation as follows: We define $S_{-i} \equiv S_1 \times S_2 \times \cdots \times S_{i-1} \times S_{i+1} \times \cdots \times S_n$ as the set of all the strategy sets of *all players who are not* player i. We then define $s_{-i} \in S_{-i}$ as a particular possible profile of strategies for all players who are not i. Hence we can rewrite the payoff of player i from strategy s as $v_i(s_i, s_{-i})$, where $s = (s_i, s_{-i})$.

4.1.1 Dominated Strategies

The Prisoner's Dilemma was easy to analyze: each of the two players has an action that is best *regardless* of what his opponent chooses. Suggesting that each player will choose this action seems natural because it is consistent with the basic concept of rationality. If we assume that the players are rational, then we should expect them to choose whatever they deem to be best for them. If it turns out that a player's best strategy does not depend on the strategies of his opponents then we should be all the more confident that he will choose it.

It is not too often that we will find ourselves in situations in which we have a best action that does not depend on the actions of our opponents. We begin, therefore, with a less demanding concept that follows from rationality. In particular consider the strategy mum in the Prisoner's Dilemma:

		Player 2	
		M	F
Player 1	M	−2, −2	−5, −1
	F	−1, −5	−4, −4

As we argued earlier, playing M is worse than playing F for each player *regardless* of what the player's opponent does. What makes it unappealing is that there is another strategy, F, that is better than M regardless of what one's opponent chooses. We say that such a strategy is *dominated*. Formally we have

Definition 4.1 Let $s_i \in S_i$ and $s_i' \in S_i$ be possible strategies for player i. We say that s_i' is **strictly dominated** by s_i if for any possible combination of the other players' strategies, $s_{-i} \in S_{-i}$, player i's payoff from s_i' is strictly less than that from s_i. That is,

$$v_i(s_i, s_{-i}) > v_i(s_i', s_{-i}) \quad \text{for all } s_{-i} \in S_{-i}.$$

We will write $s_i \succ_i s_i'$ to denote that s_i' is strictly dominated by s_i.

Now that we have a precise definition for a dominated strategy, it is straightforward to draw an obvious conclusion:

Claim 4.1 *A rational player will never play a strictly dominated strategy.*

This claim is obvious. If a player plays a dominated strategy then he cannot be playing optimally because, by the definition of a dominated strategy, the player has another strategy that will yield him a higher payoff *regardless of the strategies of his opponents*. Hence knowledge of the game implies that a player should recognize dominated strategies, and rationality implies that these strategies will be avoided.

When we apply the notion of a dominated strategy to the Prisoner's Dilemma we argue that each of the two players has one dominated strategy that he should never use, and hence each player is left with one strategy that is not dominated. Therefore, for the Prisoner's Dilemma, rationality alone is enough to offer a prediction about which outcome will prevail: (F, F) is this outcome.

Many games, however, will not be as special as the Prisoner's Dilemma, and rationality alone will not suggest a clear-cut, unique prediction. As an example, consider the following advertising game. Two competing brands can choose one of three marketing campaigns—low (L), medium (M), and high (H)—with payoffs given by the following matrix:

		Player 2		
		L	M	H
	L	6, 6	2, 8	0, 4
Player 1	M	8, 2	4, 4	1, 3
	H	4, 0	3, 1	2, 2

It is easy to see that each player has one dominated strategy, which is L. However, neither M nor H is dominated. For example, if player 2 plays M then player 1 should also play M, while if player 2 plays H then player 1 should also play H. Hence rationality alone does not offer a unique prediction. It is nonetheless worth spending some time on the extreme cases in which it does.

4.1.2 Dominant Strategy Equilibrium

Because a strictly dominated strategy is one to avoid at all costs,[1] there is a counterpart strategy, represented by F in the Prisoner's Dilemma, that would be desirable. This is a strategy that is *always* the best thing you can do, regardless of what your opponents choose. Formally we have

Definition 4.2 $s_i \in S_i$ is a **strictly dominant strategy** for i if every other strategy of i is strictly dominated by it, that is,

$$v_i(s_i, s_{-i}) > v_i(s_i', s_{-i}) \quad \text{for all } s_i' \in S_i, \quad s_i' \neq s_i, \quad \text{and all } s_{-i} \in S_{-i}.$$

If, as in the Prisoner's Dilemma, every player had such a wonderful dominant strategy, then it would be a very sensible predictor of behavior because it follows from rationality alone. We can introduce this simple idea as our first solution concept:

Definition 4.3   The strategy profile $s^D \in S$ is a **strict dominant strategy equilibrium** if $s_i^D \in S_i$ is a strict dominant strategy for all $i \in N$.

This gives a formal definition for the outcome "both players fink," or (F, F), in the Prisoner's Dilemma: it is a dominant strategy equilibrium. In this equilibrium the payoffs are $(-4, -4)$ for players 1 and 2, respectively.

Caveat   Be careful not to make a common error by referring to the pair of payoffs $(-4, -4)$ as the solution. The solution should always be described as the strategies that the players will choose. Strategies are a set of actions by the players, and payoffs are a result of the outcome. When we talk about predictions, or equilibria, we will always refer to *what players do* as the equilibrium, not their payoffs.

Using this solution concept for any game is not that difficult. It basically requires that we identify a strict dominant strategy for each player and then use this profile of strategies to predict or prescribe behavior. If, as in the Prisoner's Dilemma, we are lucky enough to find a dominant strategy equilibrium for other games, then this solution concept has a very appealing property:

Proposition 4.1   *If the game* $\Gamma = \langle N, \{S_i\}_{i=1}^n, \{v_i\}_{i=1}^n \rangle$ *has a strictly dominant strategy equilibrium* s^D, *then* s^D *is the unique dominant strategy equilibrium.*

This proposition is rather easy to prove, and that proof is left as exercise 4.1 at the end of the chapter.

---

1. This is a good point at which to stop and reflect on a very simple yet powerful lesson. In any situation, look first for your dominated strategies and avoid them!

4.1.3 Evaluating Dominant Strategy Equilibrium

Proposition 4.1 is very useful in addressing one of our proposed criteria for evaluating a solution concept: when it exists, the strict-dominance solution concept guarantees uniqueness. However, what do we know about existence? A quick observation will easily convince you that this is a problem.

Consider the Battle of the Sexes game introduced in Section 3.3 and described again in the following matrix:

		Chris	
		O	F
Alex	O	2, 1	0, 0
	F	0, 0	1, 2

Neither player has a dominated strategy, implying that neither has a dominant strategy either. The best strategy for Chris depends on what Alex chooses and vice versa. Thus if we stick to the solution concept of strict dominance we will encounter games, in fact many of them, for which there will be no equilibrium. This unfortunate conclusion implies that the strict-dominance solution concept will often fail to predict the choices that players ought to, or will, choose in many games.

Regarding the invariance criterion, the strict-dominance solution concept does comply. From definition 4.2, s^D is a strictly dominant strategy equilibrium if and only if $v_i(s_i^D, s_{-i}) > v_i(s_i', s_{-i})$ for all $s_i' \in S_i$ and all $s_{-i} \in S_{-i}$. Because the inequality is strict, we can find a small enough value $\varepsilon > 0$ such that if we either add or subtract ε from any payoff $v_i(s_i, s_{-i})$ the inequality will still hold.

We now turn to the Pareto criterion of equilibrium outcomes when a strict-dominance solution exists. The Prisoner's Dilemma has an equilibrium prediction using the strict-dominance solution concept, so we can evaluate the efficiency properties of the unique strictly dominant strategy equilibrium for that game. It is easy to see that the outcome prescribed by this solution is *not Pareto optimal:* both players would be better off if they could each commit to play M, yet left to their own devices they will not do this. Of course, in other games the solution may be Pareto optimal (see, for example, the "altruistic" Prisoner's Dilemma in Section 3.3).

The failure of Pareto optimality is *not* a failure of the solution concept. The assumption that players are rational causes them to fink in the Prisoner's Dilemma if we restrict attention to self-enforcing behavior. The failure of Pareto optimality implies that the players would benefit from modifying the environment in which they find themselves to create other enforcement mechanisms—for example, creating a "mafia" with norms of conduct that enforce implicit agreements so as to punish those who fink.

To explicitly see how this can work, imagine that a mafia member who finks on another member is very seriously reprimanded, which will change the payoff structure of the Prisoner's Dilemma if he is caught. If the pain from mafia punishment is equivalent to z, then we have to subtract z units of payoff for each player who finks. The "mafia-modified" Prisoner's Dilemma is represented by the following matrix:

		Player 2	
		M	F
Player 1	M	−2, −2	−5, −1 − z
	F	−1 − z, −5	−4 − z, −4 − z

If z is strictly greater than 1 then this punishment will be enough to flip our predicted equilibrium outcome of the game because then M becomes the strict dominant strategy (and (M, M) is Pareto optimal).

This example demonstrates that "institutional design," which changes the game that players play, can be very useful in affecting the well-being of players. By introducing this external enforcer, or institution, we are able to get the players to choose outcomes that make them both better off compared to what they can achieve without the additional institution. Moreover, notice that if the players believe that the mafia will enforce the code of conduct then there is no need to actually enforce it—the players choose not to fink, and the enforcement of punishments need not happen. However, we need to be suspicious of whether such an institution will be *self-enforcing*, that is, whether the mafia will indeed enforce the punishments. And, for it to be self-enforcing, we need to model the behavior of potential punishers and whether they themselves will have the selfish incentives to carry out the enforcement activities. This is something we will explore at length when we consider multistage and repeated games in Chapters Chapters 9 and 10.

Remark   A related notion is that of **weak dominance.** We say that s_i' is *weakly dominated* by s_i if, for any possible combination of the other players' strategies, player i's payoff from s_i' is weakly less than that from s_i. That is,

$$v_i(s_i, s_{-i}) \geq v_i(s_i', s_{-i}) \quad \text{for all } s_{-i} \in S_{-i}.$$

This means that for some $s_{-i} \in S_{-i}$ this weak inequality may hold strictly, while for other $s_{-i}' \in S_{-i}$ it will hold with equality. We define a strategy to be **weakly dominant** in a similar way. This is still useful because if we can find a dominant strategy for a player, be it weak or strict, this seems like the most obvious thing to prescribe. An important difference between weak and strict dominance is that if a weakly dominant equilibrium exists, it need not be unique. To show this is left as exercise 4.2.

4.2   Iterated Elimination of Strictly Dominated Pure Strategies

As we saw in the previous chapter, our requirement that players be rational implied two important conclusions:

1. A rational player will never play a dominated strategy.
2. If a rational player has a strictly dominant strategy then he will play it.

We used this second conclusion to define the solution concept of strict dominance, which is very appealing because, when it exists, it requires only rationality as its driving force. A drawback of the dominant strategy solution concept is, however, that it will often fail to exist. Hence if we wish to develop a predictive theory of behavior in games then we must consider alternative approaches that will apply to a wide variety of games.

4.2.1   Iterated Elimination and Common Knowledge of Rationality

We begin with the premise that players are rational, and we build on the first conclusion in the previous section, which claims that a rational player will never play a dominated strategy. This conclusion is by itself useful in that it rules out what players *will not do*. As a result, we conclude that rationality tells us which strategies will never be played.

Now turn to another important assumption introduced earlier: the structure of the game and the rationality of the players are common knowledge among the players. The introduction of *common knowledge of rationality* allows us to do much more than identify strategies that rational players will avoid. If indeed all the players *know* that each player will never play a strictly dominated strategy, they can effectively ignore those strictly dominated strategies that their opponents will never play, and their opponents can do the same thing. If the original game has some players with some strictly dominated strategies, then all the players know that *effectively* they are facing a "smaller" restricted game with fewer total strategies.

This logic can be taken further. Because it is *common knowledge* that all players are rational, then everyone knows that everyone knows that the game is effectively a smaller game. In this smaller restricted game, everyone knows that players will not play strictly dominated strategies. In fact we may indeed find additional strategies that are dominated in the restricted game that *were not dominated* in the original game. Because it is common knowledge that players will perform this kind of reasoning again, the process can continue until no more strategies can be eliminated in this way.

To see this idea more concretely, consider the following two-player finite game:

Player 2

		L	C	R
	U	4, 3	5, 1	6, 2
Player 1	M	2, 1	8, 4	3, 6
	D	3, 0	9, 6	2, 8

(4.1)

A quick observation reveals that there is no strictly dominant strategy, neither for player 1 nor for player 2. Also note that there is no strictly dominated strategy for player 1. There is, however, a strictly dominated strategy for player 2: the strategy C is strictly dominated by R because $2 > 1$ (row U), $6 > 4$ (row M), and $8 > 6$ (row D). Thus, because this is common knowledge, both players know that we can effectively eliminate the strategy C from player 2's strategy set, which results in the following reduced game:

Player 2

		L	R
	U	4, 3	6, 2
Player 1	M	2, 1	3, 6
	D	3, 0	2, 8

In this reduced game, both M and D are strictly dominated by U for player 1, allowing us to perform a second round of eliminating strategies, this time for player 1. Eliminating these two strategies yields the following trivial game:

Player 2

		L	R
Player 1	U	**4, 3**	6, 2

in which player 2 has a strictly dominated strategy, playing R. Thus for this example the iterated process of eliminating dominated strategies yields a unique prediction: the strategy profile we expect these players to play is (U, L), giving the players the payoffs of $(4, 3)$.

As the example demonstrates, this process of **iterated elimination of strictly dominated strategies** (IESDS) builds on the assumption of common knowledge of rationality. The first step of iterated elimination is a consequence of player 2's rationality; the second stage follows because players *know* that players are rational; the third stage follows because players *know that players know* that they are rational, and this ends in a unique prediction.

More generally we can apply this process to games in the following way. Let S_i^k denote the strategy set of player i that survives k rounds of IESDS. We begin the process by defining $S_i^0 = S_i$ for each i, the original strategy set of player i in the game.

Step 1: Define $S_i^0 = S_i$ for each i, the original strategy set of player i in the game, and set $k = 0$.

Step 2: Are there players for whom there are strategies $s_i \in S_i^k$ that are strictly dominated? If yes, go to step 3. If not, go to step 4.

Step 3: For all the players $i \in N$, remove any strategies $s_i \in S_i^k$ that are strictly dominated. Set $k = k + 1$, and define a new game with strategy sets S_i^k that *do not include* the strictly dominated strategies that have been removed. Go back to step 2.

Step 4: The remaining strategies in S_i^k are reasonable predictions for behavior.

In this chapter we refrain from giving a precise mathematical definition of the process because this requires us to consider richer behavior by the players, in particular, allowing them to choose randomly between their different pure strategies. We will revisit this approach briefly when such stochastic play, or **mixed strategies,** is introduced later.[2]

Using the process of IESDS we can define a new solution concept:

Definition 4.4   We will call any strategy profile $s^{ES} = (s_1^{ES}, \ldots, s_n^{ES})$ that survives the process of IESDS an **iterated-elimination equilibrium.**

Like the concept of a strictly dominant strategy equilibrium, the iterated-elimination equilibrium starts with the premise of rationality. However, in addition to rationality, IESDS requires a lot more: common knowledge of rationality. We will discuss the implications of this requirement later in this chapter.

4.2.2   Example: Cournot Duopoly

Recall the Cournot duopoly example we introduced in Section 3.1.2, but consider instead a simpler example of this problem in which the firms have linear rather than quadratic costs: the cost for each firm for producing quantity q_i is given by $c_i(q_i) = 10q_i$ for $i \in \{1, 2\}$. (Using economics jargon, this is a case of constant marginal cost

---

2. Just to satisfy your curiosity, think of the Battle of the Sexes, and imagine that Chris can pull out a coin and flip between the decision of opera or football. This by itself introduces a "new" strategy, and we will exploit such strategies to develop a formal definition of IESDS.

equal to 10 and no fixed costs.) Let the demand be given by $p(q) = 100 - q$, where $q = q_1 + q_2$.

Consider first the profit (payoff) function of firm 1:

$$v_1(q_1, q_2) = \overbrace{(100 - q_1 - q_2)q_1}^{\text{Revenue}} - \overbrace{10q_1}^{\text{Costs}}$$

$$= 90q_1 - q_1^2 - q_1 q_2.$$

What should firm 1 do? If it knew what quantity firm 2 will choose to produce, say some value of q_2, then the profits of firm 1 would be maximized when the first-order condition is satisfied, that is, when $90 - 2q_1 - q_2 = 0$. Thus, for any given value of q_2, firm 1 maximizes its profits when it sets its own quantity according to the function

$$q_1 = \frac{90 - q_2}{2}. \tag{4.2}$$

Though it is true that the choice of firm 1 depends on what it believes firm 2 is choosing, equation (4.2) implies that firm 1 will *never* choose to produce more than $q_1 = 45$. This follows from the simple observation that q_2 is never negative, in which case equation (4.2) implies that $q_1 \leq 45$. In fact, this is equivalent to showing that any quantity $q_1 > 45$ is strictly dominated by $q_1 = 45$. To see this, for any q_2, the profits from setting $q_1 = 45$ are given by

$$v_1(45, q_2) = (100 - 45 - q_2)45 - 450 = 2025 - 45q_2.$$

The profits from choosing any other q_1 are given by

$$v_1(q_1, q_2) = (100 - q_1 - q_2)q_1 - 10q_1 = 90q_1 - q_1 q_2 - q_1^2.$$

Thus we can subtract $v_1(q_1, q_2)$ from $v_1(45, q_2)$ and obtain

$$v_1(45, q_2) - v_1(q_1, q_2) = 2025 - 45q_2 - (90q_1 - q_1 q_2 - q_1^2)$$

$$= 2025 - q_1(90 - q_1) - q_2(45 - q_1).$$

It is easy to check that for any $q_1 > 45$ this difference is positive regardless of the value of q_2.[3] Hence we conclude that any $q_1 > 45$ is strictly dominated by $q_1 = 45$.

It is easy to see that firm 2 faces exactly the same profit function, which implies that any $q_2 > 45$ is strictly dominated by $q_2 = 45$. This observation leads to our first round of iterated elimination: a rational firm produces no more than 45 units, implying that the effective strategy space that survives one round of elimination is $q_i \in [0, 45]$ for $i \in \{1, 2\}$.

We can now turn to the second round of elimination. Because $q_2 \leq 45$, equation (4.2) implies that firm 1 will choose a quantity no less than 22.5, and a symmetric argument applies to firm 2. Hence the second round of elimination implies that the surviving strategy sets are $q_i \in [22.5, 45]$ for $i \in \{1, 2\}$.

The next step of this process will reduce the strategy set to $q_i \in [22.5, 33\frac{3}{4}]$, and the process will continue on and on. Interestingly the set of strategies that survives

---

3. This follows because if $q_1 > 45$ then $q_1(90 - q_1) < 2025$ and $q_2(45 - q_1) \leq 0$ for any $q_2 \geq 0$.

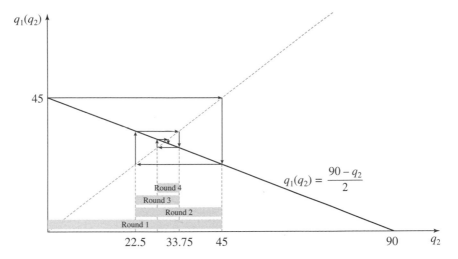

FIGURE 4.1   IESDS convergence in the Cournot game.

this process converges to a single quantity choice of $q_i = 30$. To see this, notice how we moved from one surviving interval to the next. We started by noting that $q_2 \geq 0$, and using equation (4.2) we found that $q_1 \leq 45$, creating the first-round interval of $[0, 45]$. Then, by symmetry, it follows that $q_2 \leq 45$, and using equation (4.2) again we conclude that $q_1 \geq 22.5$, creating the second-round interval $[22.5, 45]$. We can see this process graphically in Figure 4.1, where we use the upper (lower) end of the previous interval to determine the lower (upper) end of the next one. If this were to converge to an interval and not to a single point, then by the symmetry between both firms, the resulting interval for each firm would be $[q_{min}, q_{max}]$ that simultaneously satisfy two equations with two unknowns: $q_{min} = \frac{90 - q_{max}}{2}$ and $q_{max} = \frac{90 - q_{min}}{2}$. However, the only solution to these two equations is $q_{min} = q_{max} = 30$. Hence using IESDS for the Cournot game results in a unique predictor of behavior where $q_1 = q_2 = 30$, and each firm earns a profit of $v_1 = v_2 = 900$.

4.2.3   Evaluating IESDS

We turn to evaluate the IESDS solution concept using the criteria we introduced earlier. Start with existence and note that, unlike the concept of strict dominance, we can apply IESDS to *any* game by applying the algorithm just described. It does not require the existence of a strictly dominant strategy, nor does it require the existence of strictly dominated strategies. It is the latter characteristic, however, that gives this concept some bite: when strictly dominated strategies exist, the process of IESDS is able to say something about how common knowledge of rationality restricts behavior.

It is worth noting that this existence result is a consequence of assuming common knowledge of rationality. By doing so we are giving the players the ability to reason through the strategic implications of rationality, and to do so over and over again, while correctly anticipating that other players can perform the same kind of reasoning. Rationality alone does not provide this kind of reasoning.

It is indeed attractive that an IESDS solution always exists. This comes, however, at the cost of uniqueness. In the simple 3 × 3 matrix game described in (4.1) and the

Cournot duopoly game, IESDS implied the survival of a unique strategy. Consider instead the Battle of the Sexes, given by the following matrix:

Chris

		O	F
Alex	O	2, 1	0, 0
	F	0, 0	1, 2

IESDS cannot restrict the set of strategies here for the simple reason that neither O nor F is a strictly dominated strategy for each player. As we can see, this solution concept can be applied (it exists) to any game, but it will often fail to provide a unique solution. For the Battle of the Sexes game, IESDS can only conclude that "anything can happen."

After analyzing the efficiency of the outcomes that can be derived from strict dominance in Section 4.1.3, you may have anticipated the possible efficiency of IESDS equilibria. An easy illustration can be provided by the Prisoner's Dilemma. IESDS leaves (F, F) as the unique survivor, or IESDS equilibrium, after only one round of elimination. As we already demonstrated, the outcome from (F, F) is not Pareto optimal. Similarly, both previous examples (the 3×3 matrix game in (4.1) and the Cournot game) provide further evidence that Pareto optimality need not be achieved by IESDS: In the 3×3 matrix example, both strategy profiles (M, C) and (D, C) yield higher payoffs for both players—(8, 4) and (9, 6), respectively—than the unique IESDS equilibrium, which yields (4, 3). For the Cournot game, producing $q_1 = q_2 = 30$ yields profits of 900 for each firm. If instead they would both produce $q_1 = q_2 = 20$ then each would earn profits of 1000. Thus common knowledge of rationality does not mean that players can guarantee the best outcome for themselves when their own incentives dictate their behavior.

On a final note, it is interesting to observe that there is a simple and quite obvious relationship between the IESDS solution concept and the strict-dominance solution concept:

Proposition 4.2 *If for a game* $\Gamma = \langle N, \{S_i\}_{i=1}^n, \{v_i\}_{i=1}^n \rangle$ s^* *is a* strict dominant strategy equilibrium, *then* s^* *uniquely survives IESDS.*

Proof If $s^* = (s_1^*, \ldots, s_n^*)$ is a strict dominant strategy equilibrium then, by definition, for every player i all other strategies s_i' are strictly dominated by s_i^*. This implies that after one stage of elimination we will be left with a single profile of strategies, which is exactly s^*, and this concludes the proof. ■

This simple proposition is both intuitive and straightforward. Because rationality is the only requirement needed in order to eliminate all strictly dominated strategies in one round, then if all strategies but one are strictly dominated for each and every player, both IESDS and strict dominance will result in the same outcome. This shows us that whenever strict dominance results in a unique outcome, then IESDS will result in the same unique outcome after one round. However, as we saw earlier, IESDS may offer a fine prediction when strict dominance does not apply. This is exactly what the extra assumption of common knowledge of rationality delivers: a more widely applicable solution concept. However, the assumption of common knowledge of rationality is far from innocuous. It requires the players to be, in some way, extremely intelligent and to possess unusual levels of foresight. For the most part, game theory relies on this strong assumption, and hence it must be applied to the real world with caution. Remember the rule about how assumptions drive conclusions: garbage in, garbage out.

4.3 Beliefs, Best Response, and Rationalizability

Both of the solution concepts we have seen so far, strict dominance and IESDS, are based on eliminating actions that players would never play. An alternative approach is to ask: what possible strategies might players choose to play and under what conditions? When we considered eliminating strategies that no rational player would choose to play, it was by finding some strategy that is *always* better or, as we said, that dominates the eliminated strategies. A strategy that cannot be eliminated, therefore, suggests that *under some conditions* this strategy is the one that the player may like to choose. When we qualify a strategy to be the best one a player can choose *under some conditions,* these conditions must be expressed in terms that are rigorous and are related to the game that is being played.

To set the stage, think about situations in which you were puzzled about the behavior of someone you knew. To consider his choice as irrational, or simply stupid, you would have to consider whether there is a way in which he could defend his action as a good choice. A natural way to determine whether this is the case is to simply ask him, "What were you thinking?" If the response lays out a plausible situation for which his choice was a good one, then you cannot question his rationality. (You may of course question his wisdom, or even his sanity, if his thoughts seem bizarre.)

This is a type of reasoning that we will formalize and discuss in this chapter. If a strategy s_i is not strictly dominated for player i then it must be that there are combinations of strategies of player i's opponents for which the strategy s_i is player i's best choice. This reasoning will allow us to justify or *rationalize* the choice of player i.

4.3.1 The Best Response

As we discussed early in Chapter 3, what makes a game different from a single-player decision problem is that once you understand the actions, outcomes, and preferences of a decision problem, then you can choose your best or optimal action. In a game, however, your optimal decision not only depends on the structure of the game, but it will often depend on what the other players are doing.

Take the Battle of the Sexes as an example:

		Chris	
		O	*F*
Alex	*O*	2, 1	0, 0
	F	0, 0	1, 2

As the matrix demonstrates, the best choice of Alex depends on what Chris will do. If Chris goes to the opera then Alex would rather go to the opera instead of going to the football game. If, however, Chris goes to the football game then Alex's optimal action is switched around.

This simple example illustrates an important idea that will escort us throughout this book and that (one hopes) will escort you through your own decision making in strategic situations. In order for a player to be optimizing in a game, he has to choose a best strategy *as a response* to the strategies of his opponents. We therefore introduce the following formal definition:

Definition 4.5 The strategy $s_i \in S_i$ is player i's **best response** to his opponents' strategies $s_{-i} \in S_{-i}$ if

$$v_i(s_i, s_{-i}) \geq v_i(s_i', s_{-i}) \quad \forall s_i' \in S_i.$$

I can't emphasize enough how central this definition is to the concept of strategic behavior and rationality. In fact rationality implies that given any belief a player has about his opponents' behavior, he must choose an action that is best for him given his beliefs. That is,

Claim 4.2 *A rational player who believes that his opponents are playing some $s_{-i} \in S_{-i}$ will always choose a best response to s_{-i}.*

For instance, in the Battle of the Sexes, if Chris believes that Alex will go to the opera then Chris's best response is to go to the opera because

$$v_2(O, O) = 1 > 0 = v_2(O, F).$$

Similarly if Chris believes that Alex will go to the football game then Chris's best response is to go to the game as well.

There are some appealing relationships between the concept of playing a best response and the concept of dominated strategies. First, if a strategy s_i is strictly dominated, it means that some other strategy s_i' is *always* better. This leads us to the observation that the strategy s_i could not be a best response to *anything*:

Proposition 4.3 *If s_i is a strictly dominated strategy for player i, then it cannot be a best response to any $s_{-i} \in S_{-i}$.*

Proof If s_i is strictly dominated, then there exists some $s_i' \succ_i s_i$ such that $v_i(s_i', s_{-i}) > v_i(s_i, s_{-i})$ for all $s_{-i} \in S_{-i}$. But this in turn implies that there is no $s_{-i} \in S_{-i}$ for which $v_i(s_i, s_{-i}) \geq v_i(s_i', s_{-i})$, and thus that s_i cannot be a best response to any $s_{-i} \in S_{-i}$. ∎

A companion to this proposition would explore strictly dominant strategies, which are in some loose way the "opposite" of strictly dominated strategies. You should easily be able to convince yourself that if a strategy s_i^D is a strictly dominant strategy then it must be a best response to *anything* i's opponents can do. This immediately implies the next proposition, which is slightly broader than the simple intuition just provided and requires a bit more work to prove formally:

Proposition 4.4 *If in a finite normal-form game s^* is a strict dominant strategy equilibrium, or if it uniquely survives IESDS, then s_i^* is a best response to $s_{-i}^*, \forall i \in N$.*

Proof If s^* is a dominant strategy equilibrium then it uniquely survives IESDS, so it is enough to prove the proposition for strategies that uniquely survive IESDS. Suppose s^* uniquely survives IESDS, and choose some $i \in N$. Suppose in negation to the proposition that s_i^* is not a best response to s_{-i}^*. This implies that there exists an $s_i' \in S_i \setminus \{s_i^*\}$ (this is the set S_i without the strategy s_i^*) such that $v_i(s_i', s_{-i}^*) > v_i(s_i^*, s_{-i}^*)$. Let $S_i' \subset S_i$ be the set of all such s_i' for which $v_i(s_i', s_{-i}^*) > v_i(s_i^*, s_{-i}^*)$. Because s_i' was eliminated while s_{-i}^* was not (recall that s^* uniquely survives IESDS), there must be some s_i'' such that $v_i(s_i'', s_{-i}^*) > v_i(s_i', s_{-i}^*) > v_i(s_i^*, s_{-i}^*)$, implying that $s_i'' \in S_i'$. Because the game is finite, an induction argument on S_i' then implies that there exists a strategy $s_i''' \in S_i'$ that must survive IESDS. But this is a contradiction to s^* being the unique survivor of IESDS. ∎

This is a "proof by contradiction," and the intuition behind the proof may be a bit easier than the formal write-up. The logic goes as follows: If it is true that s_i^* was not a best response to s_{-i}^* then there was some other strategy s_i' that was a best response to s_{-i}^* that was eliminated at some previous round. But then it must be that there was a third strategy that was better than both of i's aforementioned strategies against s_{-i}^* in order to knock s_i' out in some earlier round. But then how could s_i^* knock out this third strategy against s_{-i}^*, which survived? This can't be true, which means that the initial negative premise in the proof, that s_i^* was not a best response to s_{-i}^*, must be false, hence the contradiction.

With the concept of a best response in hand, we need to think more seriously about the following question: to what profile of strategies should a player be playing a best response? Put differently, if my best response depends on what the other players are doing, then how should I choose between all the best responses I can possibly have? This is particularly pertinent because we are discussing static games, in which players choose their actions without knowing what their opponents are choosing. To tackle this important question, we need to give players the ability to form *conjectures* about what others are doing. We have alluded to the next step in the claim made earlier, which stated that "a rational player who believes that his opponents are playing some $s_{-i} \in S_{-i}$ will always choose a best response to s_{-i}." Thus we have to be mindful of what a player *believes* in order to draw conclusions about whether or not the player is choosing a best response.

4.3.2 Beliefs and Best-Response Correspondences

Suppose that s_i' is a best response for player i to his opponents playing s_{-i}', and assume for the moment that it is not a best response for any other profile of actions that i's opponents can choose. When would a rational player i choose to play s_i'? The answer follows directly from rationality: he will play s_i' only when his beliefs about other players' behavior justify the use of s_i', or in other words when he believes that his opponents will play s_{-i}'.

Introducing the concept of beliefs, and actions that best respond to beliefs, is central to the analysis of strategic behavior. If a player is fortunate enough to be playing in a game in which he has a strictly dominant strategy then his beliefs about the behavior of others play no role. The player's strictly dominant strategy is his best response *independent of* his opponents' play, and hence it is always a best response. But when no strictly dominant strategy exists, a player must ask himself, "What do I think my opponents will do?" The answer to this question should guide his own behavior.

To make this kind of reasoning precise we need to define what we mean by a player's belief. In a well-defined game, the only thing a player should be thinking about is what he believes his opponents are doing. Therefore we offer the following definition:

Definition 4.6 A **belief** of player i is a possible profile of his opponents' strategies, $s_{-i} \in S_{-i}$.

Given that a player has a particular belief about his opponents' strategies, he will be able to formulate a best response to that belief. The best response of a player to a certain strategy of his opponents may be unique, as in many of the games we have seen up to now.

For example, consider the Battle of the Sexes. When Chris believes that Alex is going to the opera, his unique best response is to go to the opera. Similarly, if he believes that Alex will go to the football game, then he should go to the game. For every unique belief there is a best response. Similarly recall that in the Cournot game the best choice of player 1 given any choice of player 2 solved the first-order condition of player 1's maximization problem, resulting in the function

$$q_1(q_2) = \frac{90 - q_2}{2}, \tag{4.3}$$

which assigns a unique value of q_1 to any value of q_2 for $q_2 \in [0, 90]$. Hence the function in (4.3) is the best-response function of firm 1 in the Cournot game.

We can therefore think of a rational player as having a *recipe book* that is a list of instructions as follows: "If I think my opponents are doing s_{-i}, then I should do s_i; if I think they're doing s'_{-i}, then I should do s'_i; . . . . " This list should go on until it exhausts all the possible strategies that player i's opponents can choose. If we think of this list of best responses as a plan, then this plan *maps beliefs into a choice of action,* and this choice of action must be a best response to the beliefs. We can think of this as player i's *best-response function.*

There may, however, be games in which for some beliefs a player will have more than one best-response strategy. Consider, for example, the following simple game:

			Player 2	
		L	*C*	*R*
	U	3, 3	5, 1	6, 2
Player 1	*M*	4, 1	8, 4	3, 6
	D	4, 0	9, 6	6, 8

If player 1 believes that player 2 is playing the column R then both U and D are each a best response. Similarly if player 1 believes that player 2 is playing the column L then both M and D are each a best response.

The fact that a player may have more than one best response implies that we can't think of the best-response mapping from opponents' strategies S_{-i} to an action by player i as a *function,* because by definition a function would select only *one action* as a best response (see Section 19.2 of the mathematical appendix). Thus we offer the following definition:

Definition 4.7 The **best-response correspondence** of player i selects for each $s_{-i} \in S_{-i}$ a subset $BR_i(s_{-i}) \subset S_i$ where each strategy $s_i \in BR_i(s_{-i})$ is a best response to s_{-i}.

That is, given a belief player i has about his opponents, s_{-i}, the set of all his possible strategies that are a best response to s_{-i} is denoted by $BR_i(s_{-i})$. If he has a unique best response to s_{-i} then $BR_i(s_{-i})$ will contain only one strategy from S_i.

4.3.3 Rationalizability

Equipped with the idea of beliefs and a player's best responses to his beliefs, the next natural step is to allow the players to reason about which beliefs to have about their opponents. This reasoning must take into account the rationality of all players, common knowledge of rationality, and the fact that all the players are trying to guess the behavior of their opponents.

To a large extent we employed similar reasoning when we introduced the solution concept of IESDS: instead of asking what your opponents might be doing, you asked "What would a rational player *not do*?" Then, assuming that all players follow this process by common knowledge of rationality, we were able to make some prediction about which strategies cannot be eliminated.

In what follows we introduce another way of reasoning that rules out irrational behavior with a similar iterated process that is, in many ways, the mirror image of IESDS. This next solution concept also builds on the assumption of common knowledge of rationality. However, instead of asking "What would a rational player *not do*?" our next concept asks "What *might* a rational player *do*?" A rational player will select only strategies that are a *best response to some profile of his opponents*. Thus we have

Definition 4.8 A strategy $s_i \in S_i$ is *never a best response* if there are no beliefs $s_{-i} \in S_{-i}$ for player i for which $s_i \in BR_i(s_{-i})$.

The next step, as in IESDS, is to use the common knowledge of rationality to build an iterative process that takes this reasoning to the limit. After employing this reasoning one time, we can eliminate all the strategies that are *never a best response,* resulting in a possibly "smaller" reduced game that includes only strategies that can be a best response in the original game. Then we can employ this reasoning again and again, in a similar way that we did for IESDS, in order to eliminate outcomes that should not be played by players who share a common knowledge of rationality.

The solution concept of rationalizability is defined precisely by iterating this thought process. The set of strategy profiles that survive this process is called the set of *rationalizable strategies*. (We postpone offering a definition of rationalizable strategies because the introduction of mixed strategies, in which players can play stochastic strategies, is essential for the complete definition.)

4.3.4 The Cournot Duopoly Revisited

Consider the Cournot duopoly example used to demonstrate IESDS in Section 4.2.2, with demand $p(q) = 100 - q$ and costs $c_i(q_i) = 10q_i$ for both firms. As we showed earlier, firm 1 maximizes its profits $v_1(q_1, q_2) = (100 - q_1 - q_2)q_1 - 10q_1$ by setting the first-order condition $90 - 2q_1 - q_2 = 0$. Now that we have introduced the idea of a best response, it should be clear that this firm's best response is *immediately derived* from the first-order condition. In other words, if firm 1 believes that firm 2 will choose the quantity q_2, then it should choose q_1 according to the best-response function,

$$BR_1(q_2) = \begin{cases} \frac{90-q_2}{2} & \text{if } 0 \leq q_2 < 90 \\ 0 & \text{if } q_2 \geq 90. \end{cases}$$

Notice that the best response is indeed a function. For all $0 \leq q_2 < 90$ there is a unique positive best response. For $q_2 \geq 90$ the price is guaranteed to be below 10, in which case any quantity firm 1 will choose will yield a negative profit (its costs per unit are 10), and hence the best response is to produce nothing. Similarly we can define the best-response correspondence of firm 2, which is symmetric.

Examining $BR_1(q_2)$ implies that firm 1 will choose to produce *only* quantities between 0 and 45. That is, there will be no beliefs about q_2 for which quantities above 45 are a best response. By symmetry the same is true for firm 2. Thus a first round of rationalizability implies that the only quantities that can be best-response quantities for both firms must lie in the interval [0, 45]. The next round of rationalizability for the game in which $S_i = [0, 45]$ for both firms shows that the best response of firm i is to choose any quantity $q_i \in [22.5, 45]$. Just as with IESDS, this process will continue on and on. The set of rationalizable strategies converges to a single quantity choice of $q_i = 30$ for both firms.

4.3.5 The "p-Beauty Contest"

Consider a game with n players, so that $N = \{1, \ldots, n\}$. Each player has to choose an integer between 0 and 20, so that $S_i = \{0, 1, 2, \ldots, 19, 20\}$. The winners are the players who choose an integer number that is closest to $\frac{3}{4}$ of the average. For example, if $n = 3$, and if the three players choose $s_1 = 1$, $s_2 = 5$, and $s_3 = 18$, then the average is $(1 + 5 + 18) \div 3 = 8$, and $\frac{3}{4}$ of the average is 6, so the winner is player 2. This game is called the **p-beauty contest** because, unlike in a standard beauty contest, you are not trying to guess what everyone else is guessing (beauty is what people choose it to be), but rather you are trying to guess p times the average, in this case $p = \frac{3}{4}$.[4]

Note that it is possible for more than one person to be a winner. If, for example, $s_1 = s_2 = 2$ and $s_3 = 8$, then the average is 3 and $\frac{3}{4}$ of the average is $2\frac{1}{4}$, so that both player 1 and player 2 are winners. Because more than one person can win, define the set of winners $W \subset N$ as those players who are closest to $\frac{3}{4}$ of the average, and the rest are all losers. The set of winners is defined as[5]

$$W = \left\{ i \in N : \arg\min_{i \in N} \left| s_i - \frac{3}{4} \frac{1}{n} \sum_{j=1}^{n} s_j \right| \right\}.$$

4. John Maynard Keynes (1936) described the playing of the stock market as analogous to entering a newspaper beauty-judging contest in which one selects the six prettiest faces out of a hundred photographs, with the prize going to the person whose selections are closest to those of all the other players. Keynes's depiction of a beauty contest is a situation in which you want to guess what others are guessing.

5. The notation $\arg\min_{i \in N} \left| s_i - \frac{3}{4} \frac{1}{n} \sum_{i=1}^{n} s_i \right|$ means that we are selecting all the i's for which the expression $\left| s_i - \frac{3}{4} \frac{1}{n} \sum_{j=1}^{n} s_j \right|$ is minimized. Since this expression is the absolute value of the difference between s_i and $\frac{3}{4} \frac{1}{n} \sum_{j=1}^{n} s_j$, this selection will result in the player (or players) who are closest to $\frac{3}{4}$ of the average.

To introduce payoffs, each player pays 1 to play the game, and winners split the pot equally among themselves. This implies that if there are $m \geq 1$ winners, each gets a payoff of $\frac{N-m}{m}$ (his share of the pot net of his own contribution) while losers get -1 (they lose their contribution).

Turning to rationalizable strategies, we must begin by finding strategies that are never a best response. This is not an easy task, but some simple insights can get us moving in the right direction. In particular the objective is to guess a number closest to $\frac{3}{4}$ of the average, which means that a player would want to guess a number that is generally smaller than the highest numbers that other players may be guessing.

This logic suggests that if there are strategies that are never a best response, they should be the higher numbers, and it is natural to start with 20: can choosing $s_i = 20$ be a best response? If you believe that the average is below 20, then 20 cannot be a best response—a lower number will be the best response. If the average were 20, that means that you and everyone else would be choosing 20, and you would then split the pot with all the other players. If you believe this, and instead of 20 you choose 19, then you will win the whole pot for sure, regardless of the number of players, because for any number of players n if everyone else is choosing 20 then 19 will be a unique winner.[6] This shows that 20 can never be a best response. Interestingly 19 is not the *unique* best response to the belief that all others are playing 20. (This is left as an exercise.) The important point, however, is that 20 *cannot* be a best response to *any* beliefs a player can have.

This analysis shows that only the numbers $S_i^1 = \{0, 1, \ldots, 19\}$ survive the first round of rationalizable behavior. Similarly after each additional round we will "lose" the highest number until we go through 19 rounds and are left with $S_i^{19} = \{0, 1\}$, meaning that after 19 rounds of dropping strategies that cannot be a best response, we are left with two strategies that survive: 1 and 0. If $n > 2$, we cannot reduce this set further: if, for example, player i believes that all the other players are choosing 1 then choosing 1 is a best response for him. (This is left as an exercise.) Similarly regardless of n, if he believes that everyone is choosing 0 then choosing 0 is his best response. Thus we are able to predict using rationalizability that players will not choose a number greater than 1, and if there are only two players then we will predict that both will choose 0.

Will this indeed predict behavior? Only if our *assumptions* about behavior are correct. If you were to play this game and you don't think that your opponents are doing these steps in their minds, then you may want to choose a number higher than 1. An interesting set of experiments is summarized in Camerer (2003).[7]

Remark By now you must have concluded that IESDS and rationalizability are two sides of the same coin, and you might even think that they are one and the same. This is almost true, and for two-player games it turns out that these two processes indeed result in the same outcomes. We discuss this issue briefly in Section 6.3, after we introduce the concept of mixed strategies. A more complete treatment can be found in Chapter 2 of Fudenberg and Tirole (1991).

---

6. The average, $\frac{1}{n}[(n-1)20 + 19]$, lies between 19.5 (if $n = 2$) and 20 (for $n \to \infty$), so that $\frac{3}{4}$ of the average lies between $14\frac{5}{8}$ and 15.

7. From my own experience running these games in classes of students, it is rare that the winning number is below 4.

4.3.6 Evaluating Rationalizability

In terms of existence, uniqueness, and implications for Pareto optimality, rationalizability is practically the same as IESDS. It will sometimes have bite, and may even restrict behavior quite dramatically as in the examples given. But if applied to the Battle of the Sexes, rationalizability will say "anything can happen."

4.4 Summary

- Rational players will never play a dominated strategy and will always play a dominant strategy when it exists.

- When players share common knowledge of rationality, the only strategies that are sensible are those that survive IESDS.

- Rational players will always play a best response to their beliefs. Hence any strategy for which there are no beliefs that justify its choice will never be chosen.

- Outcomes that survive IESDS, rationalizability, or strict dominance need not be Pareto optimal, implying that players may not be able to achieve desirable outcomes if they are left to their own devices.

4.5 Exercises

4.1 **Prove Proposition 4.1:** If the game $\Gamma = \left\langle N, \{S_i\}_{i=1}^n, \{v_i\}_{i=1}^n \right\rangle$ has a strictly dominant strategy equilibrium s^D, then s^D is the unique dominant strategy equilibrium.

4.2 **Weak Dominance:** We call the strategy profile $s^W \in S$ a **weakly dominant strategy equilibrium** if $s_i^W \in S_i$ is a weakly dominant strategy for all $i \in N$, that is, if $v_i(s_i, s_{-i}) \geq v_i(s_i', s_{-i})$ for all $s_i' \in S_i$ and for all $s_{-i} \in S_{-i}$.

    a. Provide an example of a game in which there is no weakly dominant strategy equilibrium.

    b. Provide an example of a game in which there is more than one weakly dominant strategy equilibrium.

4.3 **Discrete First-Price Auction:** An item is up for auction. Player 1 values the item at 3 while player 2 values the item at 5. Each player can bid either 0, 1, or 2. If player i bids more than player j then i wins the good and pays his bid, while the loser does not pay. If both players bid the same amount then a coin is tossed to determine who the winner is, and the winner gets the good and pays his bid while the loser pays nothing.

    a. Write down the game in matrix form.

    b. Does any player have a strictly dominated strategy?

    c. Which strategies survive IESDS?

4.4 **eBay's Recommendation:** It is hard to imagine that anyone is not familiar with eBay, the most popular auction web site by far. In a typical eBay auction a good is placed for sale, and each bidder places a "proxy bid," which eBay keeps in memory. If you enter a proxy bid that is lower than the current highest bid, then your bid is ignored. If, however, it is higher, then the current bid increases up to one increment (say, $0.01) above the *second highest* proxy bid. For example, imagine that three people have placed bids on a used laptop of $55, $98, and $112. The current price will be at $98.01, and if the auction ended the player who bid $112 would win at a price of $98.01. If you were to place a bid of $103.45 then the player who bid $112 would still win, but at a price of $103.46, while if your bid was $123.12 then you would win at a price of $112.01.

Now consider eBay's historical recommendation that you think hard about the value you impute to the good and that you enter your true value as your bid—no more, no less. Assume that the value of the good for each potential bidder is independent of how much other bidders value it.

    a.   Argue that bidding more than your valuation is weakly dominated by actually bidding your valuation.

    b.   Argue that bidding less than your valuation is weakly dominated by actually bidding your valuation.

    c.   Use your analysis to make sense of eBay's recommendation. Would you follow it?

4.5   **Iterated Elimination:** In the following normal-form game, which strategy profiles survive iterated elimination of strictly dominated strategies?

		Player 2		
		L	C	R
Player 1	U	6, 8	2, 6	8, 2
	M	8, 2	4, 4	9, 5
	D	8, 10	4, 6	6, 7

4.6   **Roommates:** Two roommates each need to choose to clean their apartment, and each can choose an amount of time $t_i \geq 0$ to clean. If their choices are t_i and t_j, then player i's payoff is given by $(10 - t_j)t_i - t_i^2$. (This payoff function implies that the more one roommate cleans, the less valuable is cleaning for the other roommate.)

    a.   What is the best response correspondence of each player i?

    b.   Which choices survive one round of IESDS?

    c.   Which choices survive IESDS?

4.7   **Campaigning:** Two candidates, 1 and 2, are running for office. Each has one of three choices in running his campaign: focus on the positive aspects of one's own platform (call this a positive campaign [P]), focus on the positive aspects of one's own platform while attacking one's opponent's campaign (call this a balanced campaign [B]), or finally focus only on attacking one's opponent (call this a negative campaign [N]). All a candidate cares about is the probability of winning, so assume that if a candidate expects to win with probability $\pi \in [0, 1]$, then his payoff is π. The probability that a candidate wins depends on his choice of campaign and his opponent's choice. The probabilities of winning are given as follows:

    If both choose the same campaign then each wins with probability 0.5.

    If candidate i uses a positive campaign while $j \neq i$ uses a balanced one then i loses for sure.

If candidate i uses a positive campaign while $j \neq i$ uses a negative one then i wins with probability 0.3.

If candidate i uses a negative campaign while $j \neq i$ uses a balanced one then i wins with probability 0.6.

a. Model this story as a normal-form game. (It suffices to be specific about the payoff function of one player and to explain how the other player's payoff function is different and why.)

b. Write down the game in matrix form.

c. What happens at each stage of elimination of strictly dominated strategies? Will this procedure lead to a clear prediction?

4.8 **Beauty Contest Best Responses:** Consider the p-beauty contest presented in Section 4.3.5.

a. Show that if player i believes that everyone else is choosing 20 then 19 is not the only best response for any number of players n.

b. Show that the set of best-response strategies to everyone else choosing the number 20 depends on the number of players n.

4.9 **Beauty Contest Rationalizability:** Consider the p-beauty contest presented in Section 4.3.5. Show that if the number of players $n > 2$ then the choices $\{0, 1\}$ for each player are both rationalizable, while if $n = 2$ then only the choice of $\{0\}$ by each player is rationalizable.

4.10 **Popsicle Stands:** Five lifeguard towers are lined up along a beach; the leftmost tower is number 1 and the rightmost tower is number 5. Two vendors, players 1 and 2, each have a popsicle stand that can be located next to one of five towers. There are 25 people located next to each tower, and each person will purchase a popsicle from the stand that is closest to him or her. That is, if player 1 locates his stand at tower 2 and player 2 at tower 3, then 50 people (at towers 1 and 2) will purchase from player 1, while 75 (from towers 3, 4, and 5) will purchase from vendor 2. Each purchase yields a profit of $1.

a. Specify the strategy set for each player. Are there any strictly dominated strategies?

b. Find the set of strategies that survive rationalizability.

Pinning Down Beliefs:
Nash Equilibrium

We have seen three solution concepts that offer some insights into predicting the behavior of rational players in strategic (normal-form) games. The first, strict dominance, relied only on rationality, and in some cases, like the Prisoner's Dilemma, it predicted a unique outcome, as it would in any game for which a dominant strategy equilibrium exists. However, it often fails to exist. The two sister concepts of IESDS and rationalizability relied on more than rationality by requiring common knowledge of rationality. In return a solution existed for every game, and for some games there was a unique prediction. Moreover, whenever there is a strict dominant equilibrium, it also uniquely survives IESDS and rationalizability. Even for some games for which the strict-dominance solution did not apply, like the Cournot duopoly, we obtained a unique prediction from IESDS and rationalizability.

However, when we consider a game like the Battle of the Sexes, none of these concepts had any bite. Dominant strategy equilibrium did not apply, and both IESDS and rationalizability could not restrict the set of reasonable behavior:

		Chris	
		O	*F*
Alex	*O*	2, 1	0, 0
	F	0, 0	1, 2

For example, we cannot rule out the possibility that Alex goes to the opera while Chris goes to the football game, because Alex may behave optimally given his belief that Chris is going to the opera, and Chris may behave optimally given his belief that Alex is going to the football game. Yet there is something troubling about this outcome. If we think of this pair of actions not only as actions, but as a system of actions and beliefs, then there is something of a dissonance: indeed the players are playing best responses to their beliefs, but their beliefs are wrong!

In this chapter we make a rather heroic leap that ties together beliefs and actions and results in the most central and best-known solution concept in game theory. As already mentioned, for dominant strategy equilibrium we required only that players be rational, while for IESDS and rationalizability we required common knowledge of rationality. Now we introduce a much more demanding concept, that of the Nash equilibrium, first put forth by John Nash (1950a), who received the Nobel Prize in Economics for this achievement.[1]

5.1   Nash Equilibrium in Pure Strategies

To cut to the chase, a Nash equilibrium is a system of beliefs and a profile of actions for which each player is playing a best response to his beliefs and, moreover, players have *correct beliefs*. Another common way of defining a Nash equilibrium, which does not refer to beliefs, is as a profile of strategies for which *each* player is choosing a best response to the strategies of *all other* players. Formally we have:

Definition 5.1   The pure-strategy profile $s^* = (s_1^*, s_2^*, \ldots, s_n^*) \in S$ is a **Nash equilibrium** if s_i^* is a best response to s_{-i}^*, for all $i \in N$, that is,

$$v_i(s_i^*, s_{-i}^*) \geq v_i(s_i', s_{-i}^*) \quad \text{for all } s_i' \in S_i \text{ and all } i \in N.$$

Consider as an example the following two-player discrete game, which we used to demonstrate IESDS:

		Player 2		
		L	C	R
Player 1	U	4, 3	5, 1	6, 2
	M	2, 1	8, 4	3, 6
	D	3, 0	9, 6	2, 8

In this game, the only pair of pure strategies that survived IESDS is the pair (U, L). As it turns out, this is also the only pair of strategies that constitutes a Nash equilibrium. If player 2 is playing the column L, then player 1's best response is $BR_1(L) = \{U\}$; at the same time, if player 1 is playing the row U, then player 2's best response is $BR_2(U) = \{L\}$.

What about the other games we saw? In the Prisoner's Dilemma, the unique Nash equilibrium is (F, F). This should be easy to see: if each player is playing a dominant strategy then he is by definition playing a best response to *anything* his opponent is choosing, and hence it must be a Nash equilibrium. As we will soon see in Section 5.2.3, and as you may have already anticipated, the unique Nash equilibrium in the Cournot duopoly game that we discussed earlier is $(q_1, q_2) = (30, 30)$.

The relationship between strict-dominance, IESDS, rationalizability, and Nash equilibrium outcomes in many of the examples we have analyzed is no coincidence. There is a simple relationship between the concepts we previously explored and that of Nash equilibrium, as the following proposition clearly states:

Proposition 5.1   *Consider a strategy profile $s^* = (s_1^*, s_2^*, \ldots, s_n^*)$. If s^* is either*

1. *a strict dominant strategy equilibrium,*

2. *the unique survivor of IESDS, or*

---

1. Nash's life was the subject of a very successful movie, *A Beautiful Mind,* based on the book by Sylvia Nasar (1998).

3. *the unique rationalizable strategy profile,*

then s* is the unique Nash equilibrium.

This proposition is not difficult to prove, and the proof is left as exercise 5.1 at the end of the chapter. The intuition is of course quite straightforward: we know that if there is a strict dominant strategy equilibrium then it uniquely survives IESDS and rationalizability, and this in turn must mean that each player is playing a best response to the other players' strategies.

At the risk of being repetitive, let me emphasize the requirements for a Nash equilibrium:

1. Each player is playing a *best response* to his *beliefs*.
2. The *beliefs* of the players about their opponents *are correct*.

The first requirement is a direct consequence of rationality. It is the second requirement that is very demanding and is a tremendous leap beyond the requirements we have considered so far. It is one thing to ask people to behave rationally given their beliefs (play a best response), but it is a totally different thing to ask players to predict the behavior of their opponents correctly.

Then again it may be possible to accept such a strong requirement if we allow for some reasoning that is beyond the physical structure of the game. For example, imagine, in the Battle of the Sexes game, that Alex is an influential person—people just seem to follow Alex, and this is something that Alex knows well. In this case Chris should believe, knowing that Alex is so influential, that Alex would expect Chris to go to the opera. Knowing this, Alex should believe that Chris will indeed believe that Alex is going to the opera, and so Chris will go to the opera too.

It is important to note that this argument *is not* that Chris likes to please Alex—such an argument would change the payoff of the game and increase Chris's payoff from pleasing Alex. Instead this argument is only about beliefs that are "self-fulfilling." That is, if these beliefs have some weight to them, which may be based on past experience or on some kind of deductive reasoning, then they will be self-fulfilling in that they support the behavior that players believe will occur.

Indeed (O, O) is a Nash equilibrium. However, notice that we can make the symmetric argument about Chris being an influential person: (F, F) is also a Nash equilibrium. As the external game theorist, however, we should not say more than "one of these two outcomes is what we predict." (You should be able to convince yourself that no other pair of pure strategies in the Battle of the Sexes game is a Nash equilibrium.)

5.1.1 Pure-Strategy Nash Equilibrium in a Matrix

This short section presents a simple method to find all the pure-strategy Nash equilibria in a matrix game if at least one exists. Consider the following two-person finite game in matrix form:

		Player 2		
		L	C	R
Player 1	U	7, 7	4, 2	1, 8
	M	2, 4	5, 5	2, 3
	D	8, 1	3, 2	0, 0

It is easy to see that no strategy is dominated and thus that strict dominance cannot be applied to this game. For the same reason, IESDS and rationalizability will conclude that anything can happen. However, a pure-strategy Nash equilibrium does exist, and in fact it is unique. To find it we use a simple method that builds on the fact that any Nash equilibrium must call for a pair of strategies in which each of the two players is playing a best response to his opponent's strategy. The procedure is best explained in three steps:

Step 1: For every *column,* which is a strategy of player 2, find the highest payoff entry for player 1. By definition this entry must be in the row that is a best response for the particular column being considered. Underline the pair of payoffs in this row under this column:

		Player 2		
		L	C	R
	U	7, 7	4, 2	1, 8
Player 1	M	2, 4	<u>5, 5</u>	<u>2, 3</u>
	D	<u>8, 1</u>	3, 2	0, 0

Step 1 identifies the best response of player 1 *for each of the pure strategies* (columns) of player 2. For instance, if player 2 is playing L, then player 1's best response is D, and we underline the payoffs associated with this row in column 1. After performing this step we see that there are three pairs of pure strategies at which player 1 is playing a best response: (D, L), (M, C), and (M, R).

Step 2: For every *row,* which is a strategy of player 1, find the highest payoff entry for player 2. By definition this entry must be in the column that is a best response for the particular row being considered. Overline the pair of payoffs in this entry:

		Player 2		
		L	C	R
	U	7, 7	4, 2	$\overline{1, 8}$
Player 1	M	2, 4	$\overline{5, 5}$	2, 3
	D	<u>8, 1</u>	$\overline{3, 2}$	0, 0

Step 2 similarly identifies the pairs of strategies at which player 2 is playing a best response. For instance, if player 1 is playing D, then player 2's best response is C, and we overline the payoffs associated with this column in row 3. We can continue to conclude that player 2 is playing a best response at three strategy pairs: (D, C), (M, C), and (U, R).

Step 3: If any matrix entry has both an under- and an overline, it is the outcome of a Nash equilibrium in pure strategies.

This follows immediately from the fact that both players are playing a best response at any such pair of strategies. In this example we find that (M, C) is the unique pure-strategy Nash equilibrium—it is the only pair of pure strategies for which both players are playing a best response. If you apply this approach to the Battle of the Sexes, for example, you will find both pure-strategy Nash equilibria, (O, O) and (F, F). For the Prisoner's Dilemma only (F, F) will be identified.

5.1.2 Evaluating the Nash Equilibria Solution

Considering our criteria for evaluating solution concepts, we can see from the Battle of the Sexes example that we may not have a unique Nash equilibrium. However, as alluded to in our earlier discussion of the Battle of the Sexes game, there is no reason to expect that we should. Indeed we may need to entertain other aspects of an environment in which players interact, such as social norms and historical beliefs, to make precise predictions about which of the possible Nash equilibria may result as the more likely outcome.

In Section 5.2.4 we will analyze a price competition game in which a Nash equilibrium may fail to exist. It turns out, however, that for quite general conditions games will have at least one Nash equilibrium. For the interested reader, Section 6.4 discusses some conditions that guarantee the existence of a Nash equilibrium, which was a central part of Nash's Ph.D. dissertation. This fact gives the Nash solution concept its power—like IESDS and rationalizability, the solution concept of Nash is widely applicable. It will, however, usually lead to more refined predictions than those of IESDS and rationalizability, as implied by proposition 5.1.

As with the previous solution concepts, we can easily see that Nash equilibrium does not guarantee Pareto optimality. The theme should be obvious by now: left to their own devices, people in many situations will do what is best for them, at the expense of social efficiency. This point was made quite convincingly and intuitively in Hardin's (1968) "tragedy of the commons" argument, which we explore in Section 5.2.2. This is where our focus on *self-enforcing outcomes* has its bite: our solution concepts took the game as given, and they imposed rationality and common knowledge of rationality to try to see what players would choose to do. If they each seek to maximize their individual well-being then the players may hinder their ability to achieve socially optimal outcomes.

5.2 Nash Equilibrium: Some Classic Applications

The previous section introduced the central pillar of modern noncooperative game theory, the Nash equilibrium solution concept. It has been applied widely in economics, political science, legal studies, and even biology. In what follows we demonstrate some of the best-known applications of the concept.

5.2.1 Two Kinds of Societies

The French philosopher Jean-Jacques Rousseau presented the following situation that describes a trade-off between playing it safe and relying on others to achieve a larger gain. Two hunters, players 1 and 2, can each choose to hunt a stag (S), which provides a rather large and tasty meal, or hunt a hare (H)—also tasty, but much less filling. Hunting stags is challenging and requires mutual cooperation. If either hunts a stag alone, the chance of success is negligible, while hunting hares is an individualistic enterprise that is not done in pairs. Hence hunting stags is most beneficial for society but requires "trust" between the hunters in that each believes that the other is joining forces with him. The game, often referred to as the Stag Hunt game, can be described by the following matrix:

	S	H
S	5, 5	0, 3
H	3, 0	3, 3

It is easy to see that the game has two pure-strategy equilibria: (S, S) and (H, H). However, the payoff from (S, S) Pareto dominates that from (H, H). Why then would (H, H) ever be a reasonable prediction? This is precisely the strength of the Nash equilibrium concept. If each player anticipates that the other will not join forces, then he knows that going out to hunt the stag alone is not likely to be a successful enterprise and that going after the hare will be better. This belief would result in a society of individualists who do not cooperate to achieve a better outcome. In contrast, if the players expect each other to be cooperative in going after the stag, then this anticipation is self-fulfilling and results in what can be considered a cooperative society. In the real world, societies that may look very similar in their endowments, access to technology, and physical environments have very different achievements, all because of self-fulfilling beliefs or, as they are often called, *norms* of behavior.[2]

5.2.2 The Tragedy of the Commons

The *tragedy of the commons* refers to the conflict over scarce resources that results from the tension between individual selfish interests and the common good; the concept was popularized by Hardin (1968). The central idea has proven useful for understanding how we have come to be on the brink of several environmental catastrophes.

Hardin introduces the hypothetical example of a pasture shared by local herders. Each herder wants to maximize his yield, increasing his herd size whenever possible. Each additional animal has a positive effect for its herder, but the cost of that extra animal, namely degradation of the overall quality of the pasture, is shared by all the other herders. As a consequence the individual incentive for each herder is to grow his own herd, and in the end this scenario causes tremendous losses for everyone. To those trained in economics, it is yet another example of the distortion that results from the "free-rider" problem. It should also remind you of the Prisoner's Dilemma, in which individuals driven by selfish incentives cause pain to the group.

In the course of his essay, Hardin develops the theme, drawing on examples of such latter-day commons as the atmosphere, oceans, rivers, fish stocks, national parks, advertising, and even parking meters. A major theme running throughout the essay is the growth of human populations, with the earth's resources being a global commons. (Given that this example concerns the addition of extra "animals" to the population, it is the closest to his original analogy.)

---

2. For an excellent exposition of the role that beliefs play in societies, see Greif (2006). The idea that coordinated changes are needed for developing countries to move out of poverty and into industrial growth dates back to Paul Rosenstein-Rodan's theory of the "big push," which is explored further in Murphy et al. (1989).

Let's put some game theoretic analysis behind this story. Imagine that there are n players, say firms, in the world, each choosing how much to produce. Their production activity in turn consumes some of the clean air that surrounds our planet. There is a total amount of clean air equal to K, and any consumption of clean air comes out of this common resource. Each player i chooses his own consumption of clean air for production, $k_i \geq 0$, and the amount of clean air left is therefore $K - \sum_{i=1}^{n} k_i$. The benefit of consuming an amount $k_i \geq 0$ gives player i a benefit equal to $\ln(k_i)$, and no other player benefits from i's choice. Each player also enjoys consuming the remainder of the clean air, giving each a benefit $\ln(K - \sum_{i=1}^{n} k_i)$. Hence the payoff for player i from the choice $k = (k_1, k_2, \ldots, k_n)$ is equal to

$$v_i(k_i, k_{-i}) = \ln(k_i) + \ln\left(K - \sum_{j=1}^{n} k_j \right). \tag{5.1}$$

To solve for a Nash equilibrium we can compute the best-response correspondences for each player and then find a strategy profile for which all the best-response functions are satisfied together. This is an important point that warrants further emphasis. We know that given k_{-i}, player i will want to choose an element in $BR_i(k_{-i})$. Hence if we find some profile of choices $(k_1^*, k_2^*, \ldots, k_n^*)$ for which $k_i^* = BR_i(k_{-i}^*)$ for all $i \in N$ then this must be a Nash equilibrium.

This means that if we derive all n best-response correspondences, and it turns out that they are functions (unique best responses), then we have a system of n equations, one for each player's best-response function, with n unknowns, the choices of each player. Solving this system will yield a Nash equilibrium. To get player i's best-response function (and we will verify that it is a function), we write down the first-order condition of his payoff function:

$$\frac{\partial v_i(k_i, k_{-i})}{\partial k_i} = \frac{1}{k_i} - \frac{1}{K - \sum_{j=1}^{n} k_j} = 0,$$

and this gives us player i's best response function,[3]

$$BR_i(k_{-i}) = \frac{K - \sum_{j \neq i} k_j}{2}.$$

We therefore have n such equations, one for each player, and if we substitute the choice k_i instead of $BR_i(k_{-i})$ we get the n equations with n unknowns that need to be solved.

We proceed to solve the equilibrium for two players and leave the n-player case as exercise 5.7 at the end of the chapter. Letting $k_i(k_j)$ be the best response of player i, we have two best-response equations:

$$k_1(k_2) = \frac{K - k_2}{2} \quad \text{and} \quad k_2(k_1) = \frac{K - k_1}{2}.$$

These two equations are plotted in Figure 5.1. As the figure illustrates, the more player j consumes, the less player i wants to consume. In particular if player 2 consumes nothing (effectively not existing), then player 1 will consume $k_1 = \frac{K}{2}$, and as player

---

3. Of course, we are implicitly assuming that $\sum_{j=1}^{n} k_j \leq K$.

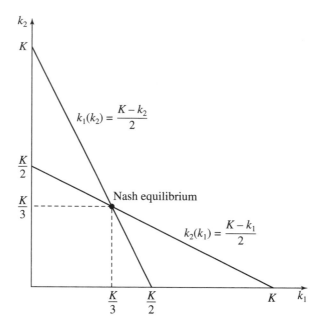

FIGURE 5.1 Best-response functions: two-player tragedy of the commons.

2's consumption increases toward K, player 1's consumption decreases toward zero. If we solve the two best-response functions simultaneously, we find the unique Nash equilibrium, which has both players playing $k_1 = k_2 = \frac{K}{3}$, as shown in Figure 5.1.

Now we can ask whether this two-player society could do better. Is consuming $\frac{K}{3}$ for each player too much or too little? The right way to answer these questions is using the Pareto criterion: can we find another consumption profile that will make everyone better off? If we can, we can compare that with the Nash equilibrium to answer this question. To find such a profile we'll use a little trick: we will maximize the sum of all the payoff functions, which we can think of as the "world's payoff function," $w(k_1, k_2)$. I won't go into the moral justification for using this approach, but it will turn out to be a useful tool.[4] The function we are maximizing is, therefore,

$$\max_{k_1, k_2} w(k_1, k_2) = \sum_{i=1}^{2} v_i(k_1, k_2) = \sum_{i=1}^{2} \ln(k_i) + 2 \ln \left(K - \sum_{i=1}^{2} k_i \right).$$

The first-order conditions for this problem are

$$\frac{\partial w(k_1, k_2)}{\partial k_1} = \frac{1}{k_1} - \frac{2}{K - k_1 - k_2} = 0$$

and

$$\frac{\partial w(k_1, k_2)}{\partial k_2} = \frac{1}{k_2} - \frac{2}{K - k_1 - k_2} = 0.$$

---

4. In general, maximizing the sum of utility functions, or maximizing total welfare, will result in a Pareto-optimal outcome, but it need not be the only one. In this example, this maximization gives us the symmetric Pareto-optimal consumption profile because the payoff function of each player is concave in his own consumption with $\frac{\partial v_i}{\partial k_i} > 0$, $\frac{\partial^2 v_i}{\partial k_i^2} < 0$, and $\lim_{k_i \to 0} \frac{\partial v_i}{\partial k_i} = \infty$.

Solving these two equations simultaneously will result in Pareto-optimal choices for k_1 and k_2. The unique solution to these two equations yields $k_1 = k_2 = \frac{K}{4}$, which means that from a social point of view the Nash equilibrium has the two players each consuming too much clean air. Indeed they would both be better off if each consumed $k_i = \frac{K}{4}$ instead of $k_i = \frac{K}{3}$.[5] In exercise 5.7 at the end of the chapter you are asked to show the consequences of having more than two players.

Thus, as Hardin puts it, giving people the freedom to make choices may make them all worse off than if those choices were somehow regulated. Of course the counterargument is whether we can trust a regulator to keep things under control; if not, the question remains which is the better of the two evils—an answer that game theory cannot offer!

5.2.3 Cournot Duopoly

Let's revisit the Cournot game with demand $P = a - bq$ and cost functions $c_i(q_i) = c_i q_i$ for firms $i \in \{1, 2\}$. The maximization problem that firm i faces when it believes that its opponent chooses quantity q_j is

$$\max_{q_i} \ v_i(q_i, q_j) = (a - bq_i - bq_j)q_i - c_i q_i.$$

Recall that the best-response function for each firm is given by the first-order condition, so that

$$BR_i(q_j) = \frac{a - bq_j - c_i}{2b}.$$

This means that each firm chooses quantities as follows:

$$q_1 = \frac{a - bq_2 - c_1}{2b} \qquad \text{and} \qquad q_2 = \frac{a - bq_1 - c_2}{2b}. \tag{5.2}$$

A pair of quantities (q_1, q_2) that are *mutual best responses* will be a Cournot-Nash equilibrium, which occurs when we solve both best-response functions (5.2) simultaneously. The best-response functions shown in Figure 5.2 depict the special case we solved earlier, in which $a = 100$, $b = 1$, and $c_1 = c_2 = 10$, in which case the unique Nash equilibrium is $q_1 = q_2 = 30$.

Notice that the Nash equilibrium coincides with the unique strategies that survive IESDS and rationalizability, which is the conclusion of proposition 5.1. An exercise that is left for you (exercise 5.8) is to explore the Cournot model with many firms.

---

5. To see this we can calculate Δv_i, the difference between a player's payoff when we maximize total surplus $\left(\text{which we solved as } k_i = \frac{K}{4}\right)$ and his Nash equilibrium payoff:

$$\Delta v_i = \ln\left(\frac{K}{4}\right) + \ln\left(\frac{K}{2}\right) - \ln\left(\frac{K}{3}\right) - \ln\left(\frac{K}{3}\right)$$

$$= \ln(K) - \ln(4) + \ln(K) - \ln(2) - \ln(K) + \ln(3) - \ln(K) + \ln(3)$$

$$= 2\ln(3) - \ln(2) - \ln(4)$$

$$= 0.051 > 0.$$

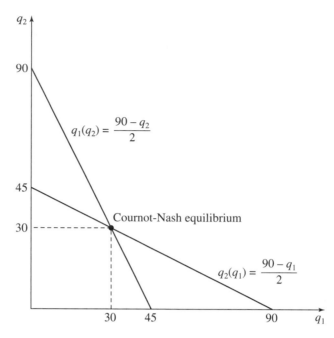

FIGURE 5.2 Cournot duopoly game: best-response functions and Nash equilibrium.

5.2.4 Bertrand Duopoly

The Cournot model assumed that the firms choose quantities and the market price adjusts to clear the demand. However, one can argue that firms often set prices and let consumers choose from where they will purchase, rather than setting quantities and waiting for the market price to equilibrate demand. We now consider the game in which each firm posts a price for otherwise identical goods. This was the situation modeled and analyzed by Joseph Bertrand (1883).

As before, assume that demand is given by $p = 100 - q$ and cost functions $c_i(q_i) = 10q_i$ for firms $i \in \{1, 2\}$. Clearly we would expect all buyers to buy from the firm whose price is the lowest. What happens if there is a tie? Let's assume that the market splits equally between the two firms. This gives us the following normal-form game:

Players: $N = \{1, 2\}$.

Strategy sets: $S_i = [0, \infty]$ for $i \in \{1, 2\}$, and firms choose prices $p_i \in S_i$.

Payoffs: To calculate payoffs we need to know what the quantities will be for each firm. Given our assumption on ties, the quantities are given by

$$q_i(p_i, p_j) = \begin{cases} 100 - p_i & \text{if } p_i < p_j \\ 0 & \text{if } p_i > p_j \\ \frac{100 - p_i}{2} & \text{if } p_i = p_j, \end{cases}$$

which in turn means that the payoff function is given by

$$v_i(p_i, p_j) = \begin{cases} (100 - p_i)(p_i - 10) & \text{if } p_i < p_j \\ 0 & \text{if } p_i > p_j \\ \frac{100 - p_i}{2}(p_i - 10) & \text{if } p_i = p_j. \end{cases}$$

Now that the description of the game is complete, we can try to calculate the best-response functions of both firms. To do this we will start with a slight modification that is motivated by reality: assume that prices cannot be *any* real number $p \geq 0$, but instead are limited to increments of some small fixed value, say $\varepsilon > 0$, which implies that the strategy (price) sets are $S_i = \{0, \varepsilon, 2\varepsilon, 3\varepsilon, \ldots\}$. For example, if we are considering cents as the price increment, so that $\varepsilon = 0.01$, then the strategy set will be $S_i = \{0, 0.01, 0.02, 0.03, \ldots\}$. We will very soon see what happens when this small denomination ε becomes very small and approaches zero.

We derive the best response of a firm by exhausting the relevant situations that it can face. It is useful to start with the situation in which only one monopolistic firm is in the market. We can calculate the monopoly price, which is the price that would maximize a single firm's profits if there were no competitors. This would be obtained by maximizing $v_i(p) = pq - 10q = (100 - p)(p - 10)$ and the first-order condition is $110 - 2p = 0$, resulting in an optimal price of $p = 55$, in a quantity of $q = 45$, and in profits equal to \$2025.

Let us now turn back to the duopoly with two firms and consider the case in which $p_j > 55$. It is easy to see that firm i can act as if there was no competition: just set the monopoly price of 55 and get the whole market. Hence we conclude that if $p_j > 55$ then the best response of firm i is to set $p_i = 55$.

It is also easy to see that in the case in which $p_j < 10$ then the best response of firm i is to set a price that is higher than that set by firm j. If it charges a price $p_i \leq p_j$ then it will sell a positive quantity at a price that is lower than its costs, causing firm i to lose money. If it charges a price $p_i > p_j$ then it sells nothing and loses nothing.

Now consider the case $55 \geq p_j \geq 10.02$. Firm i can choose one of three options: either set $p_i > p_j$ and get nothing, set $p_i = p_j$ and split the market, or set $p_i < p_j$ and get the whole market. It is not too hard to establish that firm i wants to just undercut firm j and capture the whole market, a goal that can be accomplished by setting a price of $p_i = p_j - 0.01$. To see this, observe that if $p_j > 10.01$ then by setting $p_i = p_j$ firm i gets $v_i = \frac{100 - p_j}{2}(p_j - 10)$, while if it sets $p_i = p_j - 0.01$ it will get $v_i' = (100 - (p_j - 0.01))((p_j - 0.01) - 10)$. We can calculate the difference between the two as follows:

$$\Delta v_i = v_i' - v_i = (100 - (p_j - 0.01))((p_j - 0.01) - 10) - \frac{100 - p_j}{2}(p_j - 10)$$

$$= 55.02 p_j - 0.5 p_j^2 - 501.1.$$

It is easy to check that Δv_i is positive at $p_j = 10.02$ (it is equal to 0.0002). To see that it is positive for all values of $p \in [10.02, 55]$, we show that Δv_i has a positive derivative for any $p_j \in [10.02, 55]$, which implies that this difference grows even more positive as p_j increases in this domain. That is,

$$\frac{d\Delta v_i}{dp_j} = 55.02 - p_j > 0 \quad \text{for all } p < 55.02.$$

Thus we conclude that when $p_j \in [10.02, 55]$ the best response of firm i is to charge \$0.01 less, that is, $p_i = p_j - 0.01$.

To complete the analysis, we have to explore two final cases: $p_j = 10.01$ and $p_j = 10$. The three options to consider are setting $p_i = p_j$, $p_i > p_j$, or $p_i < p_j$. When $p_j = 10.01$ then undercutting j's price means setting $p_i = 10$, which gives i zero profits and is the same as setting $p_i > p_j$. Thus the best response is setting $p_i = p_j = 10.01$ and splitting the market with very low profits. Finally, if $p_j = 10$

then any choice of price $p_i \geq p_j$ will yield firm i zero profits, whereas setting $p_i < p_j$ causes losses. Therefore any price $p_i \geq p_j$ is a best response when $p_j = 10$.

In summary we calculated:

$$BR_i(p_j) = \begin{cases} 55 & \text{if } p_j > 55 \\ p_j - 0.01 & \text{if } 55 \geq p_j \geq 10.02 \\ 10.01 & \text{if } p_j = 10.01 \\ p_i \in \{10, 10.01, 10.02, 10.03, \ldots\} & \text{if } p_j = 10. \end{cases}$$

Now given that firm j's best response is exactly symmetric, it should not be hard to see that there are two Nash equilibria that follow immediately from the form of the best-response functions: *The best response* to $p_j = 10.01$ is $BR_i(10.01) = 10.01$, and *a best response* to $p_j = 10$ is $p_i = 10$ or $10 \in BR_i(10)$. Thus the two Nash equilibria are

$$(p_1, p_2) \in \{(10, 10), (10.01, 10.01)\}.$$

It is worth pausing here for a moment to address a common point of confusion, which often arises when a player has more than one best response to a certain action of his opponents. In this example, when $p_2 = 10$, player 1 is indifferent regarding *any price at or above 10 that he chooses:* if he splits the market with $p_1 = 10$ he gets half the market with no profits, and if he sets $p_1 > 10$ he gets no customers and has no profits. One may be tempted to jump to the following conclusion: if player 2 is choosing $p_2 = 10$ then any choice of $p_1 \geq 10$ together with $p_2 = 10$ will be a Nash equilibrium. This is incorrect! It is true that player 1 is playing a best response with any one of his choices, but if $p_1 > 10$ then $p_2 = 10$ is *not* a best response of player 2 to p_1, as we can observe from the foregoing analysis.

Comparing the outcome of the Bertrand game to that of the Cournot game is an interesting exercise. Notice that when firms choose quantities (Cournot), the unique Nash equilibrium is $q_1 = q_2 = 30$. A quick calculation shows that for the aggregate quantity of $q = q_1 + q_2 = 60$ we get a demand price of $p = 40$ and each firm makes a profit of $900. When instead these firms compete on prices, the two possible equilibria have either zero profits when both choose $p_1 = p_2 = \$10$ or negligible profits (about $0.45) when they each choose $p_1 = p_2 = \$10.01$. Interestingly, for both the Cournot and Bertrand games, if we had only one player then he would maximize profits by choosing the monopoly price (or quantity) of $55 (or 45 units) and earn a profit of $2025.

The message of this analysis is quite striking: one firm may have monopoly power, but when we let one more firm compete, and they compete on prices, then the market will behave competitively—if both choose a price of $10 then price will equal marginal costs! Notice that if we add a third and a fourth firm this will not change the outcome; prices will have to be $10 (or practically the same at $10.01) for all firms in the Nash (Bertrand) equilibrium. This is not the case for Cournot competition, in which firms manage to obtain some market power as long as the number of firms is not too large.

A quick observation should lead you to realize that if we let ε be smaller than $0.01, the conclusions we reached earlier will be sustained, and we will have two Nash equilibria, one with $p_1 = p_2 = 10$ and one with $p_1 = p_2 = 10 + \varepsilon$. Clearly these two equilibria become "closer" in profits as ε becomes smaller, and they converge to each other as ε approaches zero.

It turns out that if we assume that prices can be chosen as *any* real number, we get a very "clean" result: the unique Bertrand-Nash equilibrium will have prices equal to marginal costs, implying a competitive outcome. We prove this for the more general symmetric case in which $c_i(q_i) = cq_i$ for both firms and demand is equal to $p = a - bq$ with $a > c$.[6]

Proposition 5.2 *For $\varepsilon = 0$ (prices can be any real number) there is a unique Nash equilibrium: $p_1 = p_2 = c$.*

Proof First note that in any equilibrium $p_i \geq c$ for both firms—otherwise at least one firm offering a price lower than c will lose money (pay the consumers to take its goods!). We therefore need to show that $p_i > c$ cannot be part of any equilibrium. We can see this in two steps:

1. If $p_1 = p_2 = \hat{p} > c$ then each would benefit from lowering its price to some price $\hat{p} - \varepsilon$ (ε very small) and get the whole market for almost the same price.

2. If $p_1 > p_2 \geq c$ then player 2 would want to deviate to $p_1 - \varepsilon$ (ε very small) and earn higher profits.

It is easy to see that $p_1 = p_2 = c$ is an equilibrium: Firm i's best response to $p_j = c$ is $p_i(c) \geq c$. That is, any price at or above marginal costs c is a best response to the other player charging c. Hence $p_1 = p_2 = c$ is the unique equilibrium because each is playing a best response to the other's choice and neither wants to deviate to a different price. ■

We will now see an interesting variation of the Bertrand game. Assume that $c_i(q_i) = c_i q_i$ represents the cost of each firm as before. Now, however, let $c_1 = 1$ and $c_2 = 2$ so that the two firms are not identical: firm 1 has a cost advantage. Let the demand still be $p = 100 - q$.

Now consider the case with discrete price jumps with $\varepsilon = 0.01$. The firms are not symmetric in that firm 1 has a lower marginal cost than firm 2, and unlike the example in which both had the same costs we cannot have a Nash equilibrium in which both charge the same price. To see this, imagine that $p_1 = p_2 = 2.00$. Firm 2 has no incentive to deviate, but this is no longer true for firm 1, which will be happy to cut its price by 0.01. We know that firm 2 will not be willing to sell at a price below $p = 2$, so one possible Nash equilibrium (you are asked to find more in exercise 5.12 at the end of this chapter) is

$$(p_1^*, p_2^*) = (1.99, 2.00).$$

Now we can ask ourselves what happens if $\varepsilon = 0$. If we would think of using a "limit" approach to answer this question then we may expect a result similar to the one we saw before: if we focus on the equilibrium pair $(p_1^*, p_2^*) = (2 - \varepsilon, 2)$ then as $\varepsilon \to 0$ we must get the Nash equilibrium $p_1 = p_2 = 2$. But is this really an equilibrium? Interestingly, the answer is no! To see this, consider the best response of firm 1. Its payoff function is not continuous when firm 2 offers a price of 2 (or any other positive price). The profit function of firm 1, as a function of p_1 when $p_2 = 2$, is depicted in Figure 5.3. The figure first draws the profits of firm 1 as if it were a monopolist with

---

6. The condition $a > c$ is necessary for firms to be able to produce positive quantities and not lose money.

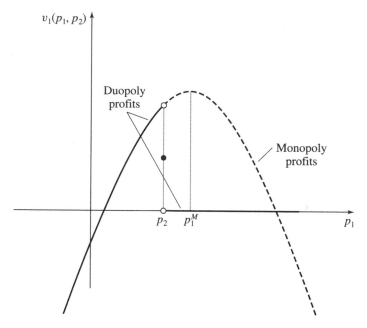

FIGURE 5.3 The profit function in the Bertrand duopoly game.

no competition (the hump-shaped curve), and if this were the case it would charge its monopoly price $p_1^M = 50.5$.[7] If firm 2 charges more than the monopoly price, this will have no impact on the choice of firm 1—it will still charge the monopoly price. If, however, firm 2 charges a price p_2 that is less than the monopoly price then there is a **discontinuity in the payoff function** of firm 1: as its price p_1 approaches p_2 from below, its profits rise. However, when it hits p_2 exactly then it will split the market and see its profits drop by half. Algebraically the profit function of firm 1 when $p_2 = 2$ is given by

$$v_i(p_1, 2) = \begin{cases} (100 - p_1)(p_1 - 1) & \text{if } p_1 < 2 \\ \frac{(100 - p_1)(p_1 - 1)}{2} & \text{if } p_1 = 2 \\ 0 & \text{if } p_1 > 2. \end{cases}$$

This discontinuity causes firm 1 to *not* have a well-defined best response correspondence when $p_2 < 50.5$. Firm 1 wants to set a price *as close to p_2 as it can,* but it does not want to reach p_2 because then it splits the market and experiences a sizable decrease in profits. Once its price goes above p_2 then firm 1's profits drop further to zero. Indeed the consequence must be that a Nash equilibrium does not exist precisely because firm 1 does not have a "well-behaved" payoff function.

To see this directly, first observe that there cannot be a Nash equilibrium with $p_i \geq p_j > 2$: firm i would want to deviate to some $p \in (2, p_j)$. Second, observe that there cannot be a Nash equilibrium with $p_i \leq p_j < 1$: firm i would want to deviate to any $p > p_j$. Hence if there is a Nash equilibrium, it must have the prices between 1 and 2. Similar to the first observation, within this range we cannot have $p_1 \geq p_2$ (firm

---

7. The maximization here is for the profit function $v_1 = (100 - p)(p - c_1)$, where $c_1 = 1$.

1 would want to deviate just slightly below p_2), and similar to the second observation, we cannot have $p_2 \leq p_1$ (firm 2 would want to deviate to any price above p_1).

This problem of payoff discontinuity is one that we will avoid in the remainder of this book precisely because it will often lead to problems of nonexistence of equilibrium. We need to remember that it is a problem that disappears when we have some discreteness in the actions of players, and if we think that our choice of continuous strategies is one of convenience (to use calculus for optimization), then we may feel comfortable enough ignoring such anomalies. We discuss the existence of Nash equilibria further in Section 6.4.

Remark   It is worth pointing out an interesting difference between the Cournot game and the Bertrand game. In the Cournot game the best-response function of each player is downward sloping. That is, the more player j produces, the lower is the best-response quantity of player i. In the Bertrand game, however, for prices between marginal costs (equal to 10 in the leading example) and the monopoly price (equal to 55), the higher the price set by player j, the higher is the best-response price of player i. These differences have received some attention in the literature. Games for which the best response of one player decreases in the choice of the other, like the Cournot game, are called games with **strategic substitutes.** Another example of a game with strategic substitutes is the tragedy of the commons. In contrast, games for which the best response of one player increases in the choice of the other, like the Bertrand game, are called games with **strategic complements.** Another example of a game with strategic complements appears in exercise 5.10 at the end of the chapter. There are several interesting insights to be derived from distinguishing between strategic substitutes and strategic complements. For a nice example see Fudenberg and Tirole (1984).

5.2.5   Political Ideology and Electoral Competition

Given a population of citizens who vote for political candidates, how should candidates position themselves along the political spectrum? One view of the world is that a politician cares only about representing his true beliefs, and that drives the campaign. Another more cynical view is that politicians care only about getting elected and hence will choose a platform that maximizes their chances. This is precisely the view taken in the seminal model introduced by Hotelling (1929).[8]

To consider a simple variant of Hotelling's original model, imagine that there are two politicians, each caring only about being elected. There are 101 citizens, each labeled by an integer $-50, -49, \ldots, 0, \ldots, +49, +50$. Each citizen has political preferences: for simplicity let's call the "-50" citizen the most "left"-leaning citizen and the "$+50$" citizen the most "right"-leaning citizen.

Each candidate i chooses his platform as a policy $a_i \in \{-50, -49, \ldots, 0, \ldots, +49, +50\}$ so that each policy is associated with the citizen for whom this policy is ideal. Each citizen chooses the candidate whose platform is closest to his political preferences. For example, if candidate 1 chooses platform $a_1 = -15$ while candidate 2 chooses platform $a_2 = +22$, then all the citizens at or above $+22$ will surely vote

8. Hotelling's main object of analysis was competition between firms. However, he did also discuss the example of electoral competition, yielding important insights into rational choice–based political science.

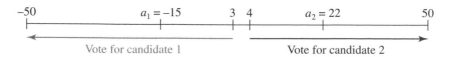

FIGURE 5.4 The Hotelling model of voting behavior.

for candidate 2, all the citizens at or below -15 will surely vote for candidate 1, and those in between will split between the candidates. In particular those citizens at or below 3 will vote for candidate 1, while those at or above 4 will vote for candidate 2. The reason is that -15 is at a distance of 18 away from citizen 3, while $+22$ is at a distance of 19 away. This is shown in Figure 5.4.

The outcome is determined by majority rule: if a majority of citizens vote for candidate i then he wins. Since there is an odd number of voters, unless someone is indifferent, one candidate will always win. In the event that a citizen is indifferent between the candidates then the citizen tosses a coin to determine for whom to vote. Assume that our candidates want to win, so that they prefer winning to a tie, and they prefer a tie to losing. Now consider the best response of player i. If player j chooses a policy $a_j > 0$ then by choosing $a_i = a_j$ or $a_i = -a_j$ there will be a tie.[9] By choosing $a_i > a_j$ or $a_i < -a_j$, player i will surely lose, while by choosing $a_i \in [-a_j + 1, a_j - 1]$, player i will win, so any platform in this interval is a best response to $a_j > 0$. Similarly a symmetric argument implies that any platform in the interval $a_i \in [a_j + 1, -a_j - 1]$ is a best response to $a_j < 0$. Observe that the best response to $a_j = 0$ is zero.[10] Thus we can write the best-response correspondence of each player as

$$BR_i(a_j) = \begin{cases} [a_j + 1, -a_j - 1] & \text{if } a_j < 0 \\ 0 & \text{if } a_j = 0 \\ [-a_j + 1, a_j - 1] & \text{if } a_j > 0. \end{cases}$$

From here it is easy to see that there is a unique Nash equilibrium, $a_1 = a_2 = 0$, implying that both candidates position their platforms smack in the middle of the political spectrum! Indeed, as Hotelling (1929, p. 54) wrote, "The competition for votes between the Republican and Democratic parties does not lead to a clear drawing of issues, and adoption of two strongly contrasted positions between which the voter may choose. Instead, each party strives to make its platform as much like the other's as possible." A fine insight in 1929, and one that is echoed frequently today.

This simple example is related to a powerful result known as the **median voter theorem.** It states that if voters are different from one another along a single-dimensional "preference" line, as in Hotelling's model, and if each prefers his own political location, with other platforms being less and less attractive the farther away

---

9. For example, if candidate 1 chooses $a_1 = -10$ and candidate 2 chooses $a_2 = 10$ then all the "positive" citizens will vote for candidate 2, all the "negative" ones will vote for candidate 1, and citizen 0 is indifferent so he may choose not to vote. In this event there will be a tie. If we force every citizen to vote one way or another then there will be no ties because we assumed an odd number of citizens.

10. Notice that the best response to $a_j = -1$ or $a_j = +1$ is also unique and equal to $a_i = 0$, because in these cases the interval previously identified collapses to 0.

they fall to either side of that location,[11] then the political platform located at the median voter will defeat *any other platform* in a simple majority vote. The theorem was first articulated by Black (1948), and it received prominence in Downs's famous 1957 book. Nevertheless one can see how the seed of the idea had been planted as far back as Hotelling's formalization of spatial competition.

Remark In the more common representation of the Hotelling competition model, the citizens are a continuum of voters, say, given by the interval $A = [\underline{a}, \overline{a}]$, with distribution $F(a)$ to determine the distribution of each political preference. We define the **median voter** as that voter a^m for which $\Pr\{a \leq a^m\} = \frac{1}{2}$. $\left(\text{For example, if } F(\cdot) \text{ is uniform then } a^m = \frac{\underline{a} + \overline{a}}{2}.\right)$ The best-response correspondence is similar to that given earlier, but for all choices of player j that are not equal to a^m player i's best response is an open interval. For example, if $a_j > a^m$, and if $a' < a^m$ is an "opposite" policy that ties with a_j, then the best response of player i is to choose any platform that lies in the open interval (a', a_j). If, however, $a_j = a^m$ then player i's unique best response is to choose $a_i = a^m$, implying that both candidates choosing a^m is the unique Nash equilibrium. You are asked to prove this in exercise 5.15.

5.3 Summary

- Any strategy profile for which players are playing mutual best responses is a Nash equilibrium, making this equilibrium concept self-enforcing.

- If a profile of strategies is the unique survivor of IESDS or is the unique rationalizable profile of strategies then it is a Nash equilibrium.

- If a profile of strategies is a Nash equilibrium then it must survive IESDS and it must be rationalizable, but not every strategy that survives IESDS or that is rationalizable is a Nash equilibrium.

- Nash equilibrium analysis can shed light on phenomena such as the tragedy of the commons and the nature of competition in markets and in politics.

5.4 Exercises

5.1 **Prove Proposition 5.1.**

5.2 **Weak Dominance:** A strategy $s^W \in S$ is a weakly dominant strategy equilibrium if $s_i^W \in S_i$ is a weakly dominant strategy for all $i \in N$, that is, if $v_i(s_i^W, s_{-i}) \geq v_i(s_i', s_{-i})$ for all $s_i' \in S_I$ and for all $s_{-i} \in S_{-i}$. Provide an example of a game for which there is a weakly dominant strategy equilibrium as well as another Nash equilibrium.

5.3 **Nash and IESDS:** Consider a two-player game with m pure strategies for each player that can be represented by an $m \times m$ matrix. Assume that for each player no two payoffs in the matrix are the same.

    a. Show that if $m = 2$ and the game has a unique pure-strategy Nash equilibrium then this is the unique strategy profile that survives IESDS.

---

11. This condition on preferences is called that of "single-peaked" preferences: if we draw the utility function of some voter with his political "bliss point" being $a \in A$, the utility is highest at a and declines in both directions, hence the single peak at a.

b. Show that if $m = 3$ and the game has a unique pure-strategy equilibrium then it may not be the only strategy profile that survives IESDS.

5.4 **Splitting Pizza:** You and a friend are in an Italian restaurant, and the owner offers both of you a free eight-slice pizza under the following condition. Each of you must simultaneously announce how many slices you would like; that is, each player $i \in \{1, 2\}$ names his desired amount of pizza, $0 \le s_i \le 8$. If $s_1 + s_2 \le 8$ then the players get their demands (and the owner eats any leftover slices). If $s_1 + s_2 > 8$, then the players get nothing. Assume that you each care only about how much pizza you individually consume, and the more the better.

   a. Write out or graph each player's best-response correspondence.
   b. What outcomes can be supported as pure-strategy Nash equilibria?

5.5 **Public Good Contribution:** Three players live in a town, and each can choose to contribute to fund a streetlamp. The value of having the streetlamp is 3 for each player, and the value of not having it is 0. The mayor asks each player to contribute either 1 or nothing. If at least two players contribute then the lamp will be erected. If one player or no players contribute then the lamp will not be erected, in which case any person who contributed will not get his money back.

   a. Write out or graph each player's best-response correspondence.
   b. What outcomes can be supported as pure-strategy Nash equilibria?

5.6 **Hawk-Dove:** The following game has been widely used in evolutionary biology to understand how fighting and display strategies by animals could coexist in a population. For a typical Hawk-Dove game there are resources to be gained (e.g., food, mates, territories), denoted as v. Each of two players can choose to be aggressive, as Hawk (H), or compromising, as Dove (D). If both players choose H then they split the resources but lose some payoff from injuries, denoted as k. Assume that $k > \frac{v}{2}$. If both choose D then they split the resources but engage in some display of power that carries a display cost d, with $d < \frac{v}{2}$. Finally, if player i chooses H while j chooses D then i gets all the resources while j leaves with no benefits and no costs.

   a. Describe this game in a matrix.
   b. Assume that $v = 10$, $k = 6$, and $d = 4$. What outcomes can be supported as pure-strategy Nash equilibria?[12]

5.7 **The n-Player Tragedy of the Commons:** Suppose there are n players in the tragedy of the commons example in Section 5.2.2.

   a. Find the Nash equilibrium of this game. How does n affect the Nash outcome?
   b. Find the socially optimal outcome with n players. How does n affect this outcome?
   c. How does the Nash equilibrium outcome compare to the socially efficient outcome as n approaches infinity?

5.8 **The n-Firm Cournot Model:** Suppose there are n firms in the Cournot oligopoly model. Let q_i denote the quantity produced by firm i, and let

---

12. In the evolutionary biology literature, the analysis performed is of a very different nature. Instead of considering the Nash equilibrium analysis of a static game, the analysis is a dynamic one in which successful strategies "replicate" in a large population. This analysis is part of a methodology called evolutionary game theory. For more on the subject see Gintis (2000).

$Q = q_i + \cdots + q_n$ denote the aggregate production. Let $P(Q)$ denote the market clearing price (when demand equals Q) and assume that the inverse demand function is given by $P(Q) = a - Q$, where $Q \leq a$. Assume that firms have no fixed cost and that the cost of producing quantity q_i is cq_i (all firms have the same marginal cost, and assume that $c < a$).

    a.   Model this as a normal-form game.

    b.   What is the Nash (Cournot) equilibrium of the game in which firms choose their quantities simultaneously?

    c.   What happens to the equilibrium price as n approaches infinity? Is this familiar?

5.9   **Tragedy of the Roommates:** You and your $n - 1$ roommates each have five hours of free time you could spend cleaning your apartment. You all dislike cleaning, but you all like having a clean apartment: each person's payoff is the total hours spent (by everyone) cleaning, minus a number c times the hours spent (individually) cleaning. That is,

$$v_i(s_1, s_2, \ldots, s_n) = -c \cdot s_i + \sum_{j=1}^{n} s_j.$$

Assume everyone chooses simultaneously how much time to spend cleaning.

    a.   Find the Nash equilibrium if $c < 1$.

    b.   Find the Nash equilibrium if $c > 1$.

    c.   Set $n = 5$ and $c = 2$. Is the Nash equilibrium Pareto efficient? If not, can you find an outcome in which everyone is better off than in the Nash equilibrium outcome?

5.10   **Synergies:** Two division managers can invest time and effort in creating a better working relationship. Each invests $e_i \geq 0$, and if both invest more then both are better off, but it is costly for each manager to invest. In particular the payoff function for player i from effort levels (e_i, e_j) is $v_i(e_i, e_j) = (a + e_j)e_i - e_i^2$.

    a.   What is the best-response correspondence of each player?

    b.   In what way are the best-response correspondences different from those in the Cournot game? Why?

    c.   Find the Nash equilibrium of this game and argue that it is unique.

5.11   **Wasteful Shipping Costs:** Consider two countries, A and B, each with a monopolist that owns the only active coal mine in the country. Let firm 1 be the firm located in country A and firm 2 the one in country B. Let q_i^j, $i \in \{1, 2\}$ and $j \in \{A, B\}$ denote the quantity that firm i sells in country j. Consequently let $q_i = q_i^A + q_i^B$ be the total quantity produced by firm $i \in \{1, 2\}$ and let $q^j = q_1^j + q_2^j$ be the total quantity sold in country $j \in \{A, B\}$. The demand for coal in countries A and B is then given respectively by

$$p^j = 90 - q^j, \quad j \in \{A, B\},$$

and the cost of production for each firm is given by

$$c_i(q_i) = 10q_i, \quad i \in \{1, 2\}.$$

    a.   Assume that the countries do not have a trade agreement and, in fact, that the importation of coal into either country is prohibited.

This implies that $q_2^A = q_1^B = 0$ is set as a political constraint. What quantities q_1^A and q_2^B will both firms produce?

Now assume that the two countries sign a free-trade agreement that allows foreign firms to sell in each country without any tariffs. There are, however, shipping costs. If firm i sells quantity q_i^j in the foreign country (i.e., firm 1 selling in B or firm 2 selling in A) then shipping costs are equal to $10q_i^j$. Assume further that *each firm* chooses a pair of quantities q_i^A, q_i^B simultaneously, $i \in \{1, 2\}$, so that a profile of actions consists of four quantity choices.

b. Model this as a normal-form game and find a Nash equilibrium of the game you described. Is it unique?

Now assume that before the game you described in (b) is played the research department of firm 1 discovers that shipping coal on the existing vessels causes the release of pollutants. If the firm would disclose this report to the World Trade Organization (WTO) then the WTO would prohibit the use of the existing ships. Instead a new shipping technology would be offered that would increase shipping costs to $40q_i^j$ (instead of $10q_i^j$ as given earlier).

c. Would firm 1 be willing to release the information to the WTO? Justify your answer with an equilibrium analysis.

5.12 **Asymmetric Bertrand:** Consider the Bertrand game with $c_1(q_1) = q_1$ and $c_2(q_2) = 2q_2$ and demand equal to $p = 100 - q$, in which firms must choose prices in increments of $0.01. We have seen in Section 5.2.4 that one possible Nash equilibrium is $(p_1^*, p_2^*) = (1.99, 2.00)$.

a. Show that there are other Nash equilibria for this game.
b. How many Nash equilibria does this game have?

5.13 **Comparative Economics:** Two high-tech firms (1 and 2) are considering a joint venture. Each firm i can invest in a novel technology and can choose a level of investment $x_i \in [0, 5]$ at a cost of $c_i(x_i) = x_i^2/4$ (think of x_i as how many hours to train employees or how much capital to spend for R&D labs). The revenue of each firm depends on both its investment and the other firm's investment. In particular if firms i and j choose x_i and x_j, respectively, then the **gross revenue** to firm i is

$$R(x_i, x_j) = \begin{cases} 0 & \text{if } x_i < 1 \\ 2 & \text{if } x_i \geq 1 \quad \text{and } x_j < 2 \\ x_i \cdot x_j & \text{if } x_i \geq 1 \quad \text{and } x_j \geq 2. \end{cases}$$

a. Write down mathematically and draw the **profit function** (gross revenue minus costs) of firm i as a function of x_i for three cases: (i) $x_j < 2$, (ii) $x_j = 2$, and (iii) $x_j = 4$.
b. What is the best-response function of firm i?
c. It turns out that there are two *identical* pairs of such firms; that is, the description applies to both pairs. One pair is in Russia, where coordination is hard to achieve and businesspeople are very cautious, and the other pair is in Germany, where coordination is common and

businesspeople expect their partners to go the extra mile. You learn that the Russian firms are earning significantly lower profits than the German firms, despite the fact that their technologies are identical. Can you use Nash equilibrium analysis to shed light on this dilemma? If so, be precise and use your previous analysis to do so.

5.14 **Negative Ad Campaigns:** Each one of two political parties can choose to buy time on commercial radio shows to broadcast negative ad campaigns against its rival. These choices are made simultaneously. Government regulations forbid a party from buying more than 2 hours of negative campaign time, so that each party cannot choose an amount of negative campaigning above 2 hours. Given a pair of choices (a_1, a_2), the payoff of party i is given by the following function: $v_i(a_1, a_2) = a_i - 2a_j + a_i a_j - (a_i)^2$.

    a. What is the normal-form representation of this game?
    b. What is the best-response function for each party?
    c. What is the pure-strategy Nash equilibrium? Is it unique?
    d. If the parties could sign a binding agreement on how much to campaign, what levels would they choose?

5.15 **Hotelling's Continuous Model:** Consider Hotelling's model, in which citizens are a continuum of voters on the interval $A = [-a, a]$, with uniform distribution $U(a)$.

    a. What is the best response of candidate i if candidate j is choosing $a_j > 0$?
    b. Show that the unique Nash equilibrium is $a_1 = a_2 = 0$.
    c. Show that for a general distribution $F(\cdot)$ over $[-a, a]$ the unique Nash equilibrium is where each candidate chooses the policy associated with the median voter.

5.16 **Hotelling's Price Competition:** Imagine a continuum of potential buyers, located on the line segment $[0, 1]$, with uniform distribution. (Hence the "mass" or quantity of buyers in the interval $[a, b]$ is equal to $b - a$.) Imagine two firms, players 1 and 2, who are located at each end of the interval (player 1 at the 0 point and player 2 at the 1 point). Each player i can choose its price p_i, and each customer goes to the vendor who offers him the highest value. However, price alone does not determine the value; distance is important as well. In particular each buyer who buys the product from player i has a net value of $v - p_i - d_i$, where d_i is the distance between the buyer and vendor i and represents the transportation costs of buying from vendor i. Thus buyer $a \in [0, 1]$ buys from 1 and not 2 if $v - p_1 - d_1 > v - p_2 - d_2$ and if buying is better than getting zero. (Here $d_1 = a$ and $d_2 = 1 - a$. The buying choice would be reversed if the inequality were reversed.) Finally, assume that the cost of production is zero.

    a. Assume that v is very large so that all the customers will be served by at least one firm, and that some customer $x^* \in [0, 1]$ is indifferent between the two firms. What is the best-response function of each player?
    b. Assume that $v = 1$. What is the Nash equilibrium? Is it unique?
    c. Assume that $v = 1$. Now assume that the transportation costs are $\frac{1}{2}d_i$, so that a buyer buys from 1 if and only if $v - p_1 - \frac{1}{2}d_1 > v - p_2 - \frac{1}{2}d_2$. Write down the best-response function of each player and solve for the Nash equilibrium.

d. Assume that $v = 1$. Following your analysis in (c), imagine that transportation costs are αd_i, with $\alpha \in [0, \frac{1}{2}]$. What happens to the Nash equilibrium as $\alpha \to 0$? What is the intuition for this result?

5.17 **To Vote or Not to Vote:** Two candidates, D and R, are running for mayor in a town with n residents. A total of $0 < d < n$ residents support candidate D, while the remainder, $r = n - d$, support candidate R. The value for each resident for having his candidate win is 4, for having him tie is 2, and for having him lose is 0. Going to vote costs each resident 1.

a. Let $n = 2$ and $d = 1$. Write down this game as a matrix and solve for the Nash equilibrium.
b. Let $n > 2$ be an even number and let $d = r = \frac{n}{2}$. Find all the Nash equilibria.
c. Assume now that the cost of voting is equal to 3. How does your answer to (a) and (b) change?

5.18 **Political Campaigning:** Two candidates are competing in a political race. Each candidate i can spend $s_i \geq 0$ on ads that reach out to voters, which in turn increases the probability that candidate i wins the race. Given a pair of spending choices (s_1, s_2), the probability that candidate i wins is given by $\frac{s_i}{s_1+s_2}$. If neither spends any resources then each wins with probability $\frac{1}{2}$. Each candidate values winning at a payoff of $v > 0$, and the cost of spending s_i is just s_i.

a. Given two spend levels (s_1, s_2), write down the expected payoff of a candidate i.
b. What is the function that represents each player's best-response function?
c. Find the unique Nash equilibrium.
d. What happens to the Nash equilibrium spending levels if v increases?
e. What happens to the Nash equilibrium levels if player 1 still values winning at v but player 2 values winning at kv, where $k > 1$?

6

Mixed Strategies

In the previous chapters we restricted players to using pure strategies and we postponed discussing the option that a player may choose to randomize between several of his pure strategies. You may wonder why anyone would wish to randomize between actions. This turns out to be an important type of behavior to consider, with interesting implications and interpretations. In fact, as we will now see, there are many games for which there will be no equilibrium predictions if we do not consider the players' ability to choose stochastic strategies.

Consider the following classic zero-sum game called Matching Pennies.[1] Players 1 and 2 each put a penny on a table simultaneously. If the two pennies come up the same side (heads or tails) then player 1 gets both; otherwise player 2 does. We can represent this in the following matrix:

| | | Player 2 | |
		H	T
Player 1	H	1, −1	−1, 1
	T	−1, 1	1, −1

The matrix also includes the best-response choices of each player using the method we introduced in Section 5.1.1 to find pure-strategy Nash equilibria. As you can see, this method does not work: Given a belief that player 1 has about player 2's choice, he always wants to match it. In contrast, given a belief that player 2 has about player 1's choice, he would like to choose the opposite orientation for his penny. Does this mean that a Nash equilibrium fails to exist? We will soon see that a Nash equilibrium will indeed exist *if we allow players to choose random strategies,* and there will be an intuitive appeal to the proposed equilibrium.

Matching Pennies is not the only simple game that fails to have a pure-strategy Nash equilibrium. Recall the child's game rock-paper-scissors, in which rock beats

---

1. A *zero-sum game* is one in which the gains of one player are the losses of another, hence their payoffs always sum to zero. The class of zero-sum games was the main subject of analysis before Nash introduced his solution concept in the 1950s. These games have some very nice mathematical properties and were a central object of analysis in von Neumann and Morgenstern's (1944) seminal book.

scissors, scissors beats paper, and paper beats rock. If winning gives the player a payoff of 1 and the loser a payoff of -1, and if we assume that a tie is worth 0, then we can describe this game by the following matrix:

		Player 2		
		R	P	S
	R	0, 0	$-1, 1$	$1, -1$
Player 1	P	$1, -1$	0, 0	$-1, 1$
	S	$-1, 1$	$1, -1$	0, 0

It is rather straightforward to write down the best-response correspondence for player 1 when he believes that player 2 will play one of his pure strategies as follows:

$$s_1(s_2) = \begin{cases} P & \text{when } s_2 = R \\ S & \text{when } s_2 = P \\ R & \text{when } s_2 = S, \end{cases}$$

and a similar (symmetric) list would be the best-response correspondence of player 2. Examining the two best-response correspondences immediately implies that there is no pure-strategy equilibrium, just like in the Matching Pennies game. The reason is that, starting with any pair of pure strategies, at least one player is not playing a best response and will want to change his strategy in response.

6.1 Strategies, Beliefs, and Expected Payoffs

We now introduce the possibility that players choose stochastic strategies, such as flipping a coin or rolling a die to determine what they will choose to do. This approach will turn out to offer us several important advances over that followed so far. Aside from giving the players a richer set of actions from which to choose, it will more importantly give them a richer set of possible beliefs that capture an uncertain world. If player i can believe that his opponents are choosing stochastic strategies, then this puts player i in the same kind of situation as a decision maker who faces a decision problem with probabilistic uncertainty. If you are not familiar with such settings, you are encouraged to review Chapter 2, which lays out the simple decision problem with random events.

6.1.1 Finite Strategy Sets

We start with the basic definition of random play when players have finite strategy sets S_i:

Definition 6.1 Let $S_i = \{s_{i1}, s_{i2}, \ldots, s_{im}\}$ be player i's finite set of pure strategies. Define $\triangle S_i$ as the **simplex** of S_i, which is the set of all probability distributions over S_i. A **mixed strategy** for player i is an element $\sigma_i \in \triangle S_i$, so that $\sigma_i = \{\sigma_i(s_{i1}), \sigma_i(s_{i2}), \ldots, \sigma_i(s_{im}))$ is a probability distribution over S_i, where $\sigma_i(s_i)$ is the probability that player i plays s_i.

That is, a mixed strategy for player i is just a probability distribution over his pure strategies. Recall that any probability distribution $\sigma_i(\cdot)$ over a finite set of elements (a finite state space), in our case S_i, must satisfy two conditions:

1. $\sigma_i(s_i) \geq 0$ for all $s_i \in S_i$, and
2. $\sum_{s_i \in S_i} \sigma_i(s_i) = 1$.

That is, the probability of any event happening must be nonnegative, and the sum of the probabilities of all the possible events must add up to one.[2] Notice that every pure strategy is a mixed strategy with a *degenerate* distribution that picks a single pure strategy with probability one and all other pure strategies with probability zero.

As an example, consider the Matching Pennies game described earlier, with the matrix

		Player 2	
		H	T
Player 1	H	1, −1	−1, 1
	T	−1, 1	1, −1

For each player i, $S_i = \{H, T\}$, and the simplex, which is the set of mixed strategies, can be written as

$$\triangle S_i = \{(\sigma_i(H), \sigma_i(T)) : \sigma_i(H) \geq 0, \sigma_i(T) \geq 0, \sigma_i(H) + \sigma_i(T) = 1\}.$$

We read this as follows: the set of mixed strategies is the set of all pairs $(\sigma_i(H), \sigma_i(T))$ such that both are nonnegative numbers, and they both sum to one.[3] We use the notation $\sigma_i(H)$ to represent the probability that player i plays H and $\sigma_i(T)$ to represent the probability that player i plays T.

Now consider the example of the rock-paper-scissors game, in which $S_i = \{R, P, S\}$ (for rock, paper, and scissors, respectively). We can define the simplex as

$$\triangle S_i = \{(\sigma_i(R), \sigma_i(P), \sigma_i(S)) : \sigma_i(R), \sigma_i(P), \sigma_i(S) \geq 0, \sigma_i(R) + \sigma_i(P) + \sigma_i(S) = 1\},$$

which is now three numbers, each defining the probability that the player plays one of his pure strategies. As mentioned earlier, a pure strategy is just a special case of a mixed strategy. For example, in this game we can represent the pure strategy of playing R with the degenerate mixed strategy: $\sigma(R) = 1$, $\sigma(P) = \sigma(S) = 0$.

From our definition it is clear that when a player uses a mixed strategy, he may choose not to use all of his pure strategies in the mix; that is, he may have some pure strategies that are not selected with positive probability. Given a player's

---

2. The notation $\sum_{s_i \in S_i} \sigma(s_i)$ means the sum of $\sigma(s_i)$ over all the $s_i \in S_i$. If S_i has m elements, as in the definition, we could write this as $\sum_{k=1}^{m} \sigma_i(s_{ik})$.

3. The simplex of this two-element strategy set can be represented by a single number $p \in [0, 1]$, where p is the probability that player i plays H and $1 - p$ is the probability that player i plays T. This follows from the definition of a probability distribution over a two-element set. In general the simplex of a strategy set with m pure strategies will be in an $(m - 1)$-dimensional space, where each of the $m - 1$ numbers is in $[0, 1]$, and will represent the probability of the first $m - 1$ pure strategies. All sum to a number equal to or less than one so that the remainder is the probability of the mth pure strategy.

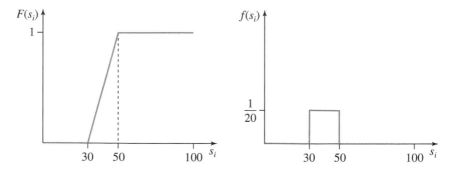

FIGURE 6.1 A continuous mixed strategy in the Cournot game.

mixed strategy $\sigma_i(\cdot)$, it will be useful to distinguish between pure strategies that are chosen with a positive probability and those that are not. We offer the following definition:

Definition 6.2 Given a mixed strategy $\sigma_i(\cdot)$ for player i, we will say that a pure strategy $s_i \in S_i$ is **in the support of** $\sigma_i(\cdot)$ if and only if it occurs with positive probability, that is, $\sigma_i(s_i) > 0$.

For example, in the game of rock-paper-scissors, a player can choose rock or paper, each with equal probability, and not choose scissors. In this case $\sigma_i(R) = \sigma_i(P) = 0.5$ and $\sigma_i(S) = 0$. We will then say that R and P are in the support of $\sigma_i(\cdot)$, but S is not.

6.1.2 Continuous Strategy Sets

As we have seen with the Cournot and Bertrand duopoly examples, or the tragedy of the commons example in Section 5.2.2, the pure-strategy sets that players have need not be finite. In the case in which the pure-strategy sets are well-defined intervals, a mixed strategy will be given by a cumulative distribution function:

Definition 6.3 Let S_i be player i's pure-strategy set and assume that S_i is an interval. A **mixed strategy** for player i is a cumulative distribution function $F_i : S_i \to [0, 1]$, where $F_i(x) = \Pr\{s_i \leq x\}$. If $F_i(\cdot)$ is differentiable with density $f_i(\cdot)$ then we say that $s_i \in S_i$ is in the support of $F_i(\cdot)$ if $f_i(s_i) > 0$.

As an example, consider the Cournot duopoly game with a capacity constraint of 100 units of production, so that $S_i = [0, 100]$ for $i \in \{1, 2\}$. Consider the mixed strategy in which player i chooses a quantity between 30 and 50 using a uniform distribution. That is,

$$F_i(s_i) = \begin{cases} 0 & \text{for } s_i < 30 \\ \frac{s_i - 30}{20} & \text{for } s_i \in [30, 50] \\ 1 & \text{for } s_i > 50 \end{cases} \quad \text{and} \quad f_i(s_i) = \begin{cases} 0 & \text{for } s_i < 30 \\ \frac{1}{20} & \text{for } s_i \in [30, 50] \\ 0 & \text{for } s_i > 50. \end{cases}$$

These two functions are depicted in Figure 6.1.

We will typically focus on games with finite strategy sets to illustrate most of the examples with mixed strategies, but some interesting examples will have infinite strategy sets and will require the use of cumulative distributions and densities to explore behavior in mixed strategies.

6.1.3 Beliefs and Mixed Strategies

As we discussed earlier, introducing probability distributions not only enriches the set of actions from which a player can choose but also allows us to enrich the beliefs that players can have. Consider, for example, player i, who plays against opponents $-i$. It may be that player i is uncertain about the behavior of his opponents for many reasons. For example, he may believe that his opponents are indeed choosing mixed strategies, which immediately implies that their behavior is not fixed but rather random. An alternative interpretation is the situation in which player i is playing a game against an opponent that he does not know, whose background will determine how he will play. This interpretation will be revisited in Section 12.5, and it is a very appealing justification for beliefs that are random and behavior that is consistent with these beliefs.

To introduce beliefs about mixed strategies formally we define them as follows:

Definition 6.4 A **belief** for player i is given by a *probability distribution* $\pi_i \in \Delta S_{-i}$ over the strategies of his opponents. We denote by $\pi_i(s_{-i})$ the probability player i assigns to his opponents playing $s_{-i} \in S_{-i}$.

Thus a *belief* for player i is a probability distribution over the strategies of his opponents. Notice that the belief of player i lies in the same set that represents the profiles of mixed strategies of player i's opponents. For example, in the rock-paper-scissors game, we can represent the beliefs of player 1 as a triplet, $(\pi_1(R), \pi_1(P), \pi_1(S))$, where by definition $\pi_1(R), \pi_1(P), \pi_1(S) \geq 0$ and $\pi_1(R) + \pi_1(P) + \pi_1(S) = 1$. The interpretation of $\pi_1(s_2)$ is the probability that player 1 assigns to player 2 playing some particular $s_2 \in S_2$. Recall that the strategy of player 2 is a triplet $\sigma_2(R), \sigma_2(P), \sigma_2(S) \geq 0$, with $\sigma_2(R) + \sigma_2(P) + \sigma_2(S) = 1$, so we can clearly see the analogy between π and σ.

6.1.4 Expected Payoffs

Consider the Matching Pennies game described previously, and assume for the moment that player 2 chooses the mixed strategy $\sigma_2(H) = \frac{1}{3}$ and $\sigma_2(T) = \frac{2}{3}$. If player 1 plays H then he will win and get 1 with probability $\frac{1}{3}$ while he will lose and get -1 with probability $\frac{2}{3}$. If, however, he plays T then he will win and get 1 with probability $\frac{2}{3}$ while he will lose and get -1 with probability $\frac{1}{3}$. Thus by choosing different actions player 1 will face different lotteries, as described in Chapter 2.

To evaluate these lotteries we will resort to the notion of expected payoff over lotteries as presented in Section 2.2. Thus we define the expected payoff of a player as follows:

Definition 6.5 The **expected payoff** of player i when he chooses the pure strategy $s_i \in S_i$ and his opponents play the mixed strategy $\sigma_{-i} \in \Delta S_{-i}$ is

$$v_i(s_i, \sigma_{-i}) = \sum_{s_{-i} \in S_{-i}} \sigma_{-i}(s_{-i}) v_i(s_i, s_{-i}).$$

Similarly the expected payoff of player i when he chooses the mixed strategy $\sigma_i \in \Delta S_i$ and his opponents play the mixed strategy $\sigma_{-i} \in \Delta S_{-i}$ is

$$v_1(\sigma_i, \sigma_{-i}) = \sum_{s_i \in S_i} \sigma_i(s_i) v_i(s_i, \sigma_{-i}) = \sum_{s_i \in S_i} \left(\sum_{s_{-i} \in S_{-i}} \sigma_i(s_i) \sigma_{-i}(s_{-i}) v_i(s_i, s_{-i}) \right).$$

The idea is a straightforward adaptation of definition 2.3 in Section 2.2.1. The randomness that player i faces if he chooses some $s_i \in S_i$ is created by the random selection of $s_{-i} \in S_{-i}$ that is described by the probability distribution $\sigma_{-i}(\cdot)$. Clearly the definition we just presented is well defined only for finite strategy sets S_i. The analog to interval strategy sets is a straightforward adaptation of the second part of definition 2.3.[4]

As an example, recall the rock-paper-scissors game:

		Player 2		
		R	P	S
Player 1	R	0, 0	−1, 1	1, −1
	P	1, −1	0, 0	−1, 1
	S	−1, 1	1, −1	0, 0

and assume that player 2 plays $\sigma_2(R) = \sigma_2(P) = \frac{1}{2}; \sigma_2(S) = 0$. We can now calculate the expected payoff for player 1 from any of his pure strategies,

$$v_1(R, \sigma_2) = \frac{1}{2} \times 0 + \frac{1}{2} \times (-1) + 0 \times 1 = -\frac{1}{2}$$

$$v_1(P, \sigma_2) = \frac{1}{2} \times 1 + \frac{1}{2} \times 0 + 0 \times (-1) = \frac{1}{2}$$

$$v_1(S, \sigma_2) = \frac{1}{2} \times (-1) + \frac{1}{2} \times 1 + 0 \times 0 = 0.$$

It is easy to see that player 1 has a unique best response to this mixed strategy of player 2. If he plays P, he wins or ties with equal probability, while his other two pure strategies are worse: with R he either loses or ties and with S he either loses or wins. Clearly if his beliefs about the strategy of his opponent are different then player 1 is likely to have a different best response.

It is useful to consider an example in which the players have strategy sets that are intervals. Consider the following game, known as an **all-pay auction,** in which two players can bid for a dollar. Each can submit a bid that is a real number (we are not restricted to penny increments), so that $S_i = [0, \infty)$, $i \in \{1, 2\}$. The person with the higher bid gets the dollar, but the twist is that both bidders have to pay their bids (hence the name of the game). If there is a tie then both pay and the dollar is awarded to each player with an equal probability of 0.5. Thus if player i bids s_i and player $j \neq i$ bids s_j then player i's payoff is

---

4. Consider a game in which each player has a strategy set given by the interval $S_i = [\underline{s}_i, \overline{s}_i]$. If player 1 is playing s_1 and his opponents, players $j = 2, 3, \ldots, n$, are using the mixed strategies given by the density function $f_j(\cdot)$ then the expected payoff of player 1 is given by

$$\int_{\underline{s}_2}^{\overline{s}_2} \int_{\underline{s}_3}^{\overline{s}_3} \cdots \int_{\underline{s}_n}^{\overline{s}_n} v_i(s_i, s_{-i}) f_2(s_2) f_3(s_3) \cdots f_n(s_n) ds_2 ds_3 \cdots ds_n.$$

For more on this topic see Section 19.4.4.

$$v_i(s_i, s_{-i}) = \begin{cases} -s_i & \text{if } s_i < s_j \\ \frac{1}{2} - s_i & \text{if } s_i = s_j \\ 1 - s_i & \text{if } s_i > s_j. \end{cases}$$

Now imagine that player 2 is playing a mixed strategy in which he is *uniformly* choosing a bid between 0 and 1. That is, player 2's mixed strategy σ_2 is a uniform distribution over the interval 0 and 1, which is represented by the cumulative distribution function and density

$$F_2(s_2) = \begin{cases} s_2 & \text{for } s_2 \in [0, 1] \\ 1 & \text{for } s_2 > 1 \end{cases} \quad \text{and} \quad f_2(s_2) = \begin{cases} 1 & \text{for } s_2 \in [0, 1] \\ 0 & \text{for } s_2 > 1. \end{cases}$$

The expected payoff of player 1 from offering a bid $s_i > 1$ is $1 - s_i < 0$ because he will win for sure, but this would not be wise. The expected payoff from bidding $s_i < 1$ is[5]

$$v_1(s_1, \sigma_2) = \Pr\{s_1 < s_2\}(-s_1) + \Pr\{s_1 = s_2\}\left(\frac{1}{2} - s_1\right) + \Pr\{s_1 > s_2\}\left(1 - s_1\right)$$

$$= (1 - F_2(s_1))(-s_1) + 0\left(\frac{1}{2} - s_1\right) + F_2(s_1)(1 - s_1)$$

$$= 0.$$

Thus when player 2 is using a uniform distribution between 0 and 1 for his bid, then player 1 cannot get any positive *expected* payoff from any bid he offers: any bid less than 1 offers an expected payoff of 0, and any bid above 1 guarantees getting the dollar at an inflated price. This game is one to which we will return later, as it has several interesting features and twists.

6.2   Mixed-Strategy Nash Equilibrium

Now that we are equipped with a richer space for both strategies and beliefs, we are ready to restate the definition of a Nash equilibrium for this more general setup as follows:

Definition 6.6   The mixed-strategy profile $\sigma^* = (\sigma_1^*, \sigma_2^*, \ldots, \sigma_n^*)$ is a *Nash equilibrium* if for each player σ_i^* is a best response to σ_{-i}^*. That is, for all $i \in N$,

$$v_i(\sigma_i^*, \sigma_{-i}^*) \geq v_i(\sigma_i, \sigma_{-i}^*) \ \forall \, \sigma_i \in \Delta S_i.$$

This definition is the natural generalization of definition 5.1. We require that each player be choosing a strategy $\sigma_i^* \in \Delta S_i$ that is (one of) the best choice(s) he can make when his opponents are choosing some profile $\sigma_{-i}^* \in \Delta S_{-i}$.

As we discussed previously, there is another interesting interpretation of the definition of a Nash equilibrium. We can think of σ_{-i}^* as the belief of player i about his opponents, π_i, which captures the idea that player i is uncertain of his opponents' behavior. The profile of mixed strategies σ_{-i}^* thus captures this uncertain belief over all of the pure strategies that player i's opponents can play. Clearly rationality requires

---

5. If player 2 is using a uniform distribution over $[0, 1]$ then $\Pr\{s_1 = s_2\} = 0$ for any $s_1 \in [0, 1]$.

that a player play a best response given his beliefs (and this now extends the notion of rationalizability to allow for uncertain beliefs). A Nash equilibrium requires that these beliefs be correct.

Recall that we defined a pure strategy $s_i \in S_i$ to be *in the support of* σ_i if $\sigma_i(s_i) > 0$, that is, if s_i is played with positive probability (see definition 6.2). Now imagine that in the Nash equilibrium profile σ^* the support of i's mixed strategy σ_i^* contains more than one pure strategy—say s_i and s_i' are both in the support of σ_i^*.

What must we conclude about a rational player i if σ_i^* is indeed part of a Nash equilibrium $(\sigma_i^*, \sigma_{-i}^*)$? By definition σ_i^* is a best response against σ_{-i}^*, which means that given σ_{-i}^* player i cannot do better than to randomize between more than one of his pure strategies, in this case, s_i and s_i'. But when would a player be willing to randomize between two alternative pure strategies? The answer is predictable:

Proposition 6.1 *If σ^* is a Nash equilibrium, and both s_i and s_i' are in the support of σ_i^*, then*

$$v_i(s_i, \sigma_{-i}^*) = v_i(s_i', \sigma_{-i}^*) = v_i(\sigma_i^*, \sigma_{-i}^*).$$

The proof is quite straightforward and follows from the observation that if a player is randomizing between two alternatives then he must be indifferent between them. If this were not the case, say $v_i(s_i, \sigma_{-i}^*) > v_i(s_i', \sigma_{-i}^*)$ with both s_i and s_i' in the support of σ_i^*, then by reducing the probability of playing s_i' from $\sigma_i^*(s_i')$ to zero, and increasing the probability of playing s_i from $\sigma_i^*(s_i)$ to $\sigma_i^*(s_i) + \sigma_i^*(s_i')$, player i's expected payoff must go up, implying that σ_i^* could not have been a best response to σ_{-i}^*.

This simple observation will play an important role in computing mixed-strategy Nash equilibria. In particular we know that if a player is playing a mixed strategy then he must be indifferent between the actions he is choosing with positive probability, that is, the actions that are in the support of his mixed strategy. One player's indifference will impose restrictions on the behavior of other players, and these restrictions will help us find the mixed-strategy Nash equilibrium.

For games with many players, or with two players who have many strategies, finding the set of mixed-strategy Nash equilibria is a tedious task. It is often done with the help of computer algorithms, because it generally takes on the form of a linear programming problem. Nevertheless it will be useful to see how one computes mixed-strategy Nash equilibria for simpler games.

6.2.1 Example: Matching Pennies

Consider the Matching Pennies game,

		Player 2	
		H	*T*
Player 1	*H*	1, −1	−1, 1
	T	−1, 1	1, −1

and recall that we showed that this game does not have a pure-strategy Nash equilibrium. We now ask, does it have a mixed-strategy Nash equilibrium? To answer this, we have to find mixed strategies for both players that are mutual best responses.

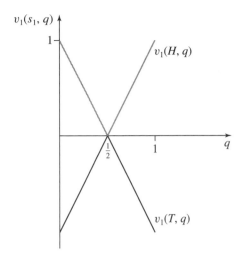

FIGURE 6.2 Expected payoffs for player 1 in the Matching Pennies game.

To simplify the notation, define mixed strategies for players 1 and 2 as follows: Let p be the probability that player 1 plays H and $1 - p$ the probability that he plays T. Similarly let q be the probability that player 2 plays H and $1 - q$ the probability that he plays T.

Using the formulas for expected payoffs in this game, we can write player 1's expected payoff from each of his two pure actions as follows:

$$v_1(H, q) = q \times 1 + (1 - q) \times (-1) = 2q - 1 \qquad (6.1)$$

$$v_1(T, q) = q \times (-1) + (1 - q) \times 1 = 1 - 2q. \qquad (6.2)$$

With these equations in hand, we can calculate the best response of player 1 for any choice q of player 2. In particular player 1 will prefer to play H over playing T if and only if $v_1(H, q) > v_1(T, q)$. Using (6.1) and (6.2), this will be true if and only if

$$2q - 1 > 1 - 2q,$$

which is equivalent to $q > \frac{1}{2}$. Similarly playing T will be strictly better than playing H for player 1 if and only if $q < \frac{1}{2}$. Finally, when $q = \frac{1}{2}$ player 1 will be indifferent between playing H or T.

It is useful to graph the expected payoff of player 1 from choosing either H or T as a function of player 2's choice of q, as shown in Figure 6.2.

The expected payoff of player 1 from playing H was given by the function $v_1(H, q) = 2q - 1$, as described in (6.1). This is the rising linear function in the figure. Similarly $v_1(T, q) = 1 - 2q$, described in (6.2), is the declining function. Now it is easy to see what determines the best response of player 1. The gray "upper envelope" of the graph will show the highest payoff that player 1 can achieve when player 2 plays any given level of q. When $q < \frac{1}{2}$ this is achieved by playing T; when $q > \frac{1}{2}$ this is achieved by playing H; and when $q = \frac{1}{2}$ both H and T are equally good for player 1, giving him an expected payoff of zero.

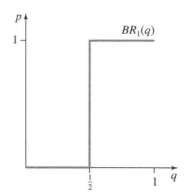

FIGURE 6.3 Player 1's best-response correspondences in the Matching Pennies game.

This simple analysis results in the best-response correspondence of player 1, which is

$$BR_1(q) = \begin{cases} p = 0 & \text{if } q < \frac{1}{2} \\ p \in [0, 1] & \text{if } q = \frac{1}{2} \\ p = 1 & \text{if } q > \frac{1}{2} \end{cases}$$

and is depicted in Figure 6.3. Notice that this is a best-response *correspondence*, and not a function, because at the value of $q = \frac{1}{2}$ any value of $p \in [0, 1]$ is a best response.

In a similar way we can calculate the payoffs of player 2 given a mixed-strategy p of player 1 to be

$$v_2(p, H) = p \times (-1) + (1 - p) \times 1 = 1 - 2p$$
$$v_2(p, T) = p \times 1 + (1 - p) \times (-1) = 2p - 1,$$

and this implies that player 2's best response is

$$BR_2(p) = \begin{cases} q = 1 & \text{if } p < \frac{1}{2} \\ q \in [0, 1] & \text{if } p = \frac{1}{2} \\ q = 0 & \text{if } p > \frac{1}{2}. \end{cases}$$

To find a Nash equilibrium we are looking for a pair of choices (p, q) for which the two best-response correspondences cross. Were we to superimpose the best response of player 2 onto Figure 6.3 then we would see that the two best-response correspondences cross at $p = q = \frac{1}{2}$. Nevertheless it is worth walking through the logic of this solution.

We know from proposition 6.1 that when player 1 is mixing between H and T, both with positive probability, then it must be the case that his payoffs from H and from T are identical. This, it turns out, imposes a restriction on the behavior of player 2, given by the choice of q. Player 1 is willing to mix between H and T if and only if $v_1(H, q) = v_1(T, q)$, which will hold if and only if $q = \frac{1}{2}$. This is the way in which *the indifference of player 1 imposes a restriction on player 2:* only when player 2 is playing $q = \frac{1}{2}$ will player 1 be willing to mix between his actions H and T. Similarly player 2 is willing to mix between H and T only when $v_2(p, H) = v_2(p, T)$, which is true only when $p = \frac{1}{2}$. We have come to the conclusion of our quest for a Nash equilibrium in this game. We can see that there is indeed a pair of mixed strategies that form a Nash equilibrium, and these are precisely when $(p, q) = \left(\frac{1}{2}, \frac{1}{2}\right)$.

There is a simple logic, which we can derive from the Matching Pennies example, that is behind the general method for finding mixed-strategy equilibria in games. The logic relies on a fact that we have already discussed: if a player is mixing several strategies then he *must be indifferent* between them. What a particular player i is willing to do depends on the strategies of his opponents. Therefore, to find out when player i is willing to mix some of his pure strategies, we must find strategies of his opponents, $-i$, that make him indifferent between some of his pure actions.

For the Matching Pennies game this can be easily illustrated as follows. First, we ask *which strategy of player 2* will make *player 1 indifferent* between playing H and T. The answer to this question (assuming it is unique) *must be player 2's strategy in equilibrium*. The reason is simple: if player 1 is to mix in equilibrium, then player 2 must be playing a strategy for which player 1's best response is mixing, and player 2's strategy must therefore make player 1 indifferent between playing H and T. Similarly we ask which strategy of player 1 will make player 2 indifferent between playing H and T, and this must be player 1's equilibrium strategy.

Remark The game of Matching Pennies is representative of situations in which one player wants to match the actions of the other, while the other wants to avoid that matching. One common example is penalty goals in soccer. The goalie wishes to jump in the direction that the kicker will kick the ball, while the kicker wishes to kick the ball in the opposite direction from the one in which the goalie chooses to jump. When they go in the same direction then the goalie wins and the kicker loses, while if they go in different directions then the opposite happens. As you can see, this is exactly the structure of the Matching Pennies game. Other common examples of such games are bosses monitoring their employees and the employees' decisions about how hard to work, or police monitoring crimes and the criminals who wish to commit them.

6.2.2 Example: Rock-Paper-Scissors

When we have games with more than two strategies for each player, then coming up with quick ways to solve mixed-strategy equilibria is a bit more involved than in 2×2 games, and it will usually involve more tedious algebra that solves several equations with several unknowns. If we consider the game of rock-paper-scissors, for example, there are many mixing combinations for each player, and we can't simply draw graphs the way we did for the Matching Pennies game.

		Player 2		
		R	P	S
	R	0, 0	−1, 1	1, −1
Player 1	P	1, −1	0, 0	−1, 1
	S	−1, 1	1, −1	0, 0

To find the Nash equilibrium of the rock-paper-scissors game we proceed in three steps. First we show that there is no Nash equilibrium in which at least one player plays a pure strategy. Then we show that there is no Nash equilibrium in which at least one player mixes only between two pure strategies. These steps will imply that in any Nash equilibrium, both players must be mixing with all three pure strategies, and this will lead to the solution.

Claim 6.1 *There can be no Nash equilibrium in which one player plays a pure strategy and the other mixes.*

To see this, suppose that player i plays a pure strategy. It's easy to see from looking at the payoff matrix that player j always receives different payoffs from each of his pure strategies whenever i plays a pure strategy. Therefore player j cannot be indifferent between any of his pure strategies, so j cannot be playing a mixed strategy if i plays a pure strategy. But we know that there are no pure-strategy equilibria, and hence we conclude that there are no Nash equilibria where either player plays a pure strategy.

Claim 6.2 *There can be no Nash equilibrium in which at least one player mixes only between two pure strategies.*

To see this, suppose that i mixes between R and P. Then j always gets a strictly higher payoff from playing P than from playing R, so no strategy requiring j to play R with positive probability can be a best response for j, and j can't play R in any Nash equilibrium. But if j doesn't play R then i gets a strictly higher payoff from S than from P, so no strategy requiring i to play P with positive probability can be a best response to j not playing R. But we assumed that i was mixing between R and P, so we've reached a contradiction. We conclude that in equilibrium i cannot mix between R and P. We can apply similar reasoning to i's other pairs of pure strategies. We conclude that in any Nash equilibrium of this game, no player can play a mixed strategy in which he only plays two pure strategies with positive probability.

If by now you've guessed that the mixed strategies $\sigma_1^* = \sigma_2^* = \left(\frac{1}{3}, \frac{1}{3}, \frac{1}{3}\right)$ form a Nash equilibrium then you are right. If player i plays σ_i^* then j will receive an expected payoff of 0 from every pure strategy, so j will be indifferent between all of his pure strategies. Therefore $BR_j(\sigma_i^*)$ includes all of j's mixed strategies and in particular $\sigma_j^* \in BR_j(\sigma_i^*)$. Similarly $\sigma_i^* \in BR_i(\sigma_j^*)$. We conclude that σ_1^* and σ_2^* form a Nash equilibrium. We will prove that (σ_1^*, σ_2^*) is *the unique Nash equilibrium*.

Suppose player i plays R with probability $\sigma_i(R) \in (0, 1)$, P with probability $\sigma_i(P) \in (0, 1)$, and S with probability $1 - \sigma_i(R) - \sigma_i(P)$. Because we proved that both players have to mix with all three pure strategies, it follows that $\sigma_i(R) + \sigma_i(P) < 1$ so that $1 - \sigma_i(R) - \sigma_i(P) \in (0, 1)$. It follows that player j receives the following payoffs from his three pure strategies:

$$v_j(R, \sigma_i) = -\sigma_i(P) + 1 - \sigma_i(R) - \sigma_i(P) = 1 - \sigma_i(R) - 2\sigma_i(P)$$

$$v_j(P, \sigma_i) = \sigma_i(R) - (1 - \sigma_i(R) - \sigma_i(P)) = 2\sigma_i(R) + \sigma_i(P) - 1$$

$$v_j(S, \sigma_i) = -\sigma_i(R) + \sigma_i(P).$$

In any Nash equilibrium in which j plays all three of his pure strategies with positive probability, he must receive the same expected payoff from all strategies. Therefore, in any equilibrium, we must have $v_j(R, \sigma_i) = v_j(P, \sigma_i) = v_j(S, \sigma_i)$. If we set these payoffs equal to each other and solve for $\sigma_i(R)$ and $\sigma_i(P)$, we get $\sigma_i(R) = \sigma_i(P) = 1 - \sigma_i(R) - \sigma_i(P) = \frac{1}{3}$. We conclude that j is willing to include all three of his pure strategies in his mixed strategy if and only if i plays $\sigma_i^* = \left(\frac{1}{3}, \frac{1}{3}, \frac{1}{3}\right)$. Similarly i will be willing to play all his pure strategies with positive probability if and only if j plays $\sigma_j^* = \left(\frac{1}{3}, \frac{1}{3}, \frac{1}{3}\right)$. Therefore there is no other Nash equilibrium in which both players play all their pure strategies with positive probability.

6.2.3 Multiple Equilibria: Pure and Mixed

In the Matching Pennies and rock-paper-scissors games, the unique Nash equilibrium was a mixed-strategy Nash equilibrium. It turns out that mixed-strategy equilibria need not be unique when they exist. In fact when a game has multiple pure-strategy Nash equilibria, it will almost always have other Nash equilibria in mixed strategies. Consider the following game:

		Player 2	
		C	R
Player 1	M	0, 0	3, 5
	D	4, 4	0, 3

It is easy to check that (M, R) and (D, C) are both pure-strategy Nash equilibria. It turns out that in 2×2 matrix games like this one, when there are two distinct pure-strategy Nash equilibria then there will almost always be a third one in mixed strategies.[6]

For this game, let player 1's mixed strategy be given by $\sigma_1 = (\sigma_1(M), \sigma_1(D))$, with $\sigma_1(M) = p$ and $\sigma_1(D) = 1 - p$, and let player 2's mixed strategy be given by $\sigma_2 = (\sigma_2(C), \sigma_2(R))$, with $\sigma_2(C) = q$ and $\sigma_2(R) = 1 - q$. Player 1 will mix when $v_1(M, q) = v_1(D, q)$, or when

$$q \times 0 + (1 - q) \times 3 = q \times 4 + (1 - q) \times 0$$

$$\Rightarrow q = \tfrac{3}{7},$$

and player 2 will mix when $v_2(p, C) = v_2(p, R)$, or when

$$p \times 0 + (1 - p) \times 4 = p \times 5 + (1 - p) \times 3$$

$$\Rightarrow p = \tfrac{1}{6}.$$

This yields our third Nash equilibrium: $(\sigma_1^*, \sigma_2^*) = \left(\left(\frac{1}{6}, \frac{5}{6}\right), \left(\frac{3}{7}, \frac{4}{7}\right)\right)$.

---

6. The statement "almost always" is not defined here, but it effectively means that if we draw numbers at random from some set of distributions to fill a game matrix, and it will result in more than one pure-strategy Nash equilibrium, then with probability 1 it will also have at least one mixed-strategy equilibrium. In fact a game will typically have an odd number of equilibria. This result is known as an *index theorem* and is far beyond the scope of this text.

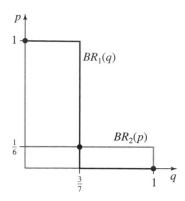

FIGURE 6.4 Best-response correspondences and Nash equilibria.

It is interesting to see that all three equilibria would show up in a careful drawing of the best-response functions. Using the payoff functions $v_1(M, q)$ and $v_1(D, q)$ we have

$$
BR_1(q) = \begin{cases} p = 1 & \text{if } q < \frac{3}{7} \\ p \in [0, 1] & \text{if } q = \frac{3}{7} \\ p = 0 & \text{if } q > \frac{3}{7}. \end{cases}
$$

Similarly using the payoff functions $v_2(p, C)$ and $v_2(p, R)$ we have

$$
BR_2(p) = \begin{cases} q = 1 & \text{if } p < \frac{1}{6} \\ q \in [0, 1] & \text{if } p = \frac{1}{6} \\ q = 0 & \text{if } p > \frac{1}{6}. \end{cases}
$$

We can draw the two best-response correspondences as they appear in Figure 6.4. Notice that all three Nash equilibria are revealed in Figure 6.4: $(p, q) \in \{(1, 0),$ $\left(\frac{1}{6}, \frac{3}{7}\right), (0, 1)\}$ are all Nash equilibria, where $(p, q) = (1, 0)$ corresponds to the pure strategy (M, R), and $(p, q) = (0, 1)$ corresponds to the pure strategy (D, C).

6.3 IESDS and Rationalizability Revisited

By introducing mixed strategies we offered two advancements: players can have richer beliefs, and players can choose a richer set of actions. This can be useful when we reconsider the concepts of IESDS and rationalizability, and in fact present them in their precise form using mixed strategies. In particular we can now state the following two definitions:

Definition 6.7 Let $\sigma_i \in \Delta S_i$ and $s_i' \in S_i$ be possible strategies for player i. We say that s_i' is **strictly dominated** by σ_i if

$$
v_i(\sigma_i, s_{-i}) > v_i(s_i', s_{-i}) \ \forall s_{-i} \in S_{-i}.
$$

Definition 6.8 A strategy $\sigma_i \in \Delta S_i$ is **never a best response** if there are no beliefs $\sigma_{-i} \in \Delta S_{-i}$ for player i for which $\sigma_i \in BR_i(\sigma_{-i})$.

That is, to consider a strategy as strictly dominated, we no longer require that some other *pure strategy* dominate it, but allow for mixed strategies to dominate it as well. The same is true for strategies that are never a best response. It turns out that this approach allows both concepts to have more bite. For example, consider the following game:

		Player 2	
	L	*C*	*R*
U	5, 1	1, 4	1, 0
M	3, 2	0, 0	3, 5
D	4, 3	4, 4	0, 3

and denote mixed strategies for players 1 and 2 as triplets, $(\sigma_1(U), \sigma_1(M), \sigma_1(D))$ and $(\sigma_2(L), \sigma_2(C), \sigma_2(R))$, respectively.

Starting with IESDS, it is easy to see that no pure strategy is strictly dominated by another pure strategy for any player. Hence if we restrict attention to pure strategies then IESDS has no bite and suggests that anything can happen in this game. However, if we allow for mixed strategies, we can find that the strategy L for player 2 is strictly dominated by a strategy that mixes between the pure strategies C and R. That is, $(\sigma_2(L), \sigma_2(C), \sigma_2(R)) = \left(0, \frac{1}{2}, \frac{1}{2}\right)$ strictly dominates choosing L for sure because this mixed strategy gives player 2 an expected payoff of 2 if player 1 chooses U, of 2.5 if player 1 chooses M, and of 3.5 if player 1 chooses D.

Effectively it is as if we are increasing the number of columns from which player 2 can choose to infinity, and one of these columns is the strategy in which player 2 mixes between C and R with equal probability, as the following diagram suggests:

	L	*C*	*R*		$\left(0, \frac{1}{2}, \frac{1}{2}\right)$
U	5, 1	1, 4	1, 0		2
M	3, 2	0, 0	3, 5	Player 2's expected payoff from mixing C and $R \Rightarrow$	2.5
D	4, 3	4, 4	0, 3		3.5

Hence we can perform the first step of IESDS with mixed strategies relying on the fact that $\left(0, \frac{1}{2}, \frac{1}{2}\right) \succ_2 L$, and now the game reduces to the following:

		Player 2	
		C	*R*
U		1, 4	1, 0
Player 1 *M*		0, 0	3, 5
D		4, 4	0, 3

$\left(0, \frac{1}{2}, \frac{1}{2}\right)$	2	1.5

In this reduced game there still are no strictly dominated pure strategies, but careful observation will reveal that the strategy U for player 1 is strictly dominated by a strategy that mixes between the pure strategies M and D. That is, $(\sigma_1(U), \sigma_1(M), \sigma_1(D))$

$= \left(0, \frac{1}{2}, \frac{1}{2}\right)$ strictly dominates choosing U for sure because this mixed strategy gives player 1 an expected payoff of 2 if player 2 chooses C and 1.5 if player 2 chooses R. We can then perform the second step of IESDS with mixed strategies relying on the fact that $\left(0, \frac{1}{2}, \frac{1}{2}\right) \succ_1 U$ in the reduced game, and now the game reduces further to the following:

	C	R
M	0, 0	3, 5
D	4, 4	0, 3

This last 2×2 game cannot be further reduced.

A question you must be asking is, how did we find these dominated strategies? Well, a good eye for numbers is what it takes—short of a computer program or brute force. Notice also that there are other mixed strategies that would work, because strict dominance implies that if we add a small $\varepsilon > 0$ to one of the probabilities, and subtract it from another, then the resulting expected payoff from the new mixed strategies can be made arbitrarily close to that of the original one; thus it too would dominate the dominated strategy.

Turning to rationalizability, in Section 4.3.3 we introduced the concept that after eliminating all the strategies that are *never a best response,* and employing this reasoning again and again in a way similar to what we did for IESDS, the strategies that remain are called the set of rationalizable strategies. If we use this concept to analyze the game we just solved with IESDS, the result will be the same. Starting with player 2, there is no belief that he can have for which playing L will be a best response. This is easy to see because either C or R will be a best response to one of player 1's pure strategies, and hence, even if player 1 mixes then the best response of player 2 will either be to play C, to play R, or to mix with both. Then after reducing the game a similar argument will work to eliminate U from player 1's strategy set.

As we mentioned briefly in Section 4.3.3, the concepts of IESDS and rationalizability are closely related. To see one obvious relation, the following fact is easy to prove:

Fact   If a strategy σ_i is strictly dominated then it is never a best response.

The reason this is obvious is because if σ_i is strictly dominated then there is some other strategy σ_i' for which $v_i(\sigma_i', \sigma_{-i}) > v_i(\sigma_i, \sigma_{-i})$ for all $\sigma_{-i} \in \Delta S_{-i}$. As a consequence, there is no belief about σ_{-i} that player i can have for which σ_i yields a payoff as good as or better than σ_i'.

This fact is useful, and it implies that the set of a player's rationalizable strategies is *no larger* than the set of a player's strategies that survive IESDS. This is true because if a strategy was eliminated using IESDS then it must have been eliminated through the process of rationalizability. Is the reverse true as well?

Proposition 6.2   *For any two-player game a strategy σ_i is strictly dominated if and only if it is never a best response.*

Hence for two-player games the set of strategies that survive IESDS is the same as the set of strategies that are rationalizable. Proving this is not that simple and is beyond the scope of this text. The eager and interested reader is encouraged to read Chapter 2 of Fudenberg and Tirole (1991), and the daring reader can refer to the original research papers by Bernheim (1984) and Pearce (1984), which simultaneously introduced the concept of rationalizability.[7]

6.4   Nash's Existence Theorem

Section 5.1.2 argued that the Nash equilibrium solution concept is powerful because on the one hand, like IESDS and rationalizability, a Nash equilibrium will exist for most games of interest and hence will be widely applicable. On the other hand, the Nash solution concept will usually lead to more refined predictions than those of IESDS and rationalizability, yet the reverse is never true (see proposition 5.1). In his seminal Ph.D. dissertation, which laid the foundations for game theory as it is used and taught today and earned him a Nobel Prize, Nash defined the solution concept that now bears his name and showed some very general conditions under which the solution concept will exist. We first state Nash's theorem:

Theorem (Nash's Existence Theorem)   *Any n-player normal-form game with finite strategy sets S_i for all players has a (Nash)* equilibrium in mixed strategies.[8]

Despite its being a bit technical, we will actually prove a restricted version of this theorem. The ideas that Nash used to prove the existence of his equilibrium concept have been widely used by game theorists, who have developed related solution concepts that refine the set of Nash equilibria, or generalize it to games that were not initially considered by Nash himself. It is illuminating to provide some basic intuition first. The central idea of Nash's proof builds on what is known in mathematics as a *fixed-point theorem*. The most basic of these theorems is known as Brouwer's fixed-point theorem:

Theorem (Brouwer's Fixed-Point Theorem)   *If $f(x)$ is a continuous function from the domain* [0, 1] *to itself then there exists at least one value $x^* \in [0, 1]$ for which* $f(x^*) = x^*$.

That is, if $f(x)$ takes values from the interval [0, 1] and generates results from this same interval (or $f : [0, 1] \rightarrow [0, 1]$) then there has to be some value x^* in the interval [0, 1] for which the operation of $f(\cdot)$ on x^* will give back the same value, $f(x^*) = x^*$. The intuition behind the proof of this theorem is actually quite simple.

First, because $f : [0, 1] \rightarrow [0, 1]$ maps the interval [0, 1] onto itself, then $0 \leq f(x) \leq 1$ for any $x \in [0, 1]$. Second, note that if $f(0) = 0$ then $x^* = 0$, while if $f(1) = 1$ then $x^* = 1$ (as shown by the function $f_1(x)$ in Figure 6.5). We need to show, therefore, that if $f(0) > 0$ and $f(1) < 1$ then when $f(x)$ is continuous there must be some value x^* for which $f(x^*) = x^*$. To see this consider the two functions, $f_2(x)$ and $f_3(x)$, depicted in Figure 6.5, both of which map the interval [0, 1] onto itself, and for which $f(0) > 0$ and $f(1) < 1$. That is, these functions start above the $45°$ line and end below it. The function $f_2(x)$ is continuous, and hence if it starts above

---

7. When there are more than two players, the set of rationalizable strategies is sometimes smaller and more refined than the set of strategies that survive IESDS. There are some conditions on the way players randomize that restore the equivalence result to many-player games, but that subject is also way beyond the scope of this text.

8. Recall that a pure strategy is a degenerate mixed strategy; hence there may be a Nash equilibrium in pure strategies.

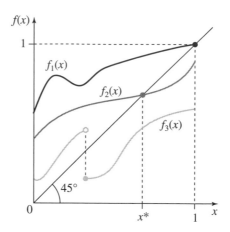

FIGURE 6.5 Brouwer's fixed-point theorem.

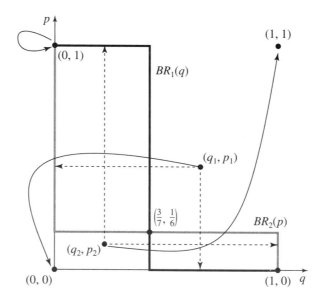

FIGURE 6.6 Mapping mixed strategies using the best-response correspondence.

the 45° line and ends below it, it must cross it at least once. In the figure, this happens at the value of x^*. To see why the continuity assumption is important, consider the function $f_3(x)$ depicted in Figure 6.5. Notice that it "jumps" down from above the 45° line to right below it, and hence this function does not cross the 45° line, in which case there is no value x for which $f(x) = x$.

You might wonder how this relates to the existence of a Nash equilibrium. What Nash showed is that something like continuity is satisfied for a mapping that uses the *best-response correspondences* of all the players at the same time to show that there must be at least one mixed-strategy profile for which each player's strategy is itself a best response to this profile of strategies. This conclusion needs some more explanation, though, because it requires a more powerful fixed-point theorem and a bit more notation and definition.

Consider the 2×2 game used in Section 6.2.3, described in the following matrix:

Player 2

		C	R
	M	0, 0	3, 5
Player 1	D	4, 4	0, 3

A mixed strategy for player 1 is to choose M with probability $p \in [0, 1]$ and for player 2 to choose C with probability $q \in [0, 1]$. The analysis in Section 6.2.3 showed that the best-response correspondences for each player are

$$BR_1(q) = \begin{cases} p = 1 & \text{if } q < \frac{3}{7} \\ p \in [0, 1] & \text{if } q = \frac{3}{7} \\ p = 0 & \text{if } q > \frac{3}{7} \end{cases} \tag{6.3}$$

and

$$BR_2(p) = \begin{cases} q = 1 & \text{if } p < \frac{1}{6} \\ q \in [0, 1] & \text{if } p = \frac{1}{6} \\ q = 0 & \text{if } p > \frac{1}{6}, \end{cases} \tag{6.4}$$

which are both depicted in Figure 6.6. We now define the **collection of best-response correspondences** as the correspondence that simultaneously represents all of the best-response correspondences of the players. This correspondence maps profiles of mixed strategies into subsets of the possible set of mixed strategies for all the players. Formally we have

Definition 6.9 The collection of best-response correspondences, $BR \equiv BR_1 \times BR_2 \times \cdots \times BR_n$, maps $\Delta S = \Delta S_1 \times \cdots \times \Delta S_n$, the set of profiles of mixed strategies, onto itself. That is, $BR : \Delta S \rightrightarrows \Delta S$ takes every element $\sigma \in \Delta S$ and converts it into a subset $BR(\sigma') \subset \Delta S$.

For a 2×2 matrix game like the one considered here, the BR correspondence can be written as[9] $BR : [0, 1]^2 \rightrightarrows [0, 1]^2$ because it takes pairs of mixed strategies of the form $(q, p) \in [0, 1]^2$ and maps them, using the best-response correspondences of the players, back to these mixed-strategy spaces, so that $BR(q, p) = (BR_2(p), BR_1(q))$. For example, consider the pair of mixed strategies (q_1, p_1) in Figure 6.6. Looking at player 1's best response, $BR_1(q) = 0$, and looking at player 2's best response, $BR_2(p) = 0$ as well. Hence $BR(q_1, p_1) = (0, 0)$, as shown by the curve that takes (q_1, p_1) and maps it onto $(0, 0)$. Similarly (q_2, p_2) is mapped onto $(1, 1)$.

Note that the point $(q, p) = (0, 1)$ is special in that $BR(0, 1) = (0, 1)$. This should be no surprise because, as we have shown in Section 6.2.3, $(q, p) = (0, 1)$ is one of the game's three Nash equilibria, so it must belong to the BR correspondence of itself. The same is true for the point $(q, p) = (1, 0)$. The third interesting point is

---

9. The space $[0, 1]^2$ is the two-dimensional square $[01] \times [01]$. It is the area in which all the action in Figure 6.6 is happening.

$(\frac{3}{7}, \frac{1}{6})$, because $BR(\frac{3}{7}, \frac{1}{6}) = ([0, 1], [0, 1])$, which means that the BR correspondence of this point is a pair of sets. This results from the fact that when player 2 mixes with probability $q = \frac{3}{7}$ then player 1 is indifferent between his two actions, causing any $p \in [0, 1]$ to be a best response, and similarly for player 2 when player 1 mixes with probability $p = \frac{1}{6}$. As a consequence, $(\frac{3}{7}, \frac{1}{6}) \in BR(\frac{3}{7}, \frac{1}{6})$, which is the reason it is the third Nash equilibrium of the game. Indeed by now you may have anticipated the following fact, which is a direct consequence of the definition of a Nash equilibrium:

Fact A mixed-strategy profile $\sigma^* \in \Delta S$ is a Nash equilibrium if and only if it is a fixed point of the collection of best-response correspondences, $\sigma^* \in BR(\sigma^*)$.

Now the connection to fixed-point theorems should be more apparent. What Nash figured out is that when the collection of best responses BR is considered, then once it is possible to prove that it has a fixed point, it immediately implies that a Nash equilibrium exists. Nash continued on to show that for games with finite strategy sets for each player it is possible to apply the following theorem:

Theorem 6.1 (Kakutani's Fixed-Point Theorem) *A correspondence $C : X \rightrightarrows X$ has a fixed point $x \in C(x)$ if four conditions are satisfied: (1) X is a non-empty, compact, and convex subset of $\mathbb{R}^n$; (2) $C(x)$ is non-empty for all x; (3) $C(x)$ is convex for all x; and (4) C has a closed graph.*

This may surely seem like a mouthful because we have not defined any of the four qualifiers required by the theorem. For the sake of completeness, we will go over them and conclude with an intuition of why the theorem is true. First, recall that a correspondence can assign more than one value to an input, whereas a function can assign only one value to any input. Now let's introduce the definitions:

- A set $X \subseteq \mathbb{R}^n$ is **convex** if for any two points $x, y \in X$ and any $\alpha \in [0, 1]$, $\alpha x + (1 - \alpha)y \in X$. That is, any point in between x and y that lies on the straight line connecting these two points lies inside the set X.

- A set $X \subseteq \mathbb{R}^n$ is **closed** if for any converging sequence $\{x_n\}_{n=1}^{\infty}$ such that $x_n \in X$ for all n and $\lim_{n \to \infty} x_n \to x^*$ then $x^* \in X$. That is, if an infinite sequence of points are all in X and this sequence converges to a point x^* then x^* must be in X. For example, the set $(0, 1]$ that does not include 0 is not closed because we can construct a sequence of points $\{\frac{1}{n}\}_{n=1}^{\infty} = \{1, \frac{1}{2}, \frac{1}{3}, \ldots\}$ that are all in the set $[0, 1)$ and that converge to the point 0, but 0 is not in $(0, 1]$.

- A set $X \subseteq \mathbb{R}^n$ is **compact** if it is both closed and bounded. That is, there is a "largest" and a "smallest" point in the set that do not involve infinity. For example, the set $[0, 1]$ is closed and bounded; the set $(0, 1]$ is bounded but not closed; and the set $[0, \infty)$ is closed but not bounded.

- The **graph** of a correspondence $C : X \rightrightarrows X$ is the set $\{(x, y) \mid x \in X, \ y \in C(x)\}$. The correspondence $C : X \rightrightarrows X$ has a **closed graph** if the graph of C is a closed set: for any sequence $\{(x_n, y_n)\}_{n=1}^{\infty}$ such that $x_n \in X$ and $y_n \in C(x_n)$ for all n, and $\lim_{n \to \infty}(x_n, y_n) = (x^*, y^*)$, then $x^* \in X$ and $y^* \in C(x^*)$. For example, if $C(x) = x^2$ then the graph is the set $\{(x, y) \mid x \in \mathbb{R}, \ y = x^2\}$, which is exactly the plot of the function. The plot of any continuous function is therefore a closed graph. (This is true whenever $C(x)$ is a real continuous

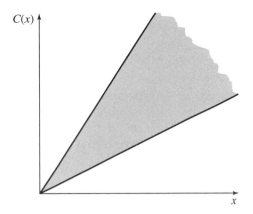

FIGURE 6.7 A correspondence with a closed graph.

function.) Another example is the correspondence $C(x) = \left[\frac{x}{2}, \frac{3x}{2}\right]$ that is depicted in Figure 6.7. In contrast the correspondence $C(x) = \left(\frac{x}{2}, \frac{3x}{2}\right)$ does not have a closed graph (it does not include the "boundaries" that are included in Figure 6.7).

The intuition for Kakutani's fixed-point theorem is somewhat similar to that for Brouwer's theorem. Brouwer's theorem was stated using two qualifiers: first, the function $f(x)$ was continuous, and second, it operated from the domain $[0, 1]$ to itself. This implied that if we draw any such function in $[0, 1]$, we will have to cross the 45° line at at least one point, which is the essence of the fixed-point theorem.

Now let's consider Kakutani's four conditions. His first condition, that X is a non-empty, compact, and convex subset of $\mathbb{R}^n$, is just the more general version of the $[0, 1]$ qualifier in Brouwer's theorem. In fact Brouwer's theorem works for $[0, 1]$ precisely because it is a non-empty, compact, and convex subset of $\mathbb{R}$.[10] His other three conditions basically guarantee that a form of "continuity" is satisfied for the correspondence $C(x)$. If we consider any continuous real function from $[0, 1]$ to itself, it satisfies all three conditions of being non-empty (it has to be well defined), convex (it is always just one point), and closed (again, just one point). Hence the four conditions identified by Kakutani guarantee that a correspondence will cross the relevant 45° line and generate at least one fixed point.

We can now show that for the 2×2 game described earlier, and in fact for any 2×2 game, the four conditions of Kakutani's theorem are satisfied:

1. $BR : [0, 1]^2 \rightrightarrows [0, 1]^2$ operates on the square $[0, 1]^2$, which is a non-empty, convex, and compact subset of $\mathbb{R}$.

---

10. If instead we consider $(0, 1)$, which is not closed and hence not compact, then the function $f(x) = \sqrt{x}$ does not have a fixed point because within the domain $(0, 1)$ it is everywhere above the 45° line. If we consider the domain $\left[0, \frac{1}{3}\right] \cup \left[\frac{2}{3}, 1\right]$, which is not convex because it is has a gap equal to $\left(\frac{1}{3}, \frac{2}{3}\right)$, then the function $f(x) = \frac{3}{4}$ for all $x \in \left[0, \frac{1}{3}\right]$ and $f(x) = \frac{1}{4}$ for all $x \in \left[\frac{2}{3}, 1\right]$ (which is continuous) will not have a fixed point precisely because of this gap.

2. $BR(\sigma)$ is non-empty for any $\sigma \in [0, 1]^2$. This is obvious for the example given earlier because both $BR_1(q)$ and $BR_2(p)$ are non-empty, as shown in (6.3) and (6.4). More generally for 2×2 games: each player's strategy is in the compact set $[0, 1]$; each player's expected payoff is a weighted average of the four possible payoffs he can achieve (weighted by the mixed strategies); and his expected payoff is therefore continuous in his strategy. As a consequence each player has at least one best response for any choice of his opponent. (This is because a continuous function operating on a compact set will achieve both a maximum and a minimum over that set. This is known as the **extreme value theorem**.)

3. $BR(\sigma)$ is convex for any $\sigma \in [0, 1]^2$. This is obvious for the example given earlier because both $BR_1(q)$ and $BR_2(p)$ are convex, as shown in (6.3) and (6.4). This follows more generally directly from proposition 6.1, which states that if a player is mixing between two pure strategies then he must be indifferent between them. This in turn implies that he is willing to mix between them in any way, and as a consequence, if two mixed strategies σ_1' and σ_1'' are in $BR_1(q)$ then any mixed strategy is in BR_1, so that BR_1 is convex. Because this argument works for both players, if any two mixed-strategy profiles σ and σ' are in BR then any convex combination of them is also in BR.

4. $BR(\sigma)$ has a closed graph. Again we can see this from (6.3) and (6.4). For each of the two players, $BR_i(\sigma_j)$ is equal to either 0, 1, or the whole interval $[0, 1]$. More generally, for any 2×2 game consider a sequence of mixed-strategy profiles $\{(q_n, p_n)\}_{n=1}^{\infty}$ and a sequence of best responses $\{(q_n', p_n')\}_{n=1}^{\infty}$, where $(q_n', p_n') \in BR((q_n, p_n))$ for all n. Let $\lim_{n \to \infty} (q_n, p_n) = (q^*, p^*)$ and $\lim_{n \to \infty} (q_n', p_n') = (q', p')$. To conclude that $BR(\sigma)$ has a closed graph we need to show that $(q', p') \in BR(q^*, p^*)$. For player 2 it must be that $v_2(q_n', p_n) \geq v_2(q, p_n)$ for any $q \in [0, 1]$ because q_n' is a best response to p_n. Because the (expected) payoff function is linear in q and p, it is continuous in both arguments and, as a consequence, we can take limits on both sides of the inequality while preserving the inequality, so that $\lim_{n \to \infty} v_2(q_n', p_n) \geq \lim_{n \to \infty} v_2(q, p_n)$ for all $q \in [0, 1]$, implying that $v_2(q', p^*) \geq v_2(q, p^*)$ for all $q \in [0, 1]$. But this implies that $q' \in BR_2(p^*)$, and a symmetric argument for player 1 implies that $p' \in BR_1(q^*)$, which together prove that $(q', p') \in BR(q^*, p^*)$.

We conclude therefore that all four conditions of Kakutani's fixed-point theorem are satisfied for the (mixed-) strategy sets and the best-response correspondence of any 2×2 game. Hence the best-response correspondence $BR(\sigma)$ has a fixed point, which in turn implies that any 2×2 game has at least one Nash equilibrium.

Recall that Nash's theorem referred to any finite n-player game and not just 2×2 games. As Nash showed, the basic application of Kakutani's fixed-point theorem to finite games holds for any finite number of pure strategies for each player. If, say, player i has a strategy set consisting of m pure strategies $\{s_{i1}, s_{i2}, \ldots, s_{im}\}$ then his set of mixed strategies is in the simplex $\Sigma_i = \{(\sigma_i(s_{i1}), \sigma_i(s_{i2}), \ldots, \sigma_i(s_{im})) \mid \sigma_i(s_{ik}) \in [0, 1]$ for all $k = 1, 2, \ldots, m$, and $\sum_{k=1}^{m} \sigma_i(s_{ik}) = 1\}$. It is easy to show that the set Σ_i is a non-empty, compact, and convex subset of $\mathbb{R}^m$, meaning that the first condition of Kakutani's theorem is satisfied. Using the same ideas as in points 1–4, it is not too difficult to show that the three other conditions of Kakutani's theorem hold, and as a result that the best-response correspondence BR has a fixed point and that any such game has a Nash equilibrium.

6.5 Summary

- Allowing for mixed strategies enriches both what players can choose and what they can believe about the choices of other players.

- In games for which players have opposing interests, like the Matching Pennies game, there will be no pure-strategy equilibrium but a mixed-strategy equilibrium will exist.

- Allowing for mixed strategies enhances the power of IESDS and of rationalizability.

- Nash proved that for finite games there will always be at least one Nash equilibrium.

6.6 Exercises

6.1 **Best Responses in the Battle of the Sexes:** Use the best-response correspondences in the Battle of the Sexes game to find all the Nash equilibria. (Follow the approach used for the example in Section 6.2.3.)

6.2 **Mixed Dominance 1:** Let σ_i be a mixed strategy of player i that puts positive weight on one strictly dominated pure strategy. Show that there exists a mixed strategy σ_i' that puts no weight on any dominated pure strategy and that dominates σ_i.

6.3 **Mixed Dominance 2:** Consider the game used in Section 6.3:

		Player 2		
		L	C	R
	U	5, 1	1, 4	1, 0
Player 1	M	3, 2	0, 0	3, 5
	D	4, 3	4, 4	0, 3

a. Find a strategy different from $(\sigma_2(L), \sigma_2(C), \sigma_2(R)) = \left(0, \frac{1}{2}, \frac{1}{2}\right)$ that strictly dominates the pure strategy L for player 2. Argue that you can find an infinite number of such strategies.

b. Find a strategy different from $(\sigma_1(U), \sigma_1(M), \sigma_1(D)) = \left(0, \frac{1}{2}, \frac{1}{2}\right)$ that strictly dominates the pure strategy U for player 1 in the game remaining after one stage of elimination. Argue that you can find an infinite number of such strategies.

6.4 **Monitoring:** An employee (player 1) who works for a boss (player 2) can either work (W) or shirk (S), while his boss can either monitor the employee (M) or ignore him (I). As in many employee-boss relationships, if the employee is working then the boss prefers not to monitor, but if the boss is not monitoring then the employee prefers to shirk. The game is represented by the following matrix:

$$
\begin{array}{ccc}
 & & \text{Player 2} \\
 & & M \qquad I \\
 & W & \boxed{1, 1 \;\; | \;\; 1, 2} \\
\text{Player 1} & & \\
 & S & \boxed{0, 2 \;\; | \;\; 2, 1}
\end{array}
$$

    a. Draw the best-response function of each player.

    b. Find the Nash equilibrium of this game. What kind of game does this game remind you of?

6.5 **Cops and Robbers:** Player 1 is a police officer who must decide whether to patrol the streets or to hang out at the coffee shop. His payoff from hanging out at the coffee shop is 10, while his payoff from patrolling the streets depends on whether he catches a robber, who is player 2. If the robber prowls the streets then the police officer will catch him and obtain a payoff of 20. If the robber stays in his hideaway then the officer's payoff is 0. The robber must choose between staying hidden or prowling the streets. If he stays hidden then his payoff is 0, while if he prowls the streets his payoff is -10 if the officer is patrolling the streets and 10 if the officer is at the coffee shop.

    a. Write down the matrix form of this game.

    b. Draw the best-response function of each player.

    c. Find the Nash equilibrium of this game. What kind of game does this game remind you of?

6.6 **Declining Industry:** Consider two competing firms in a declining industry that cannot support both firms profitably. Each firm has three possible choices, as it must decide whether or not to exit the industry immediately, at the end of this quarter, or at the end of the next quarter. If a firm chooses to exit then its payoff is 0 from that point onward. Each quarter that both firms operate yields each a loss equal to -1, and each quarter that a firm operates alone yields it a payoff of 2. For example, if firm 1 plans to exit at the end of this quarter while firm 2 plans to exit at the end of the next quarter then the payoffs are $(-1, 1)$ because both firms lose -1 in the first quarter and firm 2 gains 2 in the second. The payoff for each firm is the sum of its quarterly payoffs.

    a. Write down this game in matrix form.

    b. Are there any strictly dominated strategies? Are there any weakly dominated strategies?

    c. Find the pure-strategy Nash equilibria.

    d. Find the unique mixed-strategy Nash equilibrium. (Hint: you can use your answer to (b) to make things easier.)

6.7 **Grad School Competition:** Two students sign up to prepare an honors thesis with a professor. Each can invest time in his own project: either no time, one week, or two weeks (these are the only three options). The cost of time is 0 for no time, and each week costs 1 unit of payoff. The more time a student puts in the better his work will be, so that if one student puts in more time than the other there will be a clear "leader." If they put in the same amount of time then their thesis projects will have the same quality. The professor, however, will give out only one grade of A. If there is a clear leader then he will get the A, while if they are equally good then the professor will toss a fair coin to decide who gets the A. The other student will get a B. Since both wish to continue on to graduate school, a grade of A is worth 3 while a grade of B is worth 0.

    a.   Write down this game in matrix form.

    b.   Are there any strictly dominated strategies? Are there any weakly dominated strategies?

    c.   Find the unique mixed-strategy Nash equilibrium.

6.8   **Market Entry:** Three firms are considering entering a new market. The payoff for each firm that enters is $\frac{150}{n}$, where n is the number of firms that enter. The cost of entering is 62.

    a.   Find all the pure-strategy Nash equilibria.

    b.   Find the symmetric mixed-strategy equilibrium in which all three players enter with the same probability.

6.9   **Discrete All-Pay Auction:** In Section 6.1.4 we introduced a version of an all-pay auction that worked as follows: Each bidder submits a bid. The highest bidder gets the good, but *all bidders pay their bids.* Consider an auction in which player 1 values the item at 3 while player 2 values the item at 5. Each player can bid either 0, 1, or 2. If player i bids more than player j then i wins the good and both pay. If both players bid the same amount then a coin is tossed to determine who gets the good, but again both pay.

    a.   Write down the game in matrix form. Which strategies survive IESDS?

    b.   Find the Nash equilibria for this game.

6.10   **Continuous All-Pay Auction:** Consider an all-pay auction for a good worth 1 to each of the two bidders. Each bidder can choose to offer a bid from the unit interval so that $S_i = [0, 1]$. Players care only about the expected value they will end up with at the end of the game (i.e., if a player bids 0.4 and expects to win with probability 0.7 then his payoff is $0.7 \times 1 - 0.4$).

    a.   Model this auction as a normal-form game.

    b.   Show that this game has no pure-strategy Nash equilibrium.

    c.   Show that this game cannot have a Nash equilibrium in which each player is randomizing over a finite number of bids.

    d.   Consider mixed strategies of the following form: Each player i chooses an interval $[\underline{x}_i, \overline{x}_i]$ with $0 \leq \underline{x}_i < \overline{x}_i \leq 1$ together with a cumulative distribution $F_i(x)$ over the interval $[\underline{x}_i, \overline{x}_i]$. (Alternatively you can think of each player choosing $F_i(x)$ over the interval $[0, 1]$ with two values $\underline{x}_i$ and $\overline{x}_i$ such that $F_i(\underline{x}_i) = 0$ and $F_i(\overline{x}_i) = 1$.)

        i.   Show that if two such strategies are a mixed-strategy Nash equilibrium then it must be that $\underline{x}_1 = \underline{x}_2$ and $\overline{x}_1 = \overline{x}_2$.

        ii.   Show that if two such strategies are a mixed-strategy Nash equilibrium then it must be that $\underline{x}_1 = \underline{x}_2 = 0$.

        iii.   Using your answers to (i) and (ii), argue that if two such strategies are a mixed-strategy Nash equilibrium then both players must be getting an expected payoff of zero.

        iv.   Show that if two such strategies are a mixed-strategy Nash equilibrium then it must be that $\overline{x}_1 = \overline{x}_2 = 1$.

        v.   Show that $F_i(x)$ being uniform over $[0, 1]$ is a symmetric Nash equilibrium of this game.

6.11 **Bribes:** Two players find themselves in a legal battle over a patent. The patent is worth 20 to each player, so the winner would receive 20 and the loser 0. Given the norms of the country, it is common to bribe the judge hearing a case. Each player can offer a bribe secretly, and the one whose bribe is the highest will be awarded the patent. If both choose not to bribe, or if the bribes are the same amount, then each has an equal chance of being awarded the patent. If a player does bribe, then the bribe can be valued at either 9 or 20. Any other number is considered very unlucky, and the judge would surely rule against a party who offered a different number.

    a.   Find the unique pure-strategy Nash equilibrium for this game.

    b.   If the norm were different, so that a bribe of 15 were also acceptable, is there a pure-strategy Nash equilibrium?

    c.   Find the symmetric mixed-strategy Nash equilibrium for the game with possible bribes of 9, 15, and 20.

6.12 **The Tax Man:** A citizen (player 1) must choose whether to file taxes honestly or to cheat. The tax man (player 2) decides how much effort to invest in auditing and can choose $a \in [0, 1]$; the cost to the tax man of investing at a level a is $c(a) = 100a^2$. If the citizen is honest then he receives the benchmark payoff of 0, and the tax man pays the auditing costs without any benefit from the audit, yielding him a payoff of $-100a^2$. If the citizen cheats then his payoff depends on whether he is caught. If he is caught then his payoff is -100 and the tax man's payoff is $100 - 100a^2$. If he is not caught then his payoff is 50 while the tax man's payoff is $-100a^2$. If the citizen cheats and the tax man chooses to audit at level a then the citizen is caught with probability a and is not caught with probability $(1 - a)$.

    a.   If the tax man believes that the citizen is cheating for sure, what is his best-response level of a?

    b.   If the tax man believes that the citizen is honest for sure, what is his best-response level of a?

    c.   If the tax man believes that the citizen is honest with probability p, what is his best-response level of a as a function of p?

    d.   Is there a pure-strategy Nash equilibrium for this game? Why or why not?

    e.   Is there a mixed-strategy Nash equilibrium for this game? Why or why not?

PART III

DYNAMIC GAMES OF COMPLETE INFORMATION

Preliminaries

As we have seen in a number of examples, the normal-form representation seems like a general method of putting formal structure on strategic situations. It allowed us to analyze a variety of games and reach some conclusions about the outcomes of strategic interaction when players are rational, as well as when there is common knowledge of rationality. Setting a game within the limits of our "actions, outcomes, and preferences" language seems, therefore, to have been a useful exercise.

The simplicity and generality of the normal form notwithstanding, one of its drawbacks is its inability to capture games that unfold over time. That is, there is a sense of how players' strategy sets correspond to what they can do, and of how the combination of their actions affects each other's payoffs, but there is no way to represent situations in which the order of moves might be important.

As an example, consider the familiar Battle of the Sexes game:

		Chris	
		O	F
Alex	O	2, 1	0, 0
	F	0, 0	1, 2

but with a slight modification. Imagine that Alex finishes work at 3:00 p.m. while Chris finishes work at 5:00 p.m. This gives Alex ample time to get to either the football game or the opera and then to call Chris at 4:45 p.m. and announce "I am here." Chris then has to make a choice of where to go. Where should Chris go? If the choice is to the venue where Alex is waiting then Chris will get some payoff. (It would be 1 if Alex is at the opera and 2 if Alex is at the football game.) If Chris's choice is to go to the other venue, then he will get 0. Hence a rational Chris should go to the same venue that Alex did. Anticipating this, a rational Alex ought to choose the opera, because then Alex gets 2 instead of 1 from football.

As you can see, there is a fundamental difference between this example and the simultaneous-move Battle of the Sexes game. Here Chris makes a move *after* learning what Alex did, hence the difference is in *what Chris knows when Chris makes a move*. In other words, in the simultaneous-move Battle of the Sexes game neither player knows what the other is choosing, so each player conjectures a belief and plays a best response to this belief. Here, in contrast, when it's his turn to make a move, Chris *knows* what Alex has done, and as a result the notion of conjecturing beliefs is moot.

Furthermore Alex *knows,* by common knowledge of rationality, that Chris will choose to follow Alex because it is Chris's best response to do so. As a result Alex can get what Alex wants.

This is a simple yet convincing example of the commonly used phrase "first-mover advantage." By moving first, Alex gets to set the evening's venue. Before we draw far-reaching conclusions about the generality of the first-mover advantage, consider the Matching Pennies game. If I were to play with you, no matter what my position would be (player 1 or 2), I would be happy to give you the opportunity to show your coin first. I am sure that you can see why!

This chapter lays out a framework that allows us to represent formally such sequential strategic situations and apply strategic reasoning to these new representations. We then go a step further and introduce a solution concept that captures the important idea of **sequential rationality,** which is the general version of the dynamic reasoning offered in the Battle of the Sexes example. Earlier movers will take into account the rationality of players who move later in the game. This idea should resonate with the backward induction procedure introduced in Section 2.4.1.

7.1   The Extensive-Form Game

In this section we derive the most common representation for games that unfold over time and in which some players move after they learn the actions of other players. The innovation over the normal-form representation is to allow the knowledge of some players, when it is their turn to move, to depend on the previously made choices of other players. As with the normal form, two elements must be part of any extensive-form game's representation:

1. Set of players, N.
2. Players' payoffs as a function of outcomes, $\{v_i(\cdot)\}_{i \in N}$.

To fix these ideas, let's use the Battle of the Sexes example introduced earlier, in which Alex moves first. In the extensive form the set of players is still $N = \{1, 2\}$ (Alex is 1 and Chris is 2), and their payoffs over outcomes are given as before: $v_1(O, O) = v_2(F, F) = 2$, $v_1(F, F) = v_2(O, O) = 1$, and $v_i(O, F) = v_i(F, O) = 0$ for $i \in \{1, 2\}$.

To overcome the limitations of the normal form and capture sequential play, we need to expand the rather simplistic concept of pure-strategy sets to a more complex organization of actions. We do this by introducing two parts for actions: First, as before, *what* players can do, and second, *when* they can do it. Using our example, we need to specify that the players can choose O or F as in the normal-form game, but we also need to specify that player 1 moves first and that it is only then, after learning what player 1 has chosen, that player 2 moves. Thus in general we need two components to capture sequential play:

3. Order of moves.
4. Actions of players when they can move.

Because some players move *after* choices are made by other players, we need to be able to describe the knowledge that players have about the history of the game when it is their turn to move. Recall that we argued in Section 3.1 that the simultaneity of the normal form was illustrative of players who know nothing about their opponents' moves when they make their choices. More precisely it is not the *chronological order of play* that matters, but *what players know when they make their choices*.

In the Battle of the Sexes, for example, whether or not Chris knows what Alex chose before it is his turn to move is critical. Perhaps it is the case that Alex made a choice several hours before Chris did. However, if Chris has to make a choice *without knowing* Alex's actual choice, then it is the simultaneous-move normal-form model that correctly represents the situation. Despite moving after Alex, Chris *cannot condition* his action on the action chosen by Alex. In contrast, if Alex's choice is revealed to Chris *before* Chris makes a move, then we need to account for this circumstance in the representation of the game.

To represent the way in which information and knowledge unfold in a game, we add a fifth component to the description of an extensive-form game:

5. The knowledge that players have when they can move.

That is, we need a way to say more than "Chris moves after Alex." We must distinguish between "Chris moves after Alex and knows what Alex did" and "Chris moves after Alex but does not know what Alex did."

Finally we must account for the possibility that some random event can happen during the course of the game. Recall that in our treatment of single-person decision problems we allowed for random events. For example, in Chapter 2 we considered a firm choosing whether or not to embark on an R&D project that yields uncertain success. We can enrich the single-firm environment by adding another firm that can choose whether or not to adapt its competitive strategy given the outcome of the first firm's R&D project. Indeed this second firm may benefit from waiting to see how the first firm's R&D project develops. Because the outcome of the R&D project is not certain, we need to be able to capture these random events in order to model this kind of game correctly.

Similar to the setup described in Chapter 2, stages in a game in which some uncertainty is resolved are called *moves of Nature*. It is useful to think of Nature as a nonstrategic player that has a predetermined stochastic strategy. In the R&D story, when it is Nature's choice to move, it "chooses" whether the project succeeds or fails according to some probability distribution that is fixed and *exogenous* to the game. By "exogenous" we mean that the predetermined probability distribution of Nature's choices does not depend on the choices made by the strategic players. Thus our sixth element represents Nature as follows:

6. Probability distributions over *exogenous* events.

Finally, to be able to analyze these situations with the methods and concepts to which we have already been introduced, we add a final and familiar requirement:

7. The structure of the extensive-form game represented by 1–6 is common knowledge among all the players.

With the building blocks of 1–7 in place, one question remains: what formal notation will be used to put all this together? For this we borrow from the familiar concept of a decision tree that was introduced at the end of Section 1.1.2 and expand it to capture multiplayer strategic situations.

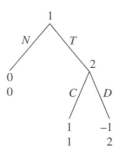

FIGURE 7.1 A trust game.

7.1.1 Game Trees

Just as decision trees offered a simple diagrammatic description of a single player's decision problem, a game tree will offer a diagrammatic description that is suitable to represent extensive-form games. Consider, for example, a game that is very common and falls under the general category of a "trust game." Player 1 first chooses whether to ask for the services of player 2. He can trust player 2 (T) or not trust him (N), the latter choice giving both players a payoff of 0. If player 1 plays T, then player 2 can choose to cooperate (C), which represents offering player 1 some fair level of service, or defect (D), by which he basically cheats player 1 with an inferior, less costly to provide service. Assume that if player 2 cooperates then both players get a payoff of 1, while if player 2 chooses to defect then player 1 gets a payoff of -1 and player 2 gets a payoff of 2.

This game represents many real-life trading situations. It could be a driver who trusts a mechanic to be honest and perform the right service for his vehicle rather than rip him off; a buyer on an auction web site like eBay who trusts a seller by paying up front and hoping that the seller will deliver the described item rather than something inferior, or nothing at all; or a local farmer who puts up a roadside produce stand with a jar for money and relies on his customers to pay for the produce they take according to the "honor system."

A simple way to represent this game is with the game tree depicted in Figure 7.1. In this figure we draw player 1 at the top with his two choices, and following the choice of N are the payoffs $(0, 0)$. Following the choice T, however, player 2 gets to move with one of his two choices, and the payoffs will be determined by the choice he makes. In this book we will use the convention of drawing the flow of the game from top to bottom, and the payoffs as a vertical list in which the top number is player 1's payoff and the bottom is player 2's. (If there are more than two players we will have more entries, still following the natural top-down order.)

This very simple structure is the most elementary form of a game tree. It includes most of the elements we described, but it still lacks the formal structure that would clearly delineate the "rules" that are used to describe such a game. Most importantly, how do we capture knowledge? For example, going back to the Battle of the Sexes game, in which Alex moved first, we may want to use the diagram in Figure 7.2 to represent it as a game tree. It starts at a node denoted x_0, at which player 1 can choose between O and F. Then, depending on the choice of player 1, player 2 gets to move at either node x_1 or x_2 and make a choice between o and f. (We're now using lowercase letters to denote the choices of player 2, a convenient way to distinguish between the

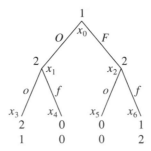

FIGURE 7.2 The sequential-move Battle of the Sexes game.

players.) The sequence of choices will result in one of the outcomes at the bottom of the tree. However, how can we distinguish between the sequential case in which player 2 *knows* the move of player 1 and the simultaneous case in which player 2 moves after player 1 but *is ignorant* about player 1's move?

To address this concern formally, and to complete the structure of a game tree, a certain amount of detail and notation needs to be introduced.[1] This is the main objective of this section; later, once the concepts are clear, we will focus our efforts on a variety of examples to master the use of game trees.

Definition 7.1 A *game tree* is a set of nodes $x \in X$ with a precedence relation $x > x'$, which means "x precedes x'." Every node in a game tree has only one predecessor. The precedence relation is *transitive* ($x > x', x' > x'' \Rightarrow x > x''$), *asymmetric* ($x > x' \Rightarrow$ not $x' > x$), and *incomplete* (not every pair of nodes x, y can be ordered). There is a special node called the *root* of the tree, denoted by x_0, that precedes any other $x \in X$. Nodes that do not precede other nodes are called *terminal nodes,* denoted by the set $Z \subset X$. Terminal nodes denote the final outcomes of the game with which payoffs are associated. Every node x that is not a terminal node is assigned either to a player, $i\,(x)$, with the action set $A_i(x)$, or to Nature.

This definition is quite a mouthful, but it formally captures the "physical" structure of a game tree, ignoring the actions of players and what they know when they move. To illustrate the definition, look back at the Battle of the Sexes game in Figure 7.2. x_0 is where the game begins, and x_0 precedes both x_1 and x_2. Each of these nodes precedes two terminal nodes, each describing a different outcome of the game. Since the terminal nodes are the game's outcomes, payoffs to the two players are noted at the terminal nodes.

Another example is given in Figure 7.3. In this game we have payoffs for four players, $N = \{1, 2, 3, 4\}$, but only players 1, 2, and 4 have actual moves. Hence we can think of player 3 as a "dummy player." The terminal nodes are $Z = \{x_4, x_5, x_6, x_7, x_8\}$, and payoffs are defined over terminal nodes: $v_i : Z \to \mathbb{R}$, where $v_i(z)$ is i's payoff if terminal node $z \in Z$ is reached. For example, if node x_5 is reached, then player 2 gets $v_2(x_5) = 7$ and player 4 gets $v_4(x_5) = -5$.

---

1. I borrow heavily from the notation in Fudenberg and Tirole (1991). This approach will serve the reader who is interested in learning about their more advanced treatment of the subject, and I hope it will not deter the reader who is not.

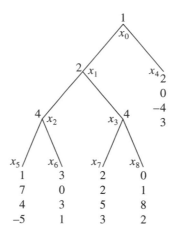

FIGURE 7.3 A game tree with a "dummy" player.

The precedence relation, together with the way in which players are assigned to nodes, describes the way in which the game unfolds. For example, in Figure 7.3 player 1 is assigned to the root, so $i(x_0) = 1$. His action set at the root, $A_1(x_0)$, includes two choices that determine whether the game will terminate at node x_4 with payoffs $(2, 0, -4, 3)$, or whether player 2 will get to play at node x_1. Player 2 then can choose whether player 4 will play at x_2 or x_3, and at each of these nodes player 4 has two choices that both end in termination of the game. Player 3 has no moves to make.[2]

There is still one missing component: how do we describe the knowledge of each player when it is his turn to move? In Figure 7.3 we see that player 2 moves after player 1. Would player 2 know what player 1 did? In this example we would think the answer should be obvious: if player 1 chose his other action, the game would end and player 2 would not have the option of choosing an action. Moving down the game tree to player 4, however, raises some questions. It seems implicit in the way we drew the game tree in Figure 7.3 that player 4 knows what happened before he moves, that is, if he is at node x_2 or at node x_3. Hence if player 4 knows where he is in the game tree then he must know what player 2 did before him.

Perhaps the game that is played actually calls on player 4 to make his move *without knowing* what player 2 did at x_1. How can we describe this situation in a game tree? Clearly we need to find a way to represent the case in which player 4 *cannot distinguish* between being at x_2 and being at x_3. That is, we need to be able to make statements like "I know that I am at either x_2 or x_3, but I don't know at which of the two I am."

We proceed to put structure on *the information that a player has* when it is his turn to move. A player can have very fine information and know exactly where he is in the game tree, or he may have coarser information and not know what has happened before his move, therefore not knowing exactly where he is in the game tree. We introduce the following definition:

---

2. Note that the moves from any nonterminal node result in a move to another node in the game. Thus we can save on notation and give moves "names" that are consistent with the nodes in which they will result. For our current example we could write $A_1(x_0) = \{x_1, x_4\}$, $A_2(x_1) = \{x_2, x_3\}$, $A_4(x_2) = \{x_5, x_6\}$, and $A_4(x_3) = \{x_7, x_8\}$.

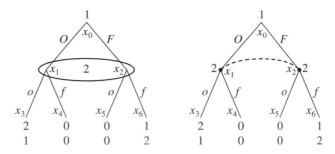

FIGURE 7.4 The simultaneous-move Battle of the Sexes game.

Definition 7.2 Every player i has a collection of information sets $h_i \in H_i$ that partition the nodes of the game at which player i moves with the following properties:

1. If h_i is a singleton that includes only x then player i who moves at x knows that he is at x.

2. If $x \neq x'$ and if both $x \in h_i$ and $x' \in h_i$ then player i who moves at x does not know whether he is at x or x'.

3. If $x \neq x'$ and if both $x \in h_i$ and $x' \in h_i$ then $A_i(x') = A_i(x)$.

The formal definition builds on a simple idea. Consider the sequential-move Battle of the Sexes game in Figure 7.2 and observe that player 2 moves at x_1. We want to describe whether or not he knows that he is at x_1. If we write $h_2 = \{x_1\}$, this means that the information set at x_1 is a singleton (it includes only the node x_1). Hence player 2 has information that says "I am at x_1," which is captured by property (1) of the definition. In this case it will follow that player 2 will have another information set, $h_2' = \{x_2\}$.

If, in contrast, we want to represent a game in which player 2 does not know whether he is at x_1 or x_2, then it must be the case that his information is "I know that I am at either x_1 or x_2, but I don't know at which of the two I am." Thus we will write $h_2 = \{x_1, x_2\}$, which exactly means that player 2 cannot tell whether he is at x_1 or x_2. This is the essence of property (2) of the definition.

Finally, property (3) is also essential to maintain the logic of information. If instead $x \in h_i$ and $x' \in h_i$ but $A_i(x') \neq A_i(x)$, then by the mere fact that player i has different actions from which to choose at each of the nodes x and x', he should be able to distinguish between these two nodes. It would therefore be illogical to assume that he cannot distinguish between them.

We are left to construct a graphical representation to show which nodes belong in the same information set. In Figure 7.4 we present the two common ways of depicting this using the extensive-form representation of the simultaneous-move Battle of the Sexes game that we have already seen and analyzed. Player 1 chooses from the action set $A_1 = \{O, F\}$, and player 2 chooses from $A_2 = \{o, f\}$, *without observing* the choice of player 1. On the left side of Figure 7.4 we use an ellipse to denote an information set, and all the nodes that are in the same ellipse belong to the same information set. In this example player 2 cannot distinguish between x_1 and x_2, so that $h_2 = \{x_1, x_2\}$. On the right side of Figure 7.4 is another common way of depicting information sets, according to which the dashed line connecting x_1 with x_2 denotes that both are in the same information set. Another example of a simultaneous-move game depicted as a

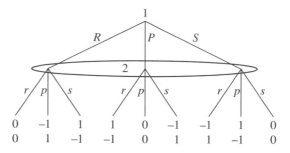

FIGURE 7.5 Game tree of rock-paper-scissors.

game tree is the game of rock-paper-scissors, depicted in Figure 7.5. In this game player 1 chooses from the set $A_1 = \{R, P, S\}$ while player 2 chooses from the set $A_2 = \{r, p, s\}$, without observing the choice made by player 1.

7.1.2 Imperfect versus Perfect Information

We defined games of *complete information* in Chapter 3 as the situation in which each player i knows the action set and the payoff function of each and every player $j \in N$, and this itself is common knowledge. This definition sufficed for the normal-form representation. For extensive-form games, however, it is useful to distinguish between two different types of complete-information games:

Definition 7.3 A game of complete information in which every information set is a singleton and there are no moves of Nature is called a **game of perfect information.** A game in which some information sets contain several nodes or in which there are moves of Nature is called a **game of imperfect information.**

In a game of perfect information every player knows exactly where he is in the game by knowing what occurred before he was called on to move. Examples would be the trust game in Figure 7.1 and the sequential-move Battle of the Sexes game in Figure 7.2. In a game of (complete but) imperfect information some players do not know where they are because some information sets include more than one node. This happens, for example, every time they move without knowing what some players have chosen previously, implying that *any simultaneous-move* game is a game of imperfect information. Examples include the simultaneous-move Battle of the Sexes game shown in Figure 7.4 and the rock-paper-scissors game depicted in Figure 7.5.

Games of imperfect information are also useful to capture the uncertainty a player may have about acts of Nature. For example, imagine the following card game: There is a large deck that includes an equal number of only kings and aces, from which player 1 pulls out a card without looking at it. The probability of getting a king is 0.5, and we can think of this as Nature's move. Hence player 1 moves after Nature and does not know if Nature chose a king or an ace. After drawing the card, player 1 can call (C) or fold (F). If he folds, he pays \$1 to player 2. If he calls, he pays \$2 to player 2 if the card is a king, while player 2 pays him \$2 if the card is an ace.

We can accurately describe this game in one of the two ways depicted in Figure 7.6. To see this, consider the game tree in the left panel. The order of appearance is loyal to the story: Nature chose K or A. Player 1 does not know what happened, but he knows that he is at either node in his information set with probability $\frac{1}{2}$. Then player 1 makes his move and the game ends. The game on the right looks different but is

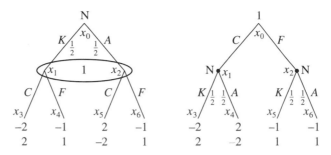

FIGURE 7.6 A card game.

strategically equivalent. Player 1 makes his move (C or F) without knowing which card was drawn, and then Nature draws the card, K or A, with equal probability. Therefore in both games we are depicting exactly the same information and moves, and hence we are getting the story right.[3]

There is an insight worth emphasizing here. A game will be one of imperfect information when a player must make a move either without knowing the move of another player or without knowing the realization of a choice of Nature. Uncertainty over the choice of Nature, or **exogenous uncertainty,** is at the heart of the single-person decision problems that were described in Chapter 2. (And the card game in Figure 7.6 is effectively a single-person decision problem because player 2 has no choices to make.) Uncertainty over the choice of another player, or **endogenous uncertainty,** is the subject of simultaneous-move games like the simultaneous-move Battle of the Sexes game in Figure 7.4. Notice, however, that both situations share a common feature: occurrences that some player does not know are captured by uncertainty over where he is in the game, be it from exogenous or endogenous uncertainty. In either case a player must form beliefs about the unobserved actions, of Nature or of other players, in order to analyze his situation.

7.2 Strategies and Nash Equilibrium

Now that we have the structure of the extensive-form game well defined, and have developed game trees to represent this structure, we move to the next important step of describing strategies. Recall that in Section 3.1 we argued that "a strategy is often defined as *a plan of action intended to accomplish a specific goal.*" In the normal-form game it was very easy to define a strategy for a player: a pure strategy was some element from his set of actions, A_i, and a mixed strategy was some probability distribution over these actions. As we will now see, a strategy is more involved in extensive-form games.

7.2.1 Pure Strategies

Consider the sequential-move Battle of the Sexes game described again in Figure 7.7. (We will now seldom include the names of nodes in our game trees because they have

---

3. This is the reason that the definition of games of perfect or imperfect information explicitly requires the reference to moves of Nature. It is sometimes possible to include them in a game tree in which all the information sets are singletons.

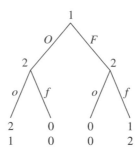

FIGURE 7.7   The sequential-move Battle of the Sexes game.

no real consequence for the issues with which we are concerned.) Player 1 has a single information set with one node, so for him a pure strategy is as simple as "play O" or "play F." For player 2, however, things are a bit more involved. Player 2 has two information sets, each associated with a different action of player 1. Hence the two simple statements "play o" and "play f" do not seem to exhaust all the possibilities for player 2. In particular, player 2 can choose the following rather attractive strategy: "If player 1 plays O then I will play o, while if player 1 plays F then I will play f."

This simple example demonstrates that when a player's move follows after the realization of previous events in the game, and if the player can distinguish between these previous events (they result in different information sets), then he can condition his behavior on the events that happened. A strategy is therefore no longer a simple statement of what a player will do, as in the normal-form simultaneous-move game. Instead we have

Pure Strategies in Extensive-Form Games   A **pure strategy** for player i is a *complete plan of play* that describes which pure action player i will choose at each of his information sets.

If we consider the simultaneous-move Battle of the Sexes game in Figure 7.8, the pure strategies for player 1 are $S_1 = \{O, F\}$, and those for player 2 are $S_2 = \{o, f\}$. Because each player has only one information set, the extensive-form game is identical to the simple normal-form game we have already encountered. In contrast, in the sequential-move Battle of the Sexes game in Figure 7.7, player 2 has two distinct information sets in which he can choose o or f, each information set resulting from the previously made choice of player 1. Therefore a "complete plan of play" must accommodate a strategy that directs what player 2 will choose for *each* choice of

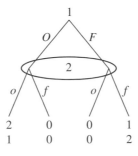

FIGURE 7.8   The simultaneous-move Battle of the Sexes game.

player 1. That is, player 2's choice of action from the set $\{o, f\}$ can be made *contingent* on what player 1 does, admitting the possibility of strategies of the form "If player 1 plays O then I will play o, while if player 1 plays F then I will play f."

For this example, we can describe the set of pure strategies for player 2 as follows:

$$S_2 = \{oo, of, fo, ff\},$$

where a pure strategy "ab" is shorthand for "I will play a if player 1 plays O and I will play b if he plays F." For player 1 the pure strategy set remains $S_1 = \{O, F\}$.

We now introduce some notation that builds on what we have already developed in order to define formally a pure strategy. Let H_i be the collection of all information sets at which player i plays, and let $h_i \in H_i$ be one of i's information sets. Let $A_i(h_i)$ be the actions that player i can take at h_i, and let A_i be the set of all actions of player i, $A_i = \cup_{h_i \in H_i} A_i(h_i)$ (i.e., the union of all the elements in all the sets $A_i(h_i)$). We can now define a pure strategy as follows:

Definition 7.4 A **pure strategy** for player i is a mapping $s_i : H_i \rightarrow A_i$ that assigns an action $s_i(h_i) \in A_i(h_i)$ for every information set $h_i \in H_i$. We denote by S_i the set of all pure-strategy mappings $s_i \in S_i$.

A final point is in order, and the sequential-move Battle of the Sexes game in Figure 7.7 is useful to illustrate it. Notice that even though player 2 has only two actions from which to choose, by moving *after observing what player 1 has chosen*, his strategy defines actions that are *conditional on his information about where he is in the game*. In this example the two actions translate into four pure strategies because he has two information sets.

This observation implies that a potentially small set of moves can translate into a much larger set of strategies when sequential moves are possible, and when players have knowledge of what preceded their play. In general assume that player i has $k > 1$ information sets, the first with m_1 actions from which to choose, the second with m_2, and so on until m_k. Letting $|S_i|$ denote the number of elements in S_i, the total number of pure strategies player i has is

$$|S_i| = m_1 \times m_2 \times \cdots \times m_k.$$

For example, a player with 3 information sets, 2 actions in the first, 3 in the second, and 4 in the third will have a total of 24 pure strategies.

7.2.2 Mixed versus Behavioral Strategies

Now that we have defined pure strategies, the definition of mixed strategies follows immediately, just like in the normal-form game:

Definition 7.5 A **mixed strategy** for player i is a probability distribution over his pure strategies $s_i \in S_i$.

How do we interpret a mixed strategy? In exactly the same way that we did for the normal form: a player randomly chooses between all his pure strategies—in this case all the complete plans of play—and once a particular plan is selected the player follows it.

You may notice that this interpretation takes away some of the dynamic flavor that we set out to capture with extensive-form games. More precisely, when a mixed strategy is used, the player selects a plan randomly *before the game is played* and then follows *a particular pure strategy.*

This description of mixed strategies was sensible for normal-form games because there it was a once-and-for-all choice to be made. In a game tree, however, the player may want to randomize at some nodes, independently of what he did at earlier nodes. In other words, the player may want to "cross the bridge when he gets there." This cannot be captured by mixed strategies as previously defined because once the randomization is over, the player is choosing a pure plan of action.

To illustrate this point, consider again the sequential-move Battle of the Sexes game in Figure 7.7. The previous definition of a mixed strategy implies that player 2 can randomize among any of his four pure strategies in the set $S_2 = \{oo, of, fo, ff\}$. It does not, however, allow him to choose strategies of the form "If player 1 plays O then I'll play f, while if he plays F then I'll mix and play f with probability $\frac{1}{3}$."

To allow for strategies that let players randomize as the game unfolds we define a new concept as follows:

Definition 7.6 A **behavioral strategy** specifies for each information set $h_i \in H_i$ an independent probability distribution over $A_i(h_i)$ and is denoted by $\sigma_i : H_i \rightarrow \triangle A_i(h_i)$, where $\sigma_i(a_i(h_i))$ is the probability that player i plays action $a_i(h_i) \in A_i(h_i)$ in information set h_i.

Arguably a behavioral strategy is more in tune with the dynamic nature of the extensive-form game. When using such a strategy, a player mixes among his actions whenever he is called to play. This differs from a mixed strategy, in which a player mixes before playing the game but then remains loyal to the selected pure strategy.

Luce and Raiffa (1957) provide a nice analogy for the different strategy types we have introduced. A pure strategy can be thought of as an instruction manual in which each page tells the player which pure action to take at a particular information set, and the number of pages is equal to the number of information sets the player has. The set S_i of pure strategies can therefore be treated like a library of such pure-strategy manuals. A mixed strategy consists of choosing one of these manuals at random and then following it precisely.

In contrast a behavioral strategy is a manual that prescribes possibly random actions on each of the pages associated with play at particular information sets. To see this, consider the sequential-move Battle of the Sexes game in Figure 7.9. Player 2 has two information sets associated with the nodes x_1 and x_2, which we denote h_2^O and h_2^F, respectively, and in each he can choose between two actions in $A_2 = \{o, f\}$. A pure strategy would be an element from $S_2 = \{oo, of, fo, ff\}$. A mixed strategy would be a probability distribution $(p_{oo}, p_{of}, p_{fo}, p_{ff})$, where $p_{s_2} \geq 0$ and $p_{oo} + p_{of} + p_{fo} + p_{ff} = 1$. Denote a behavioral strategy as four probabilities, $\sigma_2(o(h_2^O))$, $\sigma_2(f(h_2^O))$, $\sigma_2(o(h_2^F))$, and $\sigma_2(f(h_2^F))$, where $\sigma_2(o(h_2^O)) + \sigma_2(f(h_2^O)) = \sigma_2(o(h_2^F)) + \sigma_2(f(h_2^F)) = 1$. In Figure 7.9 we have used $\sigma_2(o(h_2^O)) = \frac{1}{3}$, $\sigma_2(f(h_2^O)) = \frac{2}{3}$, and $\sigma_2(o(h_2^F)) = \sigma_2(f(h_2^F)) = \frac{1}{2}$.

Having defined two kinds of nonpure strategies, the obvious question is whether we need to consider both kinds of strategies for a complete description of possible behavior, or whether it suffices to consider only one kind or the other. We can answer this question by answering two complementary questions. The first is, given a mixed (not behavioral) strategy, can we find a behavioral strategy that leads to the same outcomes? Using our example, can we replace a mixed strategy of the

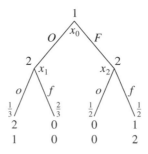

FIGURE 7.9 Behavioral strategies in the sequential-move Battle of the Sexes game.

form $(p_{oo}, p_{of}, p_{fo}, p_{ff})$ with $\sum_{s \in S_2} p_s = 1$ with a behavioral strategy of the form $(\sigma_2(o(h_2^O)), \sigma_2(f(h_2^O)), \sigma_2(o(h_2^F)), \sigma_2(f(h_2^F)))$ that leads to the same randomization over outcomes?

The answer is yes, which is quite easy to see. Conditional on reaching x_1 the probability of playing o is $\Pr\{o|O\} = p_{oo} + p_{of}$, and conditional on reaching x_1 the probability of playing f is $\Pr\{f|O\} = p_{fo} + p_{ff}$, so that $\Pr\{o|O\} + \Pr\{f|O\} = 1$. Similarly, conditional on reaching x_2 the probability of playing o is $\Pr\{o|F\} = p_{oo} + p_{fo}$, and the probability of playing f is $\Pr\{f|F\} = p_{of} + p_{ff}$, so that $\Pr\{o|F\} + \Pr\{f|F\} = 1$. Thus we can define a behavioral strategy $\sigma_2(a_2(h)) = \Pr\{a_2|h\}$ that yields the same randomization as the mixed strategy $(p_{oo}, p_{of}, p_{fo}, p_{ff})$.

The complementary question is whether, given a behavioral strategy, we can find a mixed strategy that leads to the same outcomes. Again, turn to the example as an illustration and consider the behavioral strategy shown in Figure 7.9, where $\sigma_2(o(O)) = \frac{1}{3}$, $\sigma_2(f(O)) = \frac{2}{3}$, and $\sigma_2(o(F)) = \sigma_2(f(F)) = \frac{1}{2}$. Notice that if the player uses a mixed strategy $(p_{oo}, p_{of}, p_{fo}, p_{ff})$ then *conditional on player 1 choosing O*, action o will be chosen with probability $p_{oo} + p_{of}$, and action f will be chosen with probability $p_{fo} + p_{ff}$. Similarly, *conditional on player 1 choosing F*, action o will be chosen with probability $p_{oo} + p_{fo}$, and action f will be chosen with probability $p_{of} + p_{ff}$. Thus to replicate the behavioral strategy, four equalities must be satisfied by the mixed strategy $(p_{oo}, p_{of}, p_{fo}, p_{ff})$:

$$\Pr\{o|O\} = p_{oo} + p_{of} = \tfrac{1}{3}$$

$$\Pr\{f|O\} = p_{fo} + p_{ff} = \tfrac{2}{3}$$

$$\Pr\{o|F\} = p_{oo} + p_{fo} = \tfrac{1}{2}$$

$$\Pr\{f|F\} = p_{of} + p_{ff} = \tfrac{1}{2}.$$

It may seem like we have four equations with four unknowns, but this system of equations is "underidentified" because these four equations are really only two equations. By definition it must be true that

$$p_{oo} + p_{of} + p_{fo} + p_{ff} = 1,$$

which implies that we have only three equations with four unknowns.[4] It follows that *many* values of $(p_{oo}, p_{of}, p_{fo}, p_{ff})$ will satisfy the equations, implying that this particular behavioral strategy can be generated by an infinite number of different mixed strategies. For example, $(p_{oo}, p_{of}, p_{fo}, p_{ff}) = \left(\frac{1}{3}, 0, \frac{1}{6}, \frac{1}{2}\right)$ will lead to equivalent outcomes as the behavioral strategy, but so will $(p_{oo}, p_{of}, p_{fo}, p_{ff}) = \left(\frac{1}{6}, \frac{1}{6}, \frac{1}{3}, \frac{1}{3}\right)$.

This example may suggest that any randomization over play can be represented by either mixed or behavioral strategies. As it turns out, this is true under a rather mild condition.

Definition 7.7 A game of *perfect recall* is one in which no player ever forgets information that he previously knew.

That is, a game of perfect recall is one in which, if a player is called upon to move more than once in a game, then he must remember the moves that he chose in his previous information sets. Practically all of the analysis in game theory, and in applications of game theory to the social sciences, assumes perfect recall, as will we in this text. For the class of perfect-recall games, Kuhn (1953) proved that mixed and behavioral strategies are equivalent, in the sense that given strategies of i's opponents, the same distribution over outcomes can be generated by either a mixed or a behavioral strategy of player i.

Remark (The Absent-Minded Driver) Despite our focus on games of perfect recall, it may be interesting to dwell for a moment on the example of the "absent-minded driver" that was introduced by Piccione and Rubinstein (1997) to depict a simple game (in fact, a single-person decision problem) with imperfect recall. The story goes as follows. An absent-minded driver is driving home along the highway. The first exit on the road, exit 1, takes him to a bad neighborhood, yielding him a payoff of 0. Exit 2, farther down the road, is the best route to his home, yielding him a payoff of 4. If he misses exit 2 then he will eventually get home the long way, yielding him a payoff of 1. Imperfect recall means that he cannot distinguish between exits 1 and 2 and therefore cannot remember whether he has already passed one exit or not. This decision problem is depicted in Figure 7.10. Note that the player has two pure strategies, "Exit" or "continue." Committing to the pure strategy of "Exit" implies that he exits at the first intersection and gets 0. If he chooses the pure strategy "continue," then he goes all the way on the long route and gets a payoff of 1. Hence, no mixed strategy will yield a payoff of 4 with positive probability. This, however, is not true for a behavioral strategy, which we'll call a "planning" mixed strategy. Define a "planning" mixed strategy of the player as a probability p that he will exit at any exit he passes (both nodes in his single information set)—that is, the probability that the driver will *commit* to exit when he is at an intersection. His expected payoff from this strategy is $0p + 4p(1-p) + 1(1-p)^2 = -3p^2 + 2p + 1$, which is maximized at $p = \frac{1}{3}$. As Piccione and Rubinstein argue, a puzzle emerges once the driver finds himself at an intersection. He knows that with some probability q he is at Exit 1, and with some probability $(1-q)$ he is at exit 2. The driver's payoff for choosing to exit with probability p is now, at the intersection, $q[4p(1-p) + 1(1-p)^2] + (1-q)[4p + 1(1-p)]$, which is equivalent to the planning problem only when $q = 1$,

---

4. This follows because $p_{oo} + p_{of} = \frac{1}{3}$ together with $p_{oo} + p_{of} + p_{fo} + p_{ff} = 1$ implies that $p_{fo} + p_{ff} = \frac{2}{3}$, and, at the same time, $p_{oo} + p_{fo} = \frac{1}{2}$ together with $p_{oo} + p_{of} + p_{fo} + p_{ff} = 1$ implies that $p_{of} + p_{ff} = \frac{1}{2}$. Therefore two of the five equations are redundant.

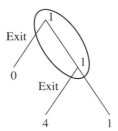

FIGURE 7.10 The absent-minded driver.

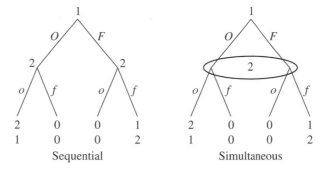

FIGURE 7.11 The Battle of the Sexes game: two versions.

that is, only when he believes for sure that he is at the first exit. This leads to an interesting dynamic inconsistency in that the driver would plan one thing in advance and then rationally change his mind once he finds himself at an intersection.

7.2.3 Normal-Form Representation of Extensive-Form Games

Consider the two variants of the Battle of the Sexes game presented in Figure 7.11. The simultaneous-move version in the right panel is one that we have seen before in its matrix form, as follows:

		Player 2	
		O	F
Player 1	O	2, 1	0, 0
	F	0, 0	1, 2

Now consider the sequential-move Battle of the Sexes game depicted in the left panel. Recall that $S_1 = \{O, F\}$ and $S_2 = \{oo, of, fo, ff\}$, where fo, for example, means that player 2 plays f after player 1 plays O, while player 2 plays o after player 1 plays F. This game can be represented by a 2×4 matrix as follows:

		Player 2			
		oo	of	fo	ff
Player 1	O	2, 1	2, 1	0, 0	0, 0
	F	0, 0	1, 2	0, 0	1, 2

As this matrix demonstrates, each of the four payoffs in the original extensive-form game is replicated twice. This happens because for any pure strategy of player 1, two of the four pure strategies of player 2 are equivalent. For example, if player 1 plays O, then only the "first component" of player 2's strategy matters (what player 2 does following player 1's choice of O). Therefore oo (player 2 playing o after O and o after F) and of (player 2 playing o after O and f after F) yield the same outcome. If, however, player 1 plays F, then only the second component of player 2's strategy matters, so that oo and fo yield the same outcome.

Any extensive-form game can be transformed into a normal-form game by using the set of pure strategies of the extensive form (see definition 7.4) as the set of pure strategies in the normal form, and the set of payoff functions is derived from how combinations of pure strategies result in the selection of terminal nodes. Furthermore every extensive-form game will have a unique normal form that represents it, which is not true for the reverse transformation (see the following remark).

Clearly this exercise of transforming extensive-form games into the normal form seems to miss the point of capturing the dynamic structure of the extensive-form game. Why then would we be interested in this exercise? The reason is that the concept of a Nash equilibrium is *static in nature,* in that the equilibrium posits that players take the strategies of others as given, and in turn they play a best response. Therefore the normal-form representation of an extensive form will suffice to find all the Nash equilibria of the game. This is particularly useful if the extensive form is a two-player game with a finite number of strategies for each player, because we can write its normal form as a matrix and solve it with the simple techniques developed earlier. As we will now see, this approach has some useful implications.

Remark Though every extensive form has a unique normal-form representation, it is not true that every normal form can be represented by a unique extensive-form game. Consider the following matrix for a normal-form game:

	C	D
N	$0, 0$	$0, 0$
E	$1, 1$	$-1, 2$

and notice that it is a consistent representation of either of the two game trees depicted in Figure 7.12. Notice that the extensive form on the left is a game of perfect information while the game on the right is one of imperfect information. However, in terms of their "strategic" character, the two extensive-form games are identical, because a game's strategic essence is captured by its normal form.

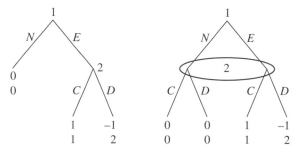

FIGURE 7.12 Two extensive forms with the same normal form.

7.3 Nash Equilibrium and Paths of Play

We are now ready to analyze the equilibrium predictions in extensive-form games using the concept of Nash equilibrium. As mentioned in the previous section, by transforming an extensive form into its normal-form representation, we are concisely capturing the strategic essence of the game and can use the normal form to find all the Nash equilibria of the game. The simultaneous-move Battle of the Sexes game in the right panel of Figure 7.11 is equivalent to the normal form that we had already analyzed. Hence we know that there are two pure-strategy Nash equilibria, (O, o) and (F, f), and a mixed-strategy equilibrium in which player 1 plays O with probability $\frac{2}{3}$ and player 2 plays o with probability $\frac{1}{3}$. By calculating (or drawing) the best-response correspondences of the two players, it is easy to show that these are the only three Nash equilibria for this game.

Now consider the sequential-move Battle of the Sexes game depicted in the left panel of Figure 7.11. It is easy to see that we can "replicate" the outcomes of the Nash equilibria we found for the simultaneous version of this game with the strategies (O, oo) and (F, ff). The first pair of strategies (O, oo) specify that player 1 chooses O and player 2 chooses "I play o after O and o after F," which yields the outcome of both players going to the opera. Similarly (F, ff) yields the outcome of both going to the football game. The question is whether there are other Nash equilibria in pure strategies. The extensive form does not offer a simple way to answer this question (though it is possible to use the extensive form to check for more equilibria). Instead consider the normal-form representation,

		\multicolumn{4}{c}{Player 2}		
		oo	of	fo	ff
Player 1	O	$\overline{2,1}$	$\overline{2,1}$	$0,0$	$0,0$
	F	$0,0$	$\overline{1,2}$	$0,0$	$\overline{1,2}$

By using the simple method described in Section 5.1.1 (underlining the best responses of player 1 to each column and overlining the best responses of player 2 to each row), we can see that a third pure-strategy Nash equilibrium exists: (O, of). This strategy yields the same outcome as (O, oo) but with the following strategies: player 1 chooses O and player 2 chooses "I play o after O and f after F."

This example demonstrates a convenient feature of the matrix-form representation of a two-player finite extensive-form game: it will immediately reveal all the pure strategies of each player, and in turn will offer an easy way to find all the pure-strategy profiles that are Nash equilibria. (This will require a bit more work for games that cannot be represented by a matrix, in which case we will have to find the best-response correspondences and use them to compute the possible Nash equilibria.)

Notice also that in the sequential-move Battle of the Sexes example there are two Nash equilibria that result in the *exact same outcome* of both players going to the opera: (O, of) and (O, oo). Is there an important difference between these two predictions? Indeed there is—the difference between these two equilibria is not in what the players actually play *in equilibrium*, but instead what player 2 plans to play in an information set that is *not reached in equilibrium.*

This observation is useful in making the following more general point. In the extensive form, every outcome of the game is associated with a unique *path of play* because there is a unique path from the root of the game to each terminal node. This

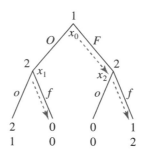

FIGURE 7.13 Equilibrium paths in the sequential-move Battle of the Sexes game.

follows from the construction that every node can be preceded by only one node. We also know that players will play a Nash equilibrium because, given their correct beliefs about the strategies of their opponents, it is in their best interest to stick with their equilibrium strategies. This implies that if a strategy profile s^* is a Nash equilibrium, each player i prefers sticking to the predicted path of play over "leaving" it and choosing some other path in the game, given his belief that the other players will stick to s^*_{-i}. It is therefore useful to define the following:

Definition 7.8 Let $\sigma^* = (\sigma_1^*, \ldots, \sigma_n^*)$ be a Nash equilibrium profile of behavioral strategies in an extensive-form game. We say that an information set is **on the equilibrium path** if given σ^* it is reached with positive probability. We say that an information set is **off the equilibrium path** if given σ^* it is never reached.

Using this definition we can reinterpret the forces that make a Nash equilibrium prediction self-enforcing. In a Nash equilibrium players choose to proceed *on the equilibrium path* because of their beliefs about what the other players are doing both *on and off the equilibrium path*. In Figure 7.13 the Nash equilibrium (F, ff) of the sequential-move Battle of the Sexes game is shown with the highlighted edges in the tree. This equilibrium is supported by player 1's *correct belief* that if he would deviate from the equilibrium path and play O then he would receive 0 because player 2 will proceed to play f in the information set x_1. In other words, the "threat" imposed by player 2's strategy of how he will proceed *off the equilibrium path* is supporting the actions of player 1 *on the equilibrium path*.

Notice, however, that the Nash equilibrium (F, ff) has a rather curious feature. The reason that player 1 is not willing to deviate from F to O is his belief that player 2 will choose f in information set x_1. However, choosing f after O is an irrational choice of player 2 given the payoffs of the game, implying that this "threat" is somewhat *incredible.* Now ask yourself: is the Nash equilibrium (F, ff) a reasonable prediction about the rational choice of player 1? By the definition of Nash equilibrium it is: F is a best response to ff and vice versa. However, this implies that if, for some unexpected reason, player 1 would suddenly choose to play O then player 2 would *not respond optimally:* his strategy ff commits him to choose f even though o would yield him a higher payoff after a choice of O by player 1. This argument sheds some light on a weakness of the normal-form representation in that it treats all choices as "once-and-for-all" simultaneous choices, and thus beliefs can never be challenged. We address this weakness in the next chapter.

7.4 Summary

- In addition to the set of players, their possible actions, and their payoffs from outcomes, the extensive-form representation captures the order in which they play as well as what each player knows when it is his turn to move. Game trees provide a useful tool to describe extensive-form games.

- If a player cannot distinguish between two or more nodes in a game tree then they belong to the same information set. Care has to be taken to specify correctly each player's information sets.

- In games of perfect information, every player knows exactly what transpired before each of his turns to move, so each information set is a singleton. If this is not the case then the players are playing a game of imperfect information.

- A pure strategy defines a player's deterministic plan of action at each of his information sets. A mixed strategy is a probability distribution over pure strategies, while a behavioral strategy is a plan of probability distributions over actions in every information set.

- Every extensive-form game has a unique normal-form representation, but the reverse is not necessarily true.

7.5 Exercises

7.1 **Strategies:** Imagine an extensive-form game in which player i has K information sets.

    a. If the player has an identical number of m possible actions in each information set, how many pure strategies does he have?

    b. If the player has m_k actions in information set $k \in \{1, 2, \ldots, K\}$, how many pure strategies does the player have?

7.2 **Strategies and Equilibrium:** Consider a two-player game in which player 1 can choose A or B. The game ends if he chooses A while it continues to player 2 if he chooses B. Player 2 can then choose C or D, with the game ending after C and continuing again with player 1 after D. Player 1 can then choose E or F, and the game ends after each of these choices.

    a. Model this as an extensive-form game tree. Is it a game of perfect or imperfect information?

    b. How many terminal nodes does the game have? How many information sets?

    c. How many pure strategies does each player have?

    d. Imagine that the payoffs following choice A by player 1 are $(2, 0)$, those following C by player 2 are $(3, 1)$, those following E by player 1 are $(0, 0)$, and those following F by player 1 are $(1, 2)$. What are the Nash equilibria of this game? Does one strike you as more "appealing" than the other? If so, explain why.

7.3 **Tic-Tac-Toe:** The extensive-form representation of a game can be cumbersome even for very simple games. Consider the game of tic-tac-toe, in which two players mark "X" or "O" in a 3×3 matrix. Player 1 moves first, then player 2, and so on. If a player gets three of his kind in a row, column, or one of the diagonals then he wins, and otherwise it is a tie. For this question assume that even after a winner is declared, the players *must completely fill* the matrix before the game ends.

    a. Is this a game of perfect or imperfect information? Why?

    b. How many information sets does player 2 have after player 1's first move?

    c. How many information sets does player 1 have after each of player 2's moves for the first time?

    d. How many information sets does each player have in total? (Hint: For this and part (e) you may want to use a spreadsheet program like Excel.)

    e. How many terminal nodes does the game have?

7.4   **Centipedes:** Imagine a two-player game that proceeds as follows. A pot of money is created with $6 in it initially. Player 1 moves first, then player 2, then player 1 again, and finally player 2 again. At each player's turn to move, he has two possible actions: grab (G) or share (S). If he grabs he gets $\frac{2}{3}$ of the current pot of money, the other player gets $\frac{1}{3}$ of the pot, and the game ends. If he shares then the size of the current pot is multiplied by $\frac{3}{2}$ and the next player gets to move. At the last stage at which player 2 moves, if he chooses S then the pot is still multiplied by $\frac{3}{2}$, player 2 gets $\frac{1}{3}$ of the pot, and player 1 gets $\frac{2}{3}$ of the pot.

    a. Model this as an extensive-form game tree. Is it a game of perfect or imperfect information?

    b. How many terminal nodes does the game have? How many information sets?

    c. How many pure strategies does each player have?

    d. Find the Nash equilibria of this game. How many outcomes can be supported in equilibrium?

    e. Now imagine that at the last stage at which player 2 moves, if he chooses to share then the pot is equally split among the players. Does your answer to part (d) change?

7.5   **Veto Power:** Two players must choose among three alternatives, a, b, and c. Player 1's preferences are given by $a \succ_1 b \succ_1 c$ while player 2's preferences are given by $c \succ_2 b \succ_2 a$. The rules are that player 1 moves first and can veto one of the three alternatives. Then player 2 chooses one of the remaining two alternatives.

    a. Model this as an extensive-form game tree (choose payoffs that represent the preferences).

    b. How many pure strategies does each player have?

    c. Find all the Nash equilibria of this game.

7.6   **Entering an Industry:** A firm (player 1) is considering entering an established industry with one incumbent firm (player 2). Player 1 must choose whether or not to enter the industry. If player 1 enters the industry then player 2 can either accommodate the entry or fight the entry by waging a price war. Player 1's most-preferred outcome is entering with player 2 not fighting, and its least-preferred outcome is entering with player 2 fighting. Player 2's most-preferred outcome is player 1 not entering, and its least-preferred outcome is player 1 entering with player 2 fighting.

     a. Model this as an extensive-form game tree (choose payoffs that represent the preferences).

     b. How many pure strategies does each player have?

     c. Find all the Nash equilibria of this game.

7.7 **Roommates Voting:** Three roommates need to vote on whether they will adopt a new rule and clean their apartment once a week or stick to the current once-a-month rule. Each votes "yes" for the new rule or "no" for the current rule. Players 1 and 2 prefer the new rule while player 3 prefers the old rule.

     a. Imagine that the players require a unanimous vote to adopt the new rule. Player 1 votes first, then player 2, and then player 3, the latter two observing the previous votes. Draw this as an extensive-form game and find the Nash equilibria.

     b. Imagine now that the players require a majority vote to adopt the new rule (at least two "yes" votes). Again player 1 votes first, then player 2, and then player 3, the latter two observing the previous votes. Draw this as an extensive-form game and find the Nash equilibria.

     c. Now imagine that the game is as in part (b), but the players put their votes into a hat—so that the votes of earlier movers are not observed by the later movers—and the votes are counted after all have voted. Draw this as an extensive-form game and find the Nash equilibria. In what way is this result different from the result in (b)?

7.8 **Brothers:** Consider the following game that proceeds in two stages: In the first stage one brother (player 2) has two $10 bills and can choose one of two options: he can give his younger brother (player 1) $20 or give him one of the $10 bills (giving nothing is inconceivable given the way they were raised). This money will then be used to buy snacks at the show they will see, and each $1 of snacks purchased yields one unit of payoff for a player who uses it. In the second stage the show they will see is determined by the following Battle of the Sexes game:

		Player 2	
		O	F
Player 1	O	16, 12	0, 0
	F	0, 0	12, 16

     a. Present the *entire* game in extensive form (a game tree).

     b. Write the (pure-) strategy sets for both players.

     c. Present the *entire* game in one matrix.

     d. Find the Nash equilibria of the *entire* game (pure and mixed strategies).

7.9 **The Dean's Dilemma:** A student stole the DVD player from the student lounge. The dean of students (player 1) suspects the student (player 2) and begins collecting evidence. However, evidence collection is a random process, and concrete evidence will be available to the dean only with probability $\frac{1}{2}$. The student knows that the evidence-gathering process is under way but does not know whether the dean has collected evidence or not. The game proceeds as follows: The dean realizes whether or not he has evidence and can then choose his action, whether to accuse the student (A) or bounce the case (B) and forget it. Once accused, the student has two options: he can either confess (*C*) or deny (*D*). Payoffs are realized as follows: If the dean bounces the case then both players get 0. If the dean accuses the student and the student confesses, the dean gains 2 and the student loses 2. If the dean accuses the student and the student denies, then payoffs depend on the evidence: If the dean has no evidence then he loses face, which gives him a loss of 4, while the student gains glory, which gives him a payoff of 4. If, however, the dean has evidence then he is triumphant and gains 4, while the student is put on probation and loses 4.

a. Draw the game tree that represents the extensive form of this game.
b. Write down the matrix that represents the normal form of the extensive form you drew in (a).
c. Solve for the Nash equilibria of the game.

8

Credibility and Sequential Rationality

Consider the simple extensive-form coordination game described in Figure 8.1, which is represented in normal form in the following matrix:

Player 2

		e	p
Player 1	E	0, 0	0, 0
	P	0, 0	1, 1

Each player can exit (E for player 1 and e for player 2), or proceed (P and p, respectively). If the players coordinate on proceeding, then they both receive a payoff of 1 while any other profile of strategies yields each a payoff of 0.

There is a good reason to assume that the players will successfully coordinate on the Nash equilibrium profile (P, p): the exit strategy is weakly dominated by the choice to proceed. However, a quick look at the matrix shows that there is another Nash equilibrium: (E, e). Using the terminology of Section 7.3, the players stick to the equilibrium path of player 1 choosing E because of player 2's threat to choose e off the equilibrium path.

There is therefore something unappealing about the logic of Nash equilibrium in extensive-form games. The concept asks only for players to act rationally

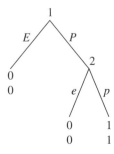

FIGURE 8.1   A coordination game.

on the equilibrium path given their beliefs about what will transpire both *on and off* the equilibrium path. Nevertheless it puts no restrictions on the beliefs of players off the equilibrium path, nor on how they should consider such beliefs. In fact the normal-form representation of a sequential game is not able to address such a requirement, which adds constraints on what we would tolerate as "rational behavior."

We would, however, expect rational players to play optimally in response to their beliefs *whenever* they are called to move. In the simple coordination game in Figure 8.1 this would require that player 2 commit to play p when it is his turn to move. Applying this reasoning to the sequential-move Battle of the Sexes game in Figure 7.11 suggests that of the three Nash equilibria of the game, two are unappealing. The equilibria (O, oo) and (F, ff) have player 2 committing to a strategy that, despite being a best response to player 1's strategy, would not have been optimal were player 1 to deviate from his strategy and cause the game to move off the equilibrium path. In what follows we will introduce a natural requirement that will result in more refined predictions for dynamic games. These will indeed rule out equilibria such as (E, e) in the coordination game and (O, oo) and (F, ff) in the sequential-move Battle of the Sexes game.

8.1 Sequential Rationality and Backward Induction

To address the critique we posed regarding the incredible nature of the equilibria (O, oo) and (F, ff) in the sequential-move Battle of the Sexes game, we will directly criticize the behavior of player 2 in the event that player 1 did not follow his prescribed strategy. We will insist that a player use strategies that are optimal at *every information set in the game tree*. We call this principle *sequential rationality,* because it implies that players are playing rationally at every stage in the sequence of the game, whether it is on or off the equilibrium path of play. Formally, we have

Definition 8.1 Given strategies $\sigma_{-i} \in \Delta S_{-i}$ of i's opponents, we say that σ_i is **sequentially rational** if and only if i is playing a best response to σ_{-i} in each of his information sets.

Using this definition we can revisit the sequential-move Battle of the Sexes game and ask: what are player 2's best responses in each of his information sets? The answer is obvious: if player 1 played O then player 2 should play o, and if player 1 played F then player 2 should play f. Any other strategy of player 2 is not a best response in at least one of these information sets, and this implies that a sequentially rational player 2 should be playing the pure strategy of.

Now move back to the root of the game where player 1 has to choose between O and F. Taking into account the sequential rationality of player 2, player 1 should conclude that playing O will result in the payoffs $(2, 1)$ while playing F will result in the payoffs $(1, 2)$. Now applying sequential rationality to player 1 requires that player 1, who is *correctly predicting* the behavior of player 2, should choose O. Hence the unique prediction from this process is the path of play that begins with player 1 choosing O followed by player 2 choosing o. Furthermore the process predicts what would happen if players *deviate from the path of play:* if player 1 chooses F then player 2 will choose f. We conclude that the Nash equilibrium (O, of) is the unique pair of strategies that survives our requirement of sequential rationality.

This type of procedure, which starts at nodes that directly precede the terminal nodes at the end of the game and then inductively moves backward through the game tree, is also known as **backward induction in games.** This is exactly the multiperson dynamic programming version of the single decision maker's process described in Section 2.4.1. As the example just given suggests, when we apply this procedure to *finite games of perfect information* it will result in a prescription of strategies for each player that are sequentially rational, as the following proposition states:

Proposition 8.1 *Any finite game of perfect information has a backward induction solution that is sequentially rational. Furthermore if no two terminal nodes prescribe the same payoffs to any player then the backward induction solution is unique.*

The proof is rather straightforward and is based on the simple idea of backward induction. By definition a finite game of perfect information has a finite set of terminal nodes. Consider the nodes that immediately precede a terminal node, and call the set of these nodes "level 1" nodes. Select for every player at a level 1 node an action that yields him his highest payoff. (There can be more than one if there are ties in the payoffs.)[1] Now similarly define "level 2" nodes as those that immediately precede a level 1 node, and let the players at these nodes choose the action that maximizes their payoff given that level 1 players will choose their action as previously specified. This process continues iteratively until we reach the root of the game (because the game has finite length) and results in a specification of moves that are sequentially rational.

Remark This proposition is often referred to as *Zermelo's Theorem,* named after the mathematician Ernst Zermelo. In 1913 Zermelo published a paper roughly stating that in games like chess, a player can guarantee at least a tie in a finite number of moves. Interestingly neither backward induction nor the notion of strategies— and surely not sequential rationality—was used in Zermelo's original paper, despite widespread belief to the contrary. For an excellent summary of the history of this line of research see Schwalbe and Walker (2001).

Notice that, by the construction of the backward induction procedure, each player will necessarily play a best response to the actions of the other players who come after him (the best responses are constructed for every information set). This results in the following corollary:

Corollary 8.1 *Any finite game of perfect information has at least one sequentially rational Nash equilibrium in pure strategies. Furthermore if no two terminal nodes prescribe the same payoffs to any player then the game has a unique sequentially rational Nash equilibrium.*

As we saw in the sequential-move Battle of the Sexes example, other pure-strategy Nash equilibria may exist.

8.2   Subgame-Perfect Nash Equilibrium: Concept

As the previous section argues, we expect rational players to play in ways that are sequentially rational. Furthermore we established that backward induction is a useful method to find a sequentially rational Nash equilibrium in finite games of

---

1. This is where the qualifier "if no two terminal nodes prescribe the same payoffs to any player" plays a role. If this is the case then there is a unique action that maximizes each player's payoff. If not, we can select any one of the maximizing actions.

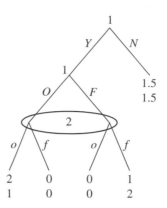

FIGURE 8.2 The voluntary Battle of the Sexes game.

perfect information. Things become a bit trickier when we try to expand our reach to suggest solutions for games of imperfect information, in which backward induction as previously defined encounters some serious problems.

Consider, for example, a game in which player 1 decides whether or not to play a Battle of the Sexes game with player 2. He can decide yes (Y), in which case they play a simultaneous-move Battle of the Sexes game, or no (N), in which case both players get a payoff of 1.5. The game is described in Figure 8.2. To try to solve this game using backward induction we need to first identify the set of "last players" that precede terminal nodes and then choose actions that would maximize their payoff at this stage. In this game, however, this is not possible, because player 2 has an information set before the terminal nodes that is not a singleton. His best response is therefore not well defined without assigning a belief to this player about what player 1 actually chose to do, and these beliefs are not part of the backward induction process.

This example shows that backward induction cannot be applied to games of imperfect information. Interestingly there is another important class of games for which we cannot apply the procedure of backward induction, and these are games that do not necessarily end in finite time. Intuitively games that do not necessarily end in a finite number of moves may not have a finite set of terminal nodes, and without such a set we cannot begin the backward induction procedure. At first thought such games may seem a bit bizarre—what kind of game will never end? As we will see in Chapters 10 and 11, such games not only are interesting in their own right but also will offer critical ways to model realistic situations. Hence our goal is to find a natural way in which to extend the concept of sequential rationality to games of imperfect information and to games that have an infinite sequence of moves.

Let's start with the problem posed by imperfect information. For example, the problem we encountered in the game depicted in Figure 8.2 is that the best response of player 2 in his information set depends on the node where he is, or on his beliefs about where he is in the information set. This of course depends on the action of player 1, which in turn depends on what player 1 believes that player 2 will do. Thus it seems that sequential rationality will have to cope with both of these decisions being interdependent. For this reason we advance the following definition:

Definition 8.2 A proper subgame G of an extensive-form game Γ consists of only a single node and all its successors in Γ with the property that if $x \in G$ and $x' \in h(x)$

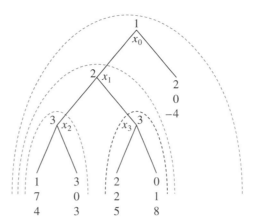

FIGURE 8.3 Subgames in a game with perfect information.

then $x' \in G$. The subgame G is itself a *game tree* with its information sets and payoffs inherited from Γ.

The idea of a proper subgame (which we will often just call a subgame) is simple and allows us to "dissect" an extensive-form game into a sequence of smaller games, an approach that in turn will allow us to apply the concept of sequential rationality to games of imperfect information. To be able to do this, however, we will require that every such smaller game be an extensive-form game in its own right, which means that it must have a unique root and follow the structure that we defined in Section 7.1.1.

In every game of perfect information, every node is a singleton and hence can be a root of a subgame. This implies that in games of perfect information every node, together with all the nodes that succeed it, forms a proper subgame. As an example, consider the game depicted in Figure 8.3. The two "smallest" subgames start at nodes x_2 and x_3. A "larger" subgame starts at x_1, and it includes the two subgames that start at x_2 and x_3. Finally the "largest" subgame starts at the original game's root, x_0, and includes all the other subgames.

Now let's return to the voluntary Battle of the Sexes game depicted again in Figure 8.4. There are only two proper subgames in this game: the whole game (which

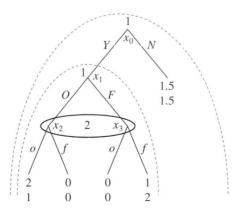

FIGURE 8.4 Proper subgames in the voluntary Battle of the Sexes game.

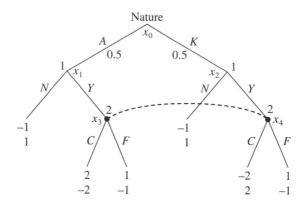

FIGURE 8.5   A game of cards.

is always, by definition, a proper subgame) and the subgame starting at the node x_1. The reason these are the only two subgames follows from definition 8.2. Note that x_2 and x_3 cannot be roots of a subgame because they belong to the same information set. From definition 8.2, if x_2 belongs to a subgame then x_3 must belong to that subgame too. In addition, from the definition we know that any subgame must begin with a single node, and thus the information set that contains x_2 and x_3 cannot begin a subgame. Hence, aside from the whole game, the only other proper subgame starts at x_1.

Another interesting example appears in Figure 8.5, which describes the following game. Players 1 and 2 put a dollar each in a pot, and player 1 pulls a card out of a deck of kings and aces, with an equal probability of getting a king (K) or an ace (A). Player 1 observes his card and then decides whether not to play the game (N), and forfeit his dollar to player 2, or proceed with the game (Y). If player 1 proceeds with the game, then without knowing which card player 1 drew (hence the information set denoted by the dashed line) player 2 can fold (F) and forfeit his dollar to player 1 or call (C), in which case each player must add another dollar to the pot. After this, if player 1 has a king then player 2 wins the pot, while if player 1 has an ace then player 1 wins the pot.

Now ask yourself: what are the proper subgames of this game? Clearly neither x_3 nor x_4 can be a root of a subgame because they belong to the same information set. Can x_1 be the root of a subgame? If it could be then x_3 must be in its subgame because x_1 precedes x_3. But by definition, if x_3 is in the subgame then $x_4 \in h(x_3)$, and therefore it too should be in the subgame (as well as all the relevant terminal nodes that follow after x_1, x_3, and x_4). But this would not be a proper subgame because x_1 does not precede x_4. A similar argument implies that x_2 cannot be the root of the subgame. We conclude that for this card game the only proper subgame is the whole game.

By now you may have realized why a subgame, which is a stand-alone game within the whole game, will prove to be useful to apply the concept of sequential rationality to extensive-form games. In particular, at any node or information set within a subgame G, a player's best response depends only on his beliefs about what the other players are doing *within* the subgame G, and not at nodes that are outside the subgame.

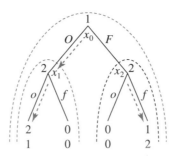

FIGURE 8.6 Subgame-perfect equilibrium in the sequential-move Battle of the Sexes game.

As demonstrated using the voluntary Battle of the Sexes game in Figure 8.4, the problem with backward induction was that we could not apply it to the end of the game because there was an information set that included two nodes that preceded terminal nodes, namely x_2 and x_3. However, what we can do is restrict attention to the *subgame* that begins at the node x_1 and require rational play in this subgame. Then we can use rational play in this "endgame" to move backward and apply sequential rationality. This is the idea behind a powerful solution concept defined as follows:

Definition 8.3 Let Γ be an n-player extensive-form game. A behavioral strategy profile $\sigma^* = (\sigma_1^*, \sigma_2^*, \ldots, \sigma_n^*)$ is a **subgame-perfect (Nash) equilibrium** if for every proper subgame G of Γ the restriction of σ^* to G is a Nash equilibrium in G.

This important concept was introduced by Reinhard Selten (1975), who was the second of the three Nobel Laureates sharing the prize in 1994 for the development of game theory. This equilibrium concept brings sequential rationality into the static Nash equilibrium solution concept. Using the terminology developed in Section 7.3, subgame perfection requires not only that a Nash equilibrium profile of strategies be a combination of best responses on the equilibrium path, which is a necessary condition of a Nash equilibrium, but also that the profile of strategies consist of mutual best responses *off the equilibrium path.* This is precisely what follows from the requirement that the restriction of the strategy profile σ^* be a Nash equilibrium in every proper subgame, including those subgames that *are not reached* in equilibrium.

Notice that, by the definition of a subgame-perfect equilibrium, every subgame-perfect equilibrium is a Nash equilibrium. However, not all Nash equilibria are necessarily subgame-perfect equilibria, implying that subgame-perfect equilibrium *refines* the set of Nash equilibria, yielding more refined predictions on behavior. To see this consider again the sequential-move Battle of the Sexes game in Figure 8.6 and its corresponding normal-form, given as:

		Player 2			
		oo	of	fo	ff
Player 1	O	$\underline{2,1}$	$\underline{2,1}$	$0,0$	$0,0$
	F	$0,0$	$1,\underline{2}$	$0,0$	$\underline{1,2}$

As we saw earlier, there are three pure-strategy Nash equilibria, (O, oo), (F, ff), and (O, of). Of these three, only (O, of) is the unique subgame-perfect equilibrium.

This follows from the fact that in the subgame beginning at x_1, the only Nash equilibrium is player 2 choosing o, because he is the only player in that subgame and he must choose a best response to his belief, which must be "I am at x_1." Similarly in the subgame beginning at x_2, the only Nash equilibrium is player 2 choosing f. Anticipating this, player 1 must choose O at x_0. Thus of the three Nash equilibria of the game, only (O, of) satisfies the condition that its restriction is a Nash equilibrium for every proper subgame of the whole game, and hence (O, oo) and (F, ff) are Nash equilibria that are not subgame perfect.

Notice that the game we just analyzed is a finite game of perfect information. For these games we have an easy and familiar way to find the subgame-perfect equilibria, which is using the procedure of backward induction because

Fact For any finite game of perfect information, the set of subgame-perfect Nash equilibria coincides with the set of Nash equilibria that survive backward induction.

This is a very useful fact because it gives us an operational procedure to find all the subgame-perfect equilibria in finite games of perfect information. However, because this procedure does not apply to games of imperfect information, we need to analyze the game using a revised version of backward induction that takes into account the proper subgames as the relevant stages in the backward induction process.

As an example, consider the voluntary Battle of the Sexes game depicted in Figure 8.4. In this game player 1 has four pure strategies. Using the previous convention, the strategy set for player 1 is $S_1 = \{YO, YF, NO, NF\}$, where, for example, YO means that player 1 plans to play Y at x_0 and O at x_1. Note that if player 1 plays either NO or NF then, regardless of player 2's strategy, the game will end after player 1's choice of N. Player 2, on the other hand, has only two strategies, $S_2 = \{o, f\}$, because he must make his choice without knowing the choice of player 1. Hence the matrix-form representation of this game is as follows:

		Player 2	
		o	f
	YO	$\underline{2, 1}$	$0, 0$
	YF	$0, 0$	$\overline{1, 2}$
Player 1	NO	$\overline{1.5, 1.5}$	$\overline{1.5, 1.5}$
	NF	$\overline{1.5, 1.5}$	$\overline{1.5, 1.5}$

As we can see, there are three pure-strategy Nash equilibria in this game, given by the set $E^{\text{Nash}} = \{(YO, o), (NO, f), (NF, f)\}$. Of these only two pairs of strategies form a subgame-perfect equilibrium, so the set of subgame-perfect equilibria profiles is $E^{\text{SPE}} = \{(YO, o), (NF, f)\}$. The reason, as shown in Figure 8.7, is that in the subgame that starts at the node x_1, the only pairs of restricted strategies that form a Nash equilibrium are (O, o) and (F, f). Hence the profile (NO, f) is not a Nash equilibrium when we restrict attention to the subgame that starts at x_1. This game has one mixed-strategy subgame-perfect equilibrium that you are asked to find in exercise 8.1.

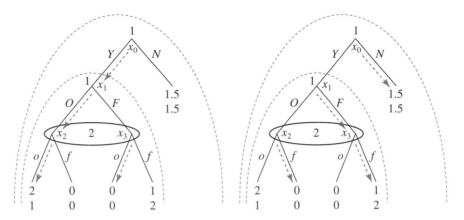

FIGURE 8.7 Subgame-perfect equilibria in the voluntary Battle of the Sexes game.

8.3 Subgame-Perfect Nash Equilibrium: Examples

This section presents some well-known examples of games and their corresponding subgame-perfect equilibria.

8.3.1 The Centipede Game

Consider the perfect-information game depicted in Figure 8.8. The game should be read from left to right as follows: Player 1 can terminate the game immediately by choosing N in his first information set or can continue by choosing C. Then player 2 faces the same choice (using lowercase letters for his choices), and if player 2 chooses to continue then the ball is back in player 1's court, who again can terminate or continue to player 2, at which stage player 2 concludes the game by choosing n or c for the second time.

It would be nice for the players to be able to continue through the game to reach the payoffs of $(3, 3)$. However, backward induction indicates that this will not happen. At his last information set, player 2 will choose n to get 4 instead of 3. Anticipating this a step earlier, player 1 will choose N to get 2 instead of 1, and the logic follows until player 1's first information set, at which he will choose N and both players will receive a payoff of 1.

Notice that this game has an interesting structure: as long as the players continue, the sum of their payoffs goes up by 1. You can easily see that we can continue with the payoffs in this way with $(2, 5)$, $(4, 4)$, $(3, 6)$, . . . and make the payoff from reaching the end of the game extremely large. Nevertheless the "curse of rationality," so to speak, predicts a unique outcome: at the last stage the last player will, by being selfish,

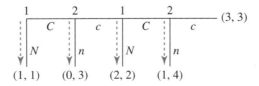

FIGURE 8.8 The Centipede Game.

prefer to stop short of the payoff that maximizes the sum of the players' payoffs, and backward induction implies that this decision is anticipated and acted on by the player before him, and so on as the actions unravel to the bleak outcome of (1, 1).

Remark The Centipede Game was first introduced by Rosenthal (1981), who used it as an example of a game in which the unique backward induction solution is extremely unappealing and goes against every intuition one might have about how the players will play. Indeed this game has been taken into the lab by letting pairs of players play it, and the experimental evidence of many studies, starting with McKelvey and Palfrey (1992), suggests that players do not play in the way predicted by backward induction. There are at least two reasons that players will not play the backward induction solution. One is that they actually care about each other's payoffs. This explanation may be less convincing for cases in which the players are anonymous, but it is not easy to rule out. The other reason is that players do not share a common knowledge of rationality. Palacios-Huerta and Volij (2009) try to put this second hypothesis to a test by taking the game to a chess tournament and having highly ranked chess players play it. Contrary to previous evidence, their results show that 69% of chess players choose to end the game immediately, and when grand masters are playing, all of them end the game immediately! This striking result suggests that when players are expected to share common knowledge of rationality they indeed play the backward induction solution.

8.3.2 Stackelberg Competition

The Stackelberg duopoly is a game of perfect information that is a sequential-moves variation on the Cournot duopoly model of competition. It was introduced and analyzed by Heinrich von Stackelberg (1934). It illustrates an important point: the order of moves might matter, and rational actors will take this into account.

Consider our familiar Cournot game with demand $p = 100 - q$, $q = q_1 + q_2$, and $c_i(q_i) = 10q_i$ for $i \in \{1, 2\}$. We have already solved for the best-response functions of each firm by maximizing their profit functions *when each takes the other's quantity as fixed* (the simultaneous-move game), so that each i solves

$$\max_{q_i}(100 - q_i - q_j)q_i - 10q_i,$$

and the best response is given by the first-order condition, which can be written as

$$q_i(q_j) = \frac{90 - q_j}{2}.$$

We now change a small but important detail of the game and assume that player 1 will choose q_1 first, and that player 2 will observe the choice made by player 1 before it makes its choice of q_2. We begin by analyzing the backward induction solution of this game.

From the fact that player 2 maximizes its profit when q_1 is already known it should be clear that player 2 will follow its best-response function, and hence sequential rationality implies that

$$q_2(q_1) = \frac{90 - q_1}{2}. \tag{8.1}$$

This follows from not only the fact that player 2 has a *belief* about q_1 but also that this belief is in reality due to the actual observation of q_1. This is precisely the backward induction conclusion of considering firm 2 first, because its moves are those that precede the terminal nodes. Notice that in this game there are infinitely many terminal nodes and infinitely many information sets preceding the terminal nodes. In particular, every different choice of q_1 is a different information set of player 2. Nevertheless, because there is perfect information, and because there is a well-defined optimal choice of player 2 for every choice of player 1, the backward induction procedure works.

Now assuming common knowledge of rationality, what should player 1 do? It would be rather naive to maximize its profits based on some fixed belief about q_2, because firm 1 knows *exactly* how a rational player 2 would respond to its choice of q_1: player 2 will choose q_2 using equation (8.1). This in turn means that a rational firm 1 would replace the "fixed" q_2 in its profit function with the best response of firm 2 and choose q_1 to solve

$$\max_{q_1} \left[100 - q_1 - \left(\frac{90 - q_1}{2} \right) \right] q_1 - 10q_1. \tag{8.2}$$

There is a fundamental difference between this maximization problem and the one we solved for the Cournot example in Section 3.1.2, and it depends on *what the firms know when they make their choices*. In the Cournot game neither firm *knew* what the other was choosing, so each set a belief and maximized its profit function. Here, in contrast, firm 2 *knows* what firm 1 has produced when it makes its choice, and as a result firm 1 *knows*, by common knowledge of rationality, that firm 2 will choose q_2 rationally, that is, by using its best-response function, given in (8.1).

The solution for firm 1 is then given by the first-order condition of (8.2), which is

$$100 - 2q_1 - 45 + q_1 - 10 = 0,$$

yielding $q_1 = 45$. Then, using firm 2's best response, we have $q_2 = 22.5$. The resulting profits are $\pi_1 = (100 - 67.5) \times 45 - 10 \times 45 = 1012.5$, and $\pi_2 = (100 - 67.5) \times 22.5 - 10 \times 22.5 = 506.25$. Recall that in the original (simultaneous-move) Cournot example the quantities and profits were $q_1 = q_2 = 30$ and $\pi_1 = \pi_2 = 900$. We conclude that when two firms are competing by setting quantities, if one firm can somehow commit to move first, it will enjoy a *first-mover advantage*.

We know already that if a profile of strategies survives backward induction then this profile is also a subgame-perfect equilibrium, and in particular a Nash equilibrium. There is a simple yet very important point worth emphasizing here. When we write down the pair of strategies as a pure-strategy Nash equilibrium *we must be careful to specify the strategies correctly:* player 2 has a continuum of information sets, each being a particular choice of q_1. This implies that the backward induction solution yields the following Nash equilibrium: $(q_1, q_2(q_1)) = \left(45, \frac{90-q_1}{2} \right)$. Writing down $(q_1, q_2) = (45, 22.5)$ is *not* identifying a Nash equilibrium because $q_1 = 45$ is not a best response to $q_2 = 22.5$. We must specify q_2 for *every information set* for it to be a well-defined strategy in the Stackelberg game, and in this case we defined it to be $q_2(q_1) = \frac{90-q_1}{2}$.

To illustrate the application of subgame perfection in this game with continuous strategy sets, consider a heuristic game tree described in Figure 8.9. Player 1 has an infinite number (a continuum) of choices, depicted by the arc, and each choice

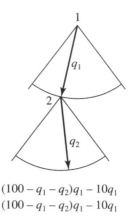

$$(100 - q_1 - q_2)q_1 - 10q_1$$
$$(100 - q_1 - q_2)q_1 - 10q_1$$

FIGURE 8.9 The Stackelberg duopoly game.

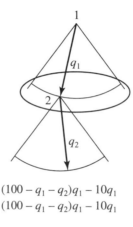

$$(100 - q_1 - q_2)q_1 - 10q_1$$
$$(100 - q_1 - q_2)q_1 - 10q_1$$

FIGURE 8.10 The Cournot duopoly game.

results in a proper subgame. Every choice q_1 of player 1 is an information set for player 2 that precedes an infinite number (a continuum) of terminal nodes associated with the specific q_1 together with player 2's choice of q_2. The payoff for player i is $(100 - q_1 - q_2)q_i - 10q_i$.

Similarly we can heuristically depict the simultaneous-move Cournot game in Figure 8.10. We draw player 2's information set to include all the choices of q_1 that player 1 can make to try to describe the simultaneous nature of the Cournot game in that player 2 makes its choice without observing the choice of player 1.

In the Cournot game we know from the analysis in Section 3.1.2 that $q_1 = q_2 = 30$ is the unique Nash equilibrium. Because every subgame-perfect equilibrium is a Nash equilibrium, and because the whole game is the only unique proper subgame in the Cournot example, then $q_1 = q_2 = 30$ is the unique subgame-perfect equilibrium in the Cournot game.

An interesting question is whether the outcome $q_1 = q_2 = 30$ could be the outcome of a Nash equilibrium in the Stackelberg game. The answer, perhaps surprisingly, is yes. To see this consider the following strategies in the Stackelberg game: $q_1 = 30$ and $q_2(q_1) = 30$ for *any* choice q_1. These two strategies are mutual best responses, implying that they constitute a Nash equilibrium. Notice two things: First, we have to specify q_2 for *every information set* for it to be a well-defined strategy in the Stackelberg game, and in this case we defined it to be $q_2(q_1) = 30$ for every q_1. Second, it is a pair of mutual best responses *only on the equilibrium path*. That is, conditional on $q_1 = 30$, $q_2 = 30$ is a best response, and conditional on $q_2(q_1) = 30$ for every q_1, $q_1 = 30$ is a best response. This requirement, mutual best responses on the equilibrium path, is all that Nash equilibrium requires.

However, as we have shown using backward induction, the unconditional strategy $q_2 = 30$ for player 2 is not sequentially rational because it is never a best response *off the equilibrium path* for $q_1 \neq 30$. The only strategy for player 2 that satisfies sequential rationality is

$$q_2(q_1) = \frac{90 - q_1}{2},$$

which in turn implies that $q_1 = 45$ and $q_2 = \frac{90-q_1}{2}$ constitute the unique subgame-perfect equilibrium of the Stackelberg game.

Remark Interestingly there are many other Nash equilibria that appear when the game is a perfect-information sequential-move game like the Stackelberg model. To see this, consider the following strategy for player 2:

$$q_2'(q_1) = \begin{cases} 35 & \text{if } q_1 = 20 \\ 100 & \text{if } q_1 \neq 20. \end{cases}$$

What would be the best response of firm 1? Clearly $q_1 = 20$ yields positive profits, while any other choice yields nonpositive profits if player 1 believes that player 2 will follow its strategy. Thus $q_1 = 20$ is the best response. Then notice that $q_2 = 40$ is a best response to $q_1 = 20$ from player 2's best-response function in (8.1), implying that $q_1 = 20$ and $q_2'(q_1)$ constitute a Nash equilibrium. At this stage you should be able to convince yourself that there are many Nash equilibria—in fact, a continuum of Nash equilibria! To see this, fix any $\tilde{q}_1 \in [0, 90]$ and replace the 35 in $q_2'(q_1)$ with player 2's best response $\tilde{q}_2 = \frac{90 - \tilde{q}_1}{2}$.

8.3.3   Mutually Assured Destruction

A very nice example for subgame-perfect equilibrium analysis in a game with imperfect information appears in Gardner (2003, p. 165), who uses a game theoretic analysis to represent the Cuban missile crisis of 1962. The crisis started with the United States' discovery of Soviet nuclear missiles in Cuba, after which the United States escalated the crisis by quarantining Cuba. The USSR then backed down, agreeing to remove its missiles from Cuba, which suggests that the United States had a credible threat along the lines of "if you don't back off we both pay dearly." Could this indeed be a credible threat?

To analyze this as a game, consider two superpowers, player 1 (the United States) and player 2 (the USSR), that have engaged in a provocative incident.[2] The game starts with player 1's choice to either ignore the incident (I), resulting in maintenance of the

---

2. The pregame phase—the USSR placing the missiles in Cuba—is assumed to be a situation that materialized without taking into account the continuing game between the superpowers.

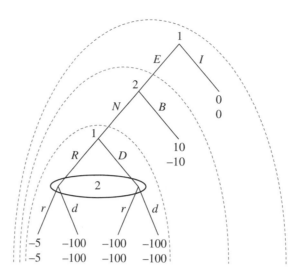

FIGURE 8.11 Mutually assured destruction.

status quo with payoffs (0, 0), or escalate the situation (E). Following escalation by player 1, player 2 can back down (B), causing it to lose face and resulting in payoffs of (10, −10), or it can choose to proceed to a nuclear confrontation (N). Upon this choice, the players play a simultaneous-move game in which they can either retreat (R for player 1, r for player 2) or choose Doomsday (D for player 1, d for player 2), in which the world is all but destroyed. If both call things off and retreat then they suffer a small loss due to the mobilization process and payoffs are (−5, −5), while if either party chooses Doomsday then the world destructs and payoffs are (−100, −100).

The extensive form of this game is depicted in Figure 8.11. Before solving for the subgame-perfect equilibria, we begin by solving for the Nash equilibria. To do this it is convenient to transform the extensive-form game into a normal-form game. Each player has two information sets, each with two actions, implying that each player has four pure strategies. Following our earlier convention, the strategy set for player 1 is $S_1 = \{IR, ID, ER, ED\}$ and that for player 2 is $S_2 = \{Br, Bd, Nr, Nd\}$ (where IR means that player 1 plays I in his first information set and R in his second one, and the other strategies are defined similarly). The normal form can be represented by the following matrix:

		Player 2			
		Br	Bd	Nr	Nd
	IR	$\overline{0, 0}$	$\overline{0, 0}$	$\mathbf{\overline{0, 0}}$	$\overline{0, 0}$
	ID	$\overline{0, 0}$	$\overline{0, 0}$	$\overline{0, 0}$	$\overline{0, 0}$
Player 1	ER	$\underline{10, -10}$	$\underline{10, -10}$	$-5, -5$	$-100, -100$
	ED	$\underline{10, -10}$	$\mathbf{\underline{10, -10}}$	$-100, -100$	$-100, -100$

In this game there are six pure-strategy Nash equilibria, and these are the profiles in the set $E^{\text{Nash}} = \{(IR, Nr), (IR, Nd), (ID, Nr), (ID, Nd), (ED, Br), (ED, Bd)\}$.

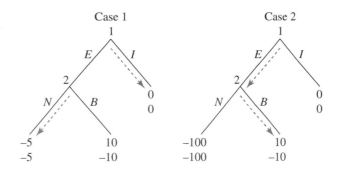

FIGURE 8.12 Reductions of the mutually assured destruction game.

(We will ignore the mixed-strategy Nash equilibria in this example. There are too many!)

To solve for the subgame-perfect equilibria, we perform backward induction using the subgames as we did earlier for the voluntary Battle of the Sexes game. There are three proper subgames depicted in Figure 8.11. The first is the whole game; the second subgame has its root at the point where player 2 chooses between B and N and includes all the moves that follow; and the third subgame starts where player 1 chooses between R and D and includes all the moves that follow.

It is easy to see that the third subgame is a simultaneous-move game, which we can call the "war stage" of the game. (Notice that we can switch the order of play in the third subgame because it is a simultaneous-move subgame.) This war-stage game can be described by the following matrix:

		Player 2	
		r	d
Player 1	R	$-5, -5$	$-100, -100$
	D	$-100, -100$	$-100, -100$

The war-stage game has two pure-strategy Nash equilibria:

NE1: (R, r) that yields a payoff of $(-5, -5)$.

NE2: (D, d) that yields a payoff of $(-100, -100)$.

We can now proceed in one of two possible ways to continue our backward induction analysis. Case 1 will perform backward induction with the belief that if the war stage is reached then the players will play NE1, and case 2 will perform backward induction with the belief that if the war stage is reached then the players will play NE2. Thus we are analyzing one of the two games depicted in Figure 8.12:

Case 1: From the left panel of Figure 8.12 we can see that for case 1 player 2 would prefer to choose N over B because losing face is worse than the consequence of mutual retreat. Moving backward another step shows that player 1 would prefer I to E because, anticipating retreat, it is better to ignore the whole situation.

Case 2: From the right panel of Figure 8.12 we can see that for case 2 player 2 would prefer to choose B over N because of the dire consequences of war, and moving backward another step shows that player 1 would prefer E to I because he anticipates that player 2 will back off.

Thus we have found two different (pure-strategy) subgame-perfect equilibria given by the set $E^{\text{SPE}} = \{(IR, Nr), (ED, Bd)\}$. We have two different subgame-perfect equilibria because in the final war-stage subgame there are two *very distinct* pure-strategy Nash equilibria, NE1 and NE2. These two Nash equilibria have different implications for the behavior of player 2 in the preceding stage of the game, which in turn affects the behavior of player 1 at the beginning of the game.

There is a nice qualitative difference between the two equilibria. The first equilibrium, (ED, Bd), represents the case in which player 1 escalates because he believes that player 2 will treat this as a signal that player 1 is willing to "go all the way." More precisely in game theoretic terms, player 1 *believes* that both will play the Doomsday equilibrium NE2 in the war stage, and in turn player 2, who shares these beliefs, will back off. The second equilibrium, (IR, Nr), is one in which player 1 chooses to ignore the incident because he believes that if he were to escalate then two things would happen: player 2 would go nuclear and they would then retreat together, resulting in a payoff of -5 instead of 0. It is interesting that in both these equilibria the war game is *off the equilibrium path*. Nonetheless it is the expected behavior in this last and final stage that dictates how players will play.

8.3.4 Time-Inconsistent Preferences

In Section 2.5.2 we analyzed the decision problem of a single player who needs to allocate a scarce resource over two periods. We assumed that the player maximizes his discounted sum of payoffs, as we defined in Section 2.4.2. We explained that a player who expects to receive a stream of payments $x_1, x_2, \ldots, x_T$ over the periods $t = 1, 2, \ldots, T$, and who evaluates per-period payments using the utility function $u(x)$, will evaluate his *discounted sum of future payoffs* in period $t = 1$ as

$$v(x_1, x_2, \ldots, x_T) = u(x_1) + \delta u(x_2) + \cdots + \delta^{T-1} u(x_T) = \sum_{t=1}^{T} \delta^{t-1} u(x_t), \quad (8.3)$$

where $\delta \in (0, 1)$ is the player's *discount factor*, which he uses to discount future payoffs.

Consider a player with $u(x) = \ln(x)$ who needs to allocate a fixed budget K across three periods. Since he will not waste any of his budget, it follows that $x_1 = K - x_2 - x_3$ and the player solves the following problem:

$$\max_{x_2, x_3} v(K - x_2 - x_3, x_2, x_3) = \ln(K - x_2 - x_3) + \delta \ln(x_2) + \delta^2 \ln(x_3).$$

To solve this problem we have to solve the two first-order conditions

$$\frac{\partial v}{\partial x_2} = -\frac{1}{K - x_2 - x_3} + \frac{\delta}{x_2} = 0$$

and

$$\frac{\partial v}{\partial x_3} = -\frac{1}{K - x_2 - x_3} + \frac{\delta^2}{x_3} = 0.$$

Solving these two equations yields the solutions

$$x_3 = K \frac{\delta^2}{1 + \delta + \delta^2} \tag{8.4}$$

and

$$x_2 = K \frac{\delta}{1 + \delta + \delta^2}, \tag{8.5}$$

and using $x_1 = K - x_2 - x_3$ gives

$$x_1 = K \frac{1}{1 + \delta + \delta^2}. \tag{8.6}$$

The player's solution is similar to the solution we saw in Section 2.5.2. The player chooses to consume more in earlier periods because he is equating the marginal utility from consumption across time, taking into account that future periods are discounted. An interesting question is the following: if the player planned out his consumption over time, would he choose to stick to his plan of action after he consumes x_1? To answer this, imagine that the player has already consumed his planned amount x_1 given in (8.6), which implies that in period $t = 2$ he is left with a budget of

$$K_2 = K - K \frac{1}{1 + \delta + \delta^2} = K \frac{\delta + \delta^2}{1 + \delta + \delta^2}. \tag{8.7}$$

Now consider the player's maximization problem in period $t = 2$, where he is left with a budget of K_2. The problem is the same two-period problem as the one we saw in Section 2.5.2:

$$\max_{x_2} \ \ln(x_2) + \delta \ln(K_2 - x_2),$$

and the solution to the player's problem will be

$$\frac{1}{x_2} = \frac{\delta}{K_2 - x_2}$$

or

$$x_2 = \frac{K_2}{1 + \delta}. \tag{8.8}$$

We can now substitute K_2 from (8.7) into (8.8) to solve for the choice of x_2 in terms of the original budget K to obtain

$$x_2 = \frac{K_2}{1 + \delta} = K \frac{\delta + \delta^2}{1 + \delta + \delta^2} \times \frac{1}{1 + \delta} = K \frac{\delta}{1 + \delta + \delta^2},$$

which is identical to the player's original choice of x_2 given by (8.5).

We call this kind of behavior *time consistent* because a player will choose to follow his original plan. The discounting rule described by (8.3) is called *exponential discounting,* and it is generally known in economic analysis that a player who uses this formula to evaluate his present-value utility will always stick to his plan of action; therefore his behavior will be time consistent. Strotz (1956) suggested that players who use discounting rules other than exponential discounting will exhibit *time-inconsistent behavior* in that, as time goes by, they will want to deviate from plans of action that they originally set.

To model this idea consider a different discounting rule known as *hyperbolic discounting*. The player uses the discount rate δ for the future periods as in exponential discounting, but he uses an *additional* discount factor $\beta \in (0, 1)$ to discount all of the future compared to present consumption. Using our three-period example, the player's modified discounted present-value problem is given by

$$\max_{x_2, x_3} v(K - x_2 - x_3, x_2, x_3) = \ln(K - x_2 - x_3) + \beta\delta \ln(x_2) + \beta\delta^2 \ln(x_3). \quad (8.9)$$

Simply put, when the player is looking toward the future, the discount factor he uses between periods $t = 1$ and $t = 2$ is stronger than the one he uses between periods $t = 2$ and $t = 3$. In particular, at any period and looking forward, the player uses $\beta\delta < \delta$ to discount between the current period and the next, while he uses only δ to discount between any two consecutive future periods.[3] As we will now see, hyperbolic discounting will cause problems of self-control, in that a player will plan to do one thing but later choose to revise his plan.

Imagine that the player is very well aware of his self-control problem—at period $t = 1$ he discounts between periods $t = 2$ and $t = 3$ at the rate of δ, but he knows that when tomorrow ($t = 2$) comes around, he will discount $t = 3$ at a rate of $\beta\delta$. In other words, the player at $t = 1$ (call him player 1) is aware that he is playing a game with his "future self" at $t = 2$ (call him player 2), who will exhibit different preferences over time. In this case a rational, forward-looking player 1 will solve his problem using backward induction; in other words he will solve for the subgame-perfect equilibrium of the implied dynamic game.

To make things simple, fix $\delta = 1$. Recall from equations (8.4)–(8.6) that a player who uses exponential discounting with $\delta = 1$ will choose to equalize his consumption over time, so that $x_t = \frac{K}{3}$ for all $t \in \{1, 2, 3\}$. Now consider a sophisticated player with hyperbolic discounting with $\delta = 1$ and $\beta = \frac{1}{2}$ (you are asked to solve for the more general problem in exercise 8.15). Looking forward to his future self, he knows that player 2, who is left with a budget of K_2, will solve the following problem (remember that $\delta = 1$ and $\beta = \frac{1}{2}$):

$$\max_{x_2} v_2(x_2, K_2 - x_2) = \ln(x_2) + \frac{1}{2} \ln(K_2 - x_2),$$

for which the first-order condition is

$$\frac{dv_2}{dx_2} = \frac{1}{x_2} - \frac{0.5}{K_2 - x_2} = 0,$$

which in turn implies that player 2's best-response function is

$$x_2(K_2) = \frac{2}{3} K_2.$$

As a consequence $x_3(K_2) = \frac{1}{3} K_2$. Hence the choices $x_2(K_2)$ and $x_3(K_2)$ made by player 2 are the solution to any of the (infinitely many) subgames that start at the

---

3. More generally the discounted present value of a player who uses hyperbolic discounting is given by $u(x_1) + \beta \sum_{t=2}^{T} \delta^{t-1} u(x_t)$.

node at which player 2 is left with a budget of K_2. Now moving up to the root of the game, where player 1 decides how much to allocate between his own consumption and that of player 2, he must solve the following problem:

$$\max_{x_1} \; v_1 \left(x_1, \tfrac{2}{3}(K - x_1), \tfrac{1}{3}(K - x_1) \right)$$

$$= \ln(x_1) + \tfrac{1}{2} \ln \left(\tfrac{2}{3}(K - x_1) \right) + \tfrac{1}{2} \ln \left(\tfrac{1}{3}(K - x_1) \right),$$

for which the first-order condition is

$$\frac{dv_1}{dx_1} = \frac{1}{x_1} - \frac{2}{3} \times \frac{1}{2} \times \frac{3}{2(K - x_1)} - \frac{1}{3} \times \frac{1}{2} \times \frac{3}{K - x_1} = 0 \,,$$

which in turn implies that player 1's best-response function is

$$x_1(K) = \frac{k}{2}.$$

This results in a very different solution than our exponential discounting bench-mark, in which (with $\delta = 1$) the player split his resources equally across the three periods. Here player 1 realizes that player 2 will choose to overconsume resources compared to what player 1 would have preferred his future self to choose. As a consequence player 1 himself overconsumes to leave a lower budget for his future self, player 2, by spending more in period 1.

Problems of self-control and time-inconsistent behavior are at the center of the field known as behavioral economics. Models of hyperbolic discounting have been used to explain insufficient savings (Laibson, 1997), procrastination (O'Donoghue and Rabin, 1999), and other related phenomena. Scholars in this literature often distinguish between players who are not aware of their self-control problem (they do not perform backward induction), referring to them as naive, and players who are aware of their self-control problem, referring to them as sophisticated. What is interesting about sophisticated players is that they would choose to constrain their own future behavior in order to achieve a higher net present value from their stream of consumption. You are asked to show this in exercise 8.16.

8.4 Summary

- Extensive-form games will often have some Nash equilibria that are not sequentially rational, yet we expect rational players to choose sequentially rational strategies.

- In games of perfect information backward induction will result in sequentially rational Nash equilibria. If there are no two payoffs at terminal nodes that are the same then there will be a unique sequentially rational Nash equilibrium.

- Subgame-perfect equilibrium is the more general construct of backward induction for games of imperfect information.

- At least one of the Nash equilibria in a game will be a subgame-perfect equilibrium.

8.5 Exercises

8.1 **Mixed-Strategy Subgame-Perfect Equilibrium:** Find the mixed-strategy subgame-perfect equilibrium of the sequential-move Battle of the Sexes game depicted in Figure 8.2.

8.2 **Mutually Assured Destruction Revisited:** Consider the game in Section 8.3.3. Now assume that the war-stage game has the following payoffs:

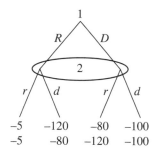

a. Find the mixed-strategy Nash equilibrium of the war-stage game and argue that it is unique.

b. What is the unique subgame-perfect equilibrium that includes the mixed strategy you found in (a)?

8.3 **Brothers Revisited:** Find all the subgame-perfect equilibria in the "Brothers" exercise (exercise 7.8) from the previous chapter.

8.4 **Industry Leader:** Three oligopolists operate in a market with inverse demand given by $P(Q) = a - Q$, where $Q = q_1 + q_2 + q_3$, and q_i is the quantity produced by firm i. Each firm has a constant marginal cost of production, c, and no fixed cost. The firms choose their quantities dynamically as follows: Firm 1, which is the industry leader, chooses $q_1 \geq 0$. Firms 2 and 3 observe q_1 and then simultaneously choose q_2 and q_3, respectively.

a. How many proper subgames does this dynamic game have? Explain briefly.

b. Is it a game of perfect or imperfect information? Explain briefly.

c. What is the subgame-perfect equilibrium of this game? Show that it is unique.

d. Find a Nash equilibrium that is not a subgame-perfect equilibrium.

8.5 **Technology Adoption:** During the adoption of a new technology a CEO (player 1) can design a new task for a division manager. The new task can be either high level (H) or low level (L). The manager simultaneously chooses to invest in good training (G) or bad training (B). The payoffs from this interaction are given by the following matrix:

		Player 2	
		G	B
Player 1	H	5, 4	−5, 2
	L	2, −2	0, 0

a. Present the game in extensive form (a game tree) and solve for all the Nash equilibria and subgame-perfect equilibria.

    b. Now assume that before the game is played the CEO can choose not to adopt this new technology, in which case the payoffs are (1, 1), or to adopt it, in which case the game is played. Present the *entire* game in extensive form. How many proper subgames does it have?

    c. Solve for all the Nash equilibria and subgame-perfect equilibria of the game described in (b).

8.6 **Investment in the Future:** Consider two firms that play a Cournot competition game with demand $p = 100 - q$ and costs for each firm given by $c_i(q_i) = 10q_i$. Imagine that before the two firms play the Cournot game firm 1 can invest in cost reduction. If it invests the costs of firm 1 will drop to $c_1(q_1) = 5q_1$. The cost of investment is $F > 0$. Firm 2 does not have this investment opportunity.

    a. Find the value F^* for which the unique subgame-perfect equilibrium involves firm 1 investing.

    b. Assume that $F > F^*$. Find a Nash equilibrium of the game that is not subgame perfect.

8.7 **Debt and Repayment:** A project costing \$100 yields a gross return of \$110. A lender (player 1) is approached by a debtor (player 2) requesting a standard loan contract to complete the project. If the lender chooses *not to offer* a loan, then both parties earn nothing. If the lender chooses to *offer* a loan of \$100, the debtor can realize the project's gains and is obliged by contract to repay \$105. For simplicity assume that money is continuous and that the debtor can choose to return any amount of money $x \leq 110$. Ignore the time value of money. Assume first that no legal system is in place that can compel the debtor to repay, so that default on the loan (less than full repayment) carries no repercussions for the debtor.

    a. Model this as an extensive-form game tree as best you can and find a subgame-perfect equilibrium of this game. Is it unique?

    b. Now assume that there is a legal system in place that allows the lender to *voluntarily* choose whether or not to sue when the debtor defaults and repays an amount $x < 105$. Furthermore assume that it is *costless* to use the legal system (it is supplied by the state), and if the lender sues a debtor that defaulted, the lender will get the \$105 repaid in full. After paying the lender, the borrower will pay a fine of \$5 to the court, above and beyond the repayment. Model this as an extensive-form game tree as best you can and find a subgame-perfect equilibrium of this game. Is it unique?

    c. Are there Nash equilibria in the game described in (b) that are not subgame-perfect equilibria?

    d. Now assume that using the legal system is *costly*: if the lender sues, he pays lawyers a *legal fee* of \$105 (this is the lawyers' price; it is unrelated to the contract). The rest of the game proceeds as before. (If the lender sues a debtor that defaulted, the lender will get repaid in full; after paying the lender, the borrower will pay a fine of \$5 above and beyond the repayment.) Model this as an extensive-form game tree as best you can and find a subgame-perfect equilibrium of this game. Is it unique?

    e. Are there Nash equilibria in the game described in (d) that are not subgame-perfect equilibria?

f. Now assume that a change in the law is proposed: upon default, if a debtor is sued he has to first repay the lender $105 and then pay the legal fees of $105 above and beyond repayment of the loan, and no extra fine is imposed. Should the lender be willing to pay to have this change in the law enacted? If so, how much?

g. If you were a social planner, would you implement the suggested law?

8.8 **Entry Deterrence 1:** NSG Corporation is considering entry into the local phone market in the San Francisco Bay Area. The incumbent, S&P Corporation, predicts that a price war will result if NSG enters. If NSG stays out, S&P earns monopoly profits valued at $10 million (net present value of profits) while NSG earns zero. If NSG enters, it must incur irreversible entry costs of $2 million. If there is a price war, each firm earns $1 million. S&P always has the option of accommodating entry (i.e., not starting a price war). In such a case both firms earn $4 million (net present value). Suppose that the timing is such that NSG first has to choose whether or not to enter the market. Then S&P decides whether to "accommodate entry" or "engage in a price war." What is the subgame-perfect equilibrium outcome to this sequential game? Set up a game tree.

8.9 **Entry Deterrence 2:** Consider the Cournot duopoly game with demand $p = 100 - (q_1 + q_2)$ and variable costs $c_i(q_i) = 0$ for $i \in \{1, 2\}$. The twist is that there is now a fixed cost of production $k > 0$ that is the same for both firms.

a. Assume first that both firms choose their quantities simultaneously. Model this as a normal-form game.

b. Write down the firm's best-response function for $k = 1000$ and solve for a pure-strategy Nash equilibrium. Is it unique?

c. Now assume that firm 1 is a "Stackelberg leader" in the sense that it moves first and chooses q_1. Then after observing q_1 firm 2 chooses q_2. Also assume that if firm 2 cannot make strictly positive profits then it will not produce at all. Model this as an extensive-form game tree as best you can and find a subgame-perfect equilibrium of this game for $k = 25$. Is it unique?

d. How does your answer in (c) change for $k = 725$?

8.10 **Playing It Safe:** Consider the following dynamic game: Player 1 can choose to play it safe (denote this choice by S), in which case both he and player 2 get a payoff of 3 each, or he can risk playing a game with player 2 (denote this choice by R). If he chooses R then they play the following simultaneous-move game:

		Player 2	
		A	B
Player 1	C	8, 0	0, 2
	D	6, 6	2, 2

a. Draw a game tree that represents this game. How many proper subgames does it have?

b. Are there other game trees that would work? Explain briefly.

c. Construct the matrix representation of the normal form of this dynamic game.

      d.   Find all the Nash and subgame-perfect equilibria of the dynamic game.

8.11   **Resident Assistant Selection with a Twist:** Two staff managers in the ΠΒΦ sorority, the house manager (player 1) and kitchen manager (player 2), must select a resident assistant from a pool of three candidates: $\{a, b, c\}$. Player 1 prefers a to b, and b to c. Player 2 prefers b to a, and a to c. The process that is imposed on them is as follows: First, the house manager vetoes one of the candidates and announces the veto to the central office for staff selection and to the kitchen manager. Next the kitchen manager vetoes one of the remaining two candidates and announces it to the central office. Finally the director of the central office assigns the remaining candidate to be a resident assistant at ΠΒΦ.

      a.   Model this as an extensive-form game (using a game tree) in which a player's most-preferred candidate gives a payoff of 2, the second most-preferred candidate gives a payoff of 1, and the last candidate gives 0.

      b.   Find the subgame-perfect equilibrium for this game. Is it unique?

      c.   Are there Nash equilibria that are not subgame-perfect equilibria?

      d.   Now assume that before the two players play the game, player 2 can send an alienating e-mail to one of the candidates, which would result in that candidate withdrawing her application. Would player 2 choose to do this, and if so, with which candidate?

8.12   **Agenda Setting:** An agenda-setting game is described as follows. The "issue space" (set of possible policies) is an interval $X = [0, 5]$. An agenda setter (player 1) proposes an alternative $x \in X$ against the status quo $q = 4$. After player 1 proposes x, the legislator (player 2) observes the proposal and selects between the proposal x and the status quo q. Player 1's most-preferred policy is 1, and for any final policy $y \in X$ his payoff is given by

$$v_1(y) = 10 - |y - 1|,$$

where $|y - 1|$ denotes the absolute value of $(y - 1)$. Player 2's most preferred policy is 3, and for any final policy $y \in X$ his payoff is given by

$$v_2(y) = 10 - |y - 3|.$$

That is, each player prefers policies that are closer to his most-preferred policy.

      a.   Write the game down as a normal form game. Is this a game of perfect or imperfect information?

      b.   Find a subgame-perfect equilibrium for this game. Is it unique?

      c.   Find a Nash equilibrium that is not subgame perfect. Is it unique? If yes, explain. If not, show all the Nash equilibria for this game.

8.13   **Junk Mail Advertising:** Suppose a single good is owned by a single seller who values it at $c > 0$ (he can consume the good and get a payoff of c). There is a single buyer who has a small transportation cost $k > 0$ to get to and back from the seller's store, and he values the good at $v > c + k$. The buyer first decides whether to make the commute or stay at home, not buy the good, and receive a payoff of 0. If the buyer commutes to the store, the seller can then make the buyer a take-it-or-leave-it price offer $p \geq 0$. The buyer can then accept the offer, pay p, and get the good, or he can walk out and not buy the good. Assume that c, v, and k are common knowledge.

    a.   As best you can, draw the extensive form of this game. What is the best response of the buyer at the node at which he decides whether to accept or reject the seller's offer?

    b.   Find the subgame-perfect equilibrium of the game and show that it is unique. Is it Pareto optimal?

    c.   Are there Nash equilibria that yield a higher payoff to both players as compared to the subgame-perfect equilibrium you found in (b)?

    d.   Now assume that before the game is played the seller can, at a small cost $\varepsilon < (v - c - k)$, send the buyer a postcard that commits the seller to a certain price at which the buyer can buy the good (e.g., "bring this coupon and get the good at a price p"). Would the seller choose to do so? Justify your answer with an equilibrium analysis.

8.14   **Hyperbolic Discounting:** Consider the three-period example of a player with hyperbolic discounting described in Section 8.3.4 with utility $\ln(x)$ in each of the three periods and with discount factors $0 < \delta < 1$ and $0 < \beta < 1$.

    a.   Solve the optimal choice of player 2, the second-period self, as a function of his budget K_2, δ, and β.

    b.   Solve the optimal choice of player 1, the first-period self, as a function of K, δ, and β.

8.15   **Time Inconsistency:** Consider the three-period example of a player with hyperbolic discounting described in Section 8.3.4 with utility $\ln(x)$ in each of the three periods, with initial budget K and with discount factors $\delta = 1$ and $\beta = \frac{1}{2}$.

    a.   Solve the optimal plan of action of a naive player 1 who does not take into account how his future self, player 2, will alter the plan. What is player 1's optimal plan x_1^*, x_2^*, and x_3^* as a function of K?

    b.   Let K_2 be the amount left from the solution to (a) after player 1 consumes his planned choice of x_1^*. Given K_2, what is the optimal plan of player 2? In what way does it differ from the optimal plan set out by player 1?

8.16   **The Value of Commitment:** Consider the three-period example of a player with hyperbolic discounting described in Section 8.3.4 with utility $\ln(x)$ in each of the three periods and with discount factors $\delta = 1$ and $\beta = \frac{1}{2}$. We solved the optimal consumption plan of a sophisticated player 1.

    a.   Imagine that an external entity can enforce any plan of action that player 1 chooses in $t = 1$ and will prevent player 2 from modifying it. What is the plan that player 1 would choose to enforce?

    b.   Assume that $K = 90$. Up to how much of his initial budget K will player 1 be willing to pay the external entity in order to enforce the plan you found in (a)?

9

Multistage Games

The normal-form game was a model of players choosing their actions simultaneously, without observing the moves of their opponents. The extensive form enriches the set of situations we can model to include those in which players learn something about the choices of other players who moved earlier, in which case a player can condition his behavior on the moves of other players in the game.

The extensive-form games analyzed so far were characterized by having payoffs delayed until the game reached a terminal node that was associated with the game's end. In reality dynamic play over time may be more complex, and it may not be correctly modeled by one "grand" game that unfolds over time with payoffs distributed at the end of the game. For instance, consider two firms that compete in one market, with the flow of profits that results from their behavior; after some time they choose to compete in another market, with a new flow of payoffs. Similarly a group of elected lawmakers may serve on a committee, on which their behavior and votes will affect the likelihood of their being reelected. Those who are reelected may sit on another committee together, with different choices and different consequences.

The broad message of these examples is that players can play one game that is followed by another, or maybe even several other games and receive some payoffs after each one of the games in this sequence is played. In such cases two natural questions emerge. First, if the players are rational and forward looking, should they not view this sequence of games as one grand game? Second, if they do view these as one grand game, should we expect that their actions in the later stages will depend on the outcomes of earlier stages? In particular, will the players be destined to play a sequence of action profiles that are Nash equilibria in each stage-game, or will they be able to use future games to support behavior in the earlier stages that is not consistent with Nash equilibrium in those early stages?

This chapter answers these questions. Before doing so, Sections 9.1–9.3 formally show how such multistage games can be modeled as extensive-form games so that we can use the tools developed in the previous chapter. It will be possible to answer the two questions just posed. First, the players should indeed anticipate future games and use them to create a richer environment for themselves. Second, they will benefit from using future play to create incentives that constrain their behavior in earlier stages. If players have the opportunity to condition future behavior on past outcomes, they may be able to create credible, self-enforcing incentives to follow behavior in earlier-stage games that would not be possible if there were no continuation games following. This is the core insight of *multistage games:* if players can condition future behavior on past outcomes then this may lead to a richer set of self-enforcing outcomes.

9.1 Preliminaries

We define a **multistage game** as a finite sequence of normal-form **stage-games,** in which each stage-game is an independent, well-defined game of complete but imperfect information (a simultaneous-move game). These stage-games are played *sequentially* by the *same players,* and the *total payoffs* from the sequence of games are evaluated using the sequence of outcomes in the games that are played. We adopt the convention that each game is played in a *distinct period,* so that game 1 is played in period $t = 1$, game 2 in period $t = 2$, and so on, up until period $t = T$, which will be the last stage in the game.[1] We will also assume that, after each stage is completed, all the players observe the outcome of that stage, and that this information structure is common knowledge.[2]

In each stage-game, players have a set of actions from which they can choose, and the profiles of actions lead to payoffs for *that specific game,* which is then followed by another game, and another, until the sequence of games is over. A sequence of normal-form games of complete information with the same players is not too hard to conceive. Consider again the examples used earlier in this chapter, such as firms in common markets or lawmakers interacting on a sequence of committees. To make things concrete, let's start with a stylized example, which is a small twist on the familiar Prisoner's Dilemma game.

Consider a two-stage game that we call the Prisoner-Revenge Game. Suppose that in a first period labeled $t = 1$ two players from different local neighborhoods play the familiar Prisoner's Dilemma with pure actions mum and fink (uppercase for player 1 and lowercase for player 2) and with the payoff matrix

		Player 2	
		m	f
Player 1	M	4, 4	−1, 5
	F	5, −1	1, 1

Now imagine that after this game is over the same two players play the following Revenge Game in a second period $t = 2$. Each player can choose whether to join his local neighborhood gang or remain a "loner." If the players remain loners then they go their separate ways, regardless of the outcome of the first game, and obtain second-period payoffs of 0 each. If both join gangs, then because of the nature of neighborhood gangs they will fight each other and suffer a loss, causing each to receive a payoff of −3. Finally if player i joins a gang while player j remains a loner then

---

1. It is possible to consider an infinite sequence of games, but we restrict attention to a finite sequence of games in this chapter. In the next chapter we will consider such infinite sequences, in which the stage-games are always the same game that repeats itself again and again.

2. It is possible to imagine games in which some stages are played before the outcomes of previous stages are learned. In the extreme case of no information being revealed, this is no different than a simultaneous-move game. If some but not all the information is revealed, the analysis becomes more complicated, and such games are beyond the scope of this text.

the loner suffers dearly since he has no gang to defend him, thus receiving a payoff of -4, while the new gang member suffers much less and receives a payoff of -1. The Revenge Game is described by the following matrix:

Player 2

		l	g
Player 1	L	$0, 0$	$-4, -1$
	G	$-1, -4$	$-3, -3$

From our previous analysis of static games of complete information, we know how to analyze these two games independently. In the first-stage game, the Prisoner's Dilemma, there is a unique Nash equilibrium, (F, f). In the second-stage game, the Revenge Game, there are two pure-strategy Nash equilibria, (L, l) and (G, g), and a mixed-strategy equilibrium in which each player plays loner with probability $\frac{1}{2}$ (verify this!). In what follows, we will see that once these two games are played in sequence, equilibrium behavior can be more interesting.

9.2 Payoffs

How should we evaluate the *total payoffs* from a sequence of outcomes in each of the sequentially played stage-games? We adopt the well-defined notion of present value, which we used to analyze the single-person decision problems of consumption over time described in Sections 2.5.2 and 8.3.4.

In particular at any period t we will collapse any sequence of payments from the games played starting at period t onward into a *single value* that can be evaluated at period t. To do this we take the payoffs derived from each of the individual games and add them up with the added assumption (as described and motivated in Section 2.5.2) that future payoffs are discounted at a rate $0 \leq \delta \leq 1$. A higher discount factor δ means that the players are more patient and will care more about future payoffs. An extreme case of impatience occurs when $\delta = 0$, which means that only the payoffs of the current stage-game matter.

Consider a multistage game in which there are T stage-games played in each of the periods $1, 2, \ldots, T$. Let v_i^t be player i's payoff from the anticipated outcome in the stage-game played in period t. We denote by v_i the *total payoff* of player i from playing the sequence of games in the multistage game and define it as

$$v_i = v_i^1 + \delta v_i^2 + \delta^2 v_i^3 + \cdots + \delta^{T-1} v_i^T = \sum_{t=1}^{T} \delta^{t-1} v_i^t,$$

which is the *discounted sum* of payoffs that the player expects to get in the sequence of games. The payoffs one period away are discounted once, those two periods away are discounted twice (hence the δ^2), and so on. For example, if in the Prisoner-Revenge Game we have the sequence of play be (F, m) followed by (L, g) then, given a discount factor δ for each player, the payoffs for players 1 and 2 are $v_1 = 5 + \delta(-4)$ and $v_2 = -1 + \delta(-1)$.

Using the concept of discounted payoffs we can write down the extensive form of the multistage Prisoner-Revenge Game to demonstrate all the possible outcomes, and the total sum of discounted payoffs, as shown in Figure 9.1. Notice that this way of

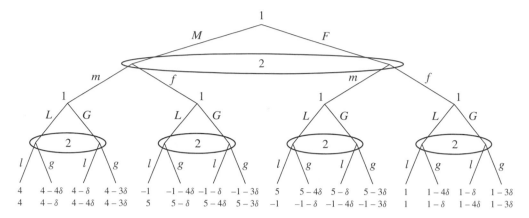

FIGURE 9.1 The multistage Prisoner-Revenge Game.

representing the game does not seem to capture the fact that payoffs are accumulated as the stages of the game proceed. That said, this is not important from a strategic point of view because we have incorporated the payoffs into the extensive form of the multistage game by accurately taking into account the sum of discounted payoffs that are a consequence of any combination of play in the two stage-games.

9.3 Strategies and Conditional Play

Imagine that the players are playing a multistage game that consists of T stage-games. If we naively treat each game independently, so that players do not make any attempt to link their behavior in a given stage-game to the outcomes of previous stage-games, then strategies would be extremely simple: each player treats each stage-game independently and commits to some action for each of the stage-games that is independent of how the games unfold over time.

More sophisticated players may benefit from using strategies that create a *strategic link* between the stage-games in ways that are no different than what children do in preschool: "If you play with the dolls this way now, I will play hide-and-seek with you later, and if you don't then I won't play with you at all." More generally players can use strategies of the form "If such-and-such happens in games $1, 2, \ldots, t - 1$ then I will choose action a_i in game t." In the Prisoner-Revenge Game, for example, player 1 can use the following strategy: "I will play F in the first game and I will play L in the second game only if player 2 played m in the first game. If player 2 played f in the first game then I will play G in the second."

A convenient way to introduce such strategies is by writing down the extensive form of the complete multistage game as we did in Figure 9.1 and focus on what the players can do at every stage of the game. It is important to determine correctly the information sets of each player. Because the outcomes of every stage are revealed before the next stage is played, the number of information sets at any stage t must be equal to the number of possible outcomes from the previously played stage-games, $1, 2, \ldots, t - 1$.

For example, in the multistage Prisoner-Revenge Game, the first-stage game has four outcomes. This implies that in the second-stage game each player should have four information sets, each associated with one of the four outcomes of the first-stage game. Note, however, that in each stage-game the players do not know what their opponent is choosing *at that stage,* which means that we must have some information sets with more than one node in them, as demonstrated in Figure 9.1. In the first stage each player chooses between mum and fink without knowing what his opponent is choosing at that stage. In the second stage he chooses between loner and gang without knowing what his opponent is choosing at that stage, but each player knows exactly how the first-stage game concluded: with either Mm, Mf, Fm, or Ff.

We can therefore define a strategy for player i in the multistage Prisoner-Revenge Game as a quintuple, $s_i = (s_i^1, s_i^2(Mm), s_i^2(Mf), s_i^2(Fm), s_i^2(Ff))$, where s_i^1 is what player i will do in the first-stage Prisoner's Dilemma game and $s_i^2(ab)$ is what player i will do in the second-stage Revenge Game, if player 1 played $a \in \{M, F\}$ and player 2 played $b \in \{m, f\}$ in the first-stage game. Notice that each element of the quintuple that represents each player's strategy can take on two values (e.g., $s_1^1 \in \{M, F\}$ and $s_1^2(Mm) \in \{L, G\}$). Hence each player has 32 pure strategies, which means that if we try to draw a matrix it will be of size 32×32 and will have 1024 pure-strategy combinations! For this reason it is most useful to use the extensive form to analyze such a game.

More generally, in a multistage game that consists of T stage-games, a pure strategy of player i will be a list of *conditional* pure strategies of the following form: $S_i = \{s_i^1, s_i^2(h_1), \ldots, s_i^t(h_{t-1}), \ldots, s_i^T(h_{T-1})\}$, where h_{t-1} is a particular outcome that occurred up to period t, not including period t, and $s_i^t(h_{t-1})$ is an action for player i from the tth stage-game. We can conveniently think of h_{t-1} as a particular *history* of events that occurred up to period t, from the set of all possible histories (or outcomes) H_{t-1}. The different histories that occurred before stage-game t are associated with different information sets in the stage-game of period t, each history being associated with a distinct sequence of outcomes from the previous stage-games in periods $1, 2, \ldots, t-1$ combined.

To solidify this idea consider a game with n firms that are choosing prices in a sequence of markets. At stage $t = 1$ they choose prices in market 1, profits in this market are determined, and this is followed by $T - 1$ more market games, each with its own demand structure. If each firm selects its price p_i^t in period t (which is stage-game t), then a pure strategy for firm i is a list of prices $p_i^1, p_i^2(h_1), \ldots, p_i^t(h_{t-1}), \ldots, p_i^T(h_{T-1})$, where $p_i^1 \geq 0$ and $p_i^t(h_{t-1}) \geq 0$ for all $t \in \{2, \ldots, T\}$. Each history h_t is the sequence of previously chosen prices: $h_1 = (p_1^1, p_2^1, \ldots, p_n^1)$, $h_2 = ((p_1^1, p_2^1, \ldots, p_n^1), (p_1^2, p_2^2, \ldots, p_n^2))$, and generally

$$h_{t-1} = ((p_1^1, p_2^1, \ldots, p_n^1), (p_1^2, p_2^2, \ldots, p_n^2), \ldots, (p_1^{t-1}, p_2^{t-1}, \ldots, p_n^{t-1})).$$

Notice that for this example a pure strategy for player i is a list of continuous functions, each from the set of continuously many histories to a single price.

Just as we have defined pure strategies for multistage games, we can define mixed behavioral strategies in which every player can mix between his or her pure actions in every information set. The formal extension is straightforward: in a multistage game that consists of T stage-games a behavioral strategy of player i will be a list of conditional randomizations of the following form: $\sigma_i = \{\sigma_i^1, \sigma_i^2(h_1), \ldots, \sigma_i^t(h_{t-1}), \ldots, \sigma_i^T(h_{T-1})\}$, where h_{t-1} is the set of outcomes that occurred up to period t, not including period t, and $\sigma_i^t(h_{t-1})$ is a randomization over player i's actions in the stage-game of period t.

In summary the definition of strategies for multistage games uses the extensive form to represent the game's structure. Furthermore, as in extensive-form games, strategies are defined as a complete list of (mixed or pure) actions for each player at each of his information sets. In multistage games, each player's information set is associated with the history of play from the previous stage-games.

9.4 Subgame-Perfect Equilibria

Because multistage games are dynamic in nature, and because past play is revealed over time, we turn to the notion of a subgame-perfect equilibrium as a solution concept. In particular, rational players should play sequentially rational strategies, which justifies the use of the subgame-perfect equilibrium concept. The question then is, how do we find a subgame-perfect equilibrium for such a game? The following result will offer a natural first step:

Proposition 9.1 *Consider a multistage game with T stages, and let σ^{t*} be a Nash equilibrium strategy profile for the tth stage-game. There exists a subgame-perfect equilibrium in the multistage game in which the equilibrium path coincides with the path generated by $\sigma^{1*}, \sigma^{2*}, \ldots, \sigma^{T*}$.*

Proof Fix a sequence of Nash equilibria for each stage-game, and let σ^{t*} denote the profile of strategies that form the Nash equilibrium for the stage-game t, where σ_i^{t*} is player i's strategy in the profile σ^{t*}. Consider the following *unconditional* strategies for each player i: in each stage $t \in \{1, 2, \ldots, T\}$ play σ_i^{t*} *regardless* of the history of plays before stage t (i.e., $\sigma_i^1 = \sigma_i^{1*}$ and $\sigma_i^t(h_{t-1}) = \sigma_i^{t*}$ for any $h_{t-1} \in H_{t-1}$). By construction, for subgames that start at stage T this is a Nash equilibrium. Now consider all the subgames that start at stage $T - 1$. The total payoffs from play in stage $T - 1$ are equal to the payoffs from the $T - 1$ stage-game plus a constant that has two components. One is the fixed payoff resulting from the previous games, and the second is the expected payoff from the game at stage T that does not depend on the outcomes of stage $T - 1$. This implies that σ_{T-1}^* followed by σ_T^* in any subgame at stage T is a Nash equilibrium in any subgame starting at time $T - 1$. Using backward induction, this argument implies that at any stage t, current play does not affect future play, and this inductive argument for a finite number of stages in turn implies that in any subgame these strategies constitute a Nash equilibrium. ∎

The proof follows from a very intuitive argument. To see this consider the Prisoner-Revenge Game, described earlier, and focus on the sequence of stage-game Nash equilibria, (F, f) followed by (L, l). Now consider the following multistage strategies for the two players:

$$s_1 = (s_1^1, s_1^2(Mm), s_1^2(Mf), s_1^2(Fm), s_1^2(Ff)) = (F, L, L, L, L)$$

$$s_2 = (s_2^1, s_2^2(Mm), s_2^2(Mf), s_2^2(Fm), s_2^2(Ff)) = (f, l, l, l, l).$$

That is, each player plays fink in the first-stage game and then plays loner in the second-stage game *regardless of what was played in the first stage*. Now observe that with this pair of strategies the players are clearly playing a Nash equilibrium in each of the four subgames in the second stage. This follows because the first-period payoffs are set in stone and do not depend on the second-period choices, and in each of these second-period subgames both players playing loner is a Nash equilibrium. Then, moving back to the subgame that begins with the first stage-game (which is the whole subgame), because the payoffs in the second period are a constant equal to 0 for each player, regardless of the play in the first stage-game, it follows that each player has a dominant strategy, which is to play fink.[3] Thus both playing fink in the first-stage game is part of the Nash equilibrium in the whole game. It is easy to see that this argument immediately implies that there is another (pure-strategy) subgame-perfect equilibrium of the multistage game in which the players play

$$s_1 = (s_1^1, s_1^2(Mm), s_1^2(Mf), s_1^2(Fm), s_1^2(Ff)) = (F, G, G, G, G)$$

$$s_2 = (s_2^1, s_2^2(Mm), s_2^2(Mf), s_2^2(Fm), s_2^2(Ff)) = (f, g, g, g, g).$$

The logic is identical because both players playing gang is a Nash equilibrium of the Revenge Game.[4]

The proposition should therefore be quite consistent with what we expect. It basically says that if we look at a sequence of plays that is a Nash equilibrium in each game *independently* then players should be content playing these stage-game strategies in each stage-game. By considering such unconditional strategies, we are removing any *strategic link* between stages, and thus each stage can be treated as if it were an independent game.

The interesting question that remains to be answered is whether we are ignoring behavior that can be supported as a subgame-perfect equilibrium if we actually use a strategic linkage between the games. In other words, if players were to condition their future play on past play in a sequentially rational manner, would we be able to support behavior in early stages that is not a Nash equilibrium in the early stage-games?

The answer is yes, and this approach yields some interesting insights, which we discuss subsequently. The trick for strategically linking the stage-games will be to *condition the behavior in later stage-games* on the *actions taken in earlier stage-games*. As we will see, this will be possible only when some of the stage-games in later periods have multiple Nash equilibria. As a first step toward understanding this requirement it is easy to establish the following proposition:

Proposition 9.2 *If σ^* is a Nash equilibrium of the multistage game consisting of stage-games $G_1, G_2, \ldots, G_T$ then the restriction of σ^* to the stage-game in period T must be a Nash equilibrium of that stage-game.*

The proposition states that for any finite stage-game of length T, in the last stage-game G_T the players *must play a Nash equilibrium* of that stage-game. The reasoning is simple: because there is no future that can depend on the actions taken at stage T, the only thing that determines what is best for any player at that stage is to play a best response to what he believes the other players are doing at that stage. This insight implies another important fact:

Proposition 9.3 *If a finite multistage game consists of stage-games that each have a unique Nash equilibrium then the multistage game has a unique subgame-perfect equilibrium.*

---

3. Note that adding a constant to the payoffs of players at every terminal node of an extensive-form game will not change the players' ranking of the outcomes and hence will not change the set of equilibria.
4. Because there is also a mixed-strategy Nash equilibrium in the Revenge Game, there is a third subgame-perfect equilibrium that can be constructed in a similar way: fink in the first-stage game followed by the mixed-strategy Nash equilibrium in the Revenge Game, which is played unconditional on the outcome of the first-stage game.

The proof of this simple observation follows immediately from backward induction. In the last stage, T, the players must play the unique Nash equilibrium of that game. In the stage before last, $T - 1$, they cannot condition future (T-stage) behavior on current ($T - 1$)-stage outcomes, so they must play the unique Nash equilibrium of the stage-game in period $T - 1$, continued by the unique Nash equilibrium of the stage-game in period T. This induction argument continues until the first stage of the game.

We now use the Prisoner-Revenge Game as an illustration of what players can do when the stage-game played at the end of a multistage game has more than one Nash equilibrium. It would be interesting, and surely appealing for the players, to try to support a subgame-perfect equilibrium in which the players will play (M, m) in the first period. This would be interesting because, as we know, in the stand-alone Prisoner's Dilemma, playing (M, m) cannot be supported as an equilibrium. However, given the appended Revenge Game, we may be able to use history-dependent strategies to support the cooperative behavior in the first stage. As a starting point, proposition 9.2 implies that as part of any pure-strategy subgame-perfect equilibrium of the multistage Prisoner-Revenge Game the players will have to play (L, l) or (G, g) in any of the subgames that include the second-stage game (or the unique mixed-strategy Nash equilibrium of the second-stage game). Intuitively sequential rationality implies that players *must play a Nash equilibrium* in the second-stage game.

To support cooperative behavior (M, m) as part of an equilibrium path of play in a subgame-perfect equilibrium, the players must find a way to give themselves incentives to stick to (M, m) and resist the temptation to deviate to fink. Notice that in the second-stage Revenge Game the two pure-strategy Nash equilibria are very different. The Nash equilibrium (L, l) yields each player a payoff of 0, which we can call the "friendly" equilibrium, and (G, g) yields each player a payoff of -3, which we can call the "gang" equilibrium. To provide the incentives for the first-stage game, the players can agree to play the friendly equilibrium of the Revenge Game if (M, m) was played in the first stage, while they can *punish themselves* by switching to the gang equilibrium of the Revenge Game if anything other than (M, m) was played in the first-stage game. That is, each player will play the following strategy in the multistage game:

Player 1: Play M in stage 1. In stage 2 play L if (M, m) was played in stage 1, and play G if anything but (M, m) was played in stage 1 (or $s_1^* = (s_1^1, s_1^2(Mm), s_1^2(Mf), s_1^2(Fm), s_1^2(Ff)) = (M, L, G, G, G)$).

Player 2: Play m in stage 1. In stage 2 play l if (M, m) was played in stage 1, and play g if anything but (M, m) was played in stage 1 (or $s_2^* = (s_2^1, s_2^2(Mm), s_2^2(Mf), s_2^2(Fm), s_2^2(Ff)) = (m, l, g, g, g)$).

We are left to check if this pair of strategies is a subgame-perfect equilibrium. It is easy to see that in each of the subgames beginning at stage 2 the players are playing a Nash equilibrium (either both play loner following (M, m) or both play gang following other histories of play). Hence to check that this is a subgame-perfect equilibrium we need to check that players would not want to deviate from mum in the first stage of the game. In other words, is mum a best response in period 1, given what each player believes about his opponent, and given the continuation play in the second-period stage-games? Consider player 1 and observe that

$$v_1(M, s_2) = 4 + 0 \times \delta \qquad \text{and} \qquad v_1(F, s_2) = 5 + (-3) \times \delta,$$

which implies that M is a best response if and only if $4 \geq 5 - 3\delta$, or

$$\delta \geq \tfrac{1}{3}.$$

Thus if the discount factor is not too small then we can support the behavior of (M, m) in the first-stage game even though it is not a Nash equilibrium in the stand-alone Prisoner's Dilemma. The fact that the Revenge Game has two different Nash equilibria, one of which is significantly better than the other, allows the players to offer a self-enforcing incentive scheme that supports cooperative behavior in the first-stage game.

This example is illuminating because it sheds light on two requirements that are crucial to supporting behavior in the first stage (or early periods in general) that is *not* a Nash equilibrium at this first stage. These requirements are as follows:

1. There must be at least two distinct equilibria in the second stage: a "stick" and a "carrot."

2. The discount factor has to be large enough for the difference in payoffs between the "stick" and the "carrot" to have enough impact in the first stage of the game.

The first requirement is necessary to offer the players the possibility of "reward and punishment" strategies, which help support first-period play that is not a Nash equilibrium in the stand-alone first-stage game. To see this, consider the following: If we are trying to support behavior in the first period that is not a Nash equilibrium in that period, it must be the case that at least one player would like to deviate in the first stage (because at least one is not playing a best response). The player would indeed deviate from the proposed play in the first period if there were no future consequences to such a deviation. In other words, in the first stage some players would benefit in the short run from deviating from the proposed path of play, and to keep them on the path of play we must guarantee that such deviations will be met with credible "punishments." For these punishments to be credible, they must take the form of moving from an equilibrium in which the player gets a high payoff to an equilibrium in which he gets a low payoff. That is, the players use *long-term losses* to deter themselves from pursuing *short-term gains*.

This is where the second requirement has bite: it must be that these long-term losses are enough to deter the players from deviating toward their short-term gains. For this to be the case it must be that the players value the payoffs that they will receive in the second period, which can only happen if they do not discount the future too much. Thus the *effective punishment* from deviating will depend first on the difference in payoffs of the two equilibria, the "reward" and the "punishment," and second on the discount factor. A more figurative way to think about this is that the friendly and gang continuation equilibria are the "carrot" and the "stick," respectively. Then a smaller discount factor makes the carrot smaller and the stick less painful, which in turn makes deterrence harder.

To see another example of a similarly constructed subgame-perfect equilibrium in the Prisoner-Revenge Game, we can support the path of play (F, m) followed by (L, l). To do this we use the following strategies for each player: Player 1 plays F in stage 1, and player 2 plays m. In stage 2 both players play loner if (F, m) was played in stage 1, while they play gang if anything but (F, m) was played in stage 1 (formally, $s_1 = (s_1^1, s_1^2(Mm), s_1^2(Mf), s_1^2(Fm), s_1^2(Ff)) = (F, G, G, L, G)$, and $s_2 = (s_2^1, s_2^2(Mm), s_2^2(Mf), s_2^2(Fm), s_2^2(Ff)) = (m, g, g, l, g)$). Clearly player 1 would not want to deviate from F in the first period because he's playing a dominant strategy in that period, followed by the best possible second-stage equilibrium. For player 2 things are different. In the first-stage game his payoffs from m and f, respectively, are:

$$v_2(s_1, m) = -1 + 0 \times \delta \qquad \text{and} \qquad v_2(s_1, f) = 1 - 3 \times \delta,$$

which implies that m is a best response for player 2 in the first-stage game if and only if $-1 \geq 1 - 3\delta$, or if

$$\delta \geq \tfrac{2}{3}.$$

As we can see, to support this outcome we need a higher discount factor compared with supporting (M, m) in the first period. Compared to the earlier path of (M, m) followed by (L, l), here the future loss of 3 is enough to deter player 2's deviation only when the future is not discounted as much, because the short-term gains from deviating are equal to 2. These are higher than the gains from deviating from (M, m), which were only 1.

The foregoing analysis of the multistage Prisoner-Revenge Game captures most of what is interesting about multistage games. For this reason it is not too useful to consider more involved games. The next section offers a result that is quite technical in nature but that is also very useful in proving which strategies are or are not part of a subgame-perfect equilibrium. To summarize: in order to check if a strategy profile $(\sigma_i^*, \sigma_{-i}^*)$ is a subgame-perfect equilibrium in a multistage (or any finite extensive-form) game, all we need to show is that for every player i, given σ_{-i}^*, player i does not have a *single information set* from which he would want to deviate. In the next chapter this result will prove to be even more useful.

9.5 The One-Stage Deviation Principle

As an illustration of multistage games, the Prisoner-Revenge Game was not hard to analyze. In particular, to check that a certain proposed strategy profile was a subgame-perfect equilibrium, we needed only to check that no player wanted to deviate in the first stage, because we constructed our candidates for subgame-perfect equilibria to consist of Nash equilibria in each of the second-stage subgames. It might seem that if we had, say, a five-stage game it would be more complicated to check that a profile of strategies constitutes a subgame-perfect equilibrium, and that in particular we would have to check that the players do not want to deviate at any single stage-game. That is, maybe they could gain by combining several deviations in separate stages of the game.

This is where the **one-stage deviation principle** comes in to simplify what seems like a rather daunting task. Interestingly the principle is based on an idea that was originally formulated by David Blackwell (1965) in the context of dynamic programming for a single-player decision problem. The relation to a single-player decision problem comes from the fact that when we want to check if a player i is playing a best response to σ_{-i} in every subgame, then all we are really doing is checking to see whether he is playing an optimal action in each of his information sets, *taking the actions of all other players σ_{-i} as given*. Thus once the strategies σ_{-i} of other players in the extensive form are considered to be fixed, player i solves a standard dynamic programming problem like the ones discussed in Section 2.4. This in turn implies that we can formulate the one-stage deviation principle within the context of a single-person decision tree.

Consider an extensive-form game tree of finite length (every path ends after a finite number of moves) with payoffs associated with each terminal node, and fix the strategies of all players that are not i. The strategy σ_i assigns to each information set h_i a probability distribution over i's actions $A_i(h_i)$, and starting from any information set h_i a strategy profile (σ_i, σ_{-i}) induces a probability distribution over paths of play and terminal nodes.

Thus if we treat every information set h_i as a node in the single-player decision tree induced by σ_{-i}, we can define $v_i(\sigma_i, h_i)$ to be the expected payoff of player i from the information set h_i onward by playing σ_i (recall that σ_{-i} is considered to be fixed). We say that a strategy σ_i is *optimal* if there is no strategy σ_i' and information set h_i such that $v_i(\sigma_i', h_i) > v_i(\sigma_i, h_i)$.

Given a strategy σ_i, define the strategy σ_i^{a,h_i} as the strategy that is identical to σ_i everywhere except at h_i, and at h_i we substitute the prescribed (possibly random) action of σ_i with the action $a \in A_i(h_i)$. We introduce the following definition:

Definition 9.1 A strategy σ_i is **one-stage unimprovable** if there is no information set h_i, action $a \in A_i(h_i)$, and corresponding strategy σ_i^{a,h_i} such that $v_i(\sigma_i^{a,h_i}, h_i) > v_i(\sigma_i, h_i)$.

Clearly an optimal strategy is one-stage unimprovable. It is the converse that is stated in the following theorem:

Theorem 9.1 *A one-stage unimprovable strategy must be optimal.*

Proof Suppose in negation that σ_i is one-stage unimprovable but it is not optimal. Then there exist σ_i' and information set h_i^1 such that $v_i(\sigma_i', h_i^1) > v_i(\sigma_i, h_i^1)$. Because stochastic strategies do not affect the highest attainable payoffs, this is equivalent to saying that there is a path starting from h_i^1 through a finite number n of i's information sets (the last being a terminal node) such that i's payoff from this path is greater than $v_i(\sigma_i, h_i^1)$. Call these n information sets that define this path $h_i^1, h_i^2, \ldots, h_i^n$. This implies that a finite number of one-stage deviations at the information sets h_i^1 through h_i^{n-1}, with σ_i used everywhere else in the tree, will offer i a higher payoff than σ_i. Define a collection of $n-1$ different strategies σ_i^t, for $t = 1, \ldots, n-1$, such that σ_i^t chooses the action $a_i \in A_i(h_i^r)$ that induces the information set h_i^{r+1}, for every r between 1 and t, and coincides with σ_i at every other information set of i. Then our earlier assertion that σ_i is not optimal implies that

$$v_i(\sigma_i^{n-1}, h_i^1) > v_i(\sigma_i, h_i^1). \tag{9.1}$$

Notice that σ_i^{n-2} fully coincides with σ_i from the information set h_i^{n-1} up to h_i^n, while σ_i^{n-1} is a one-stage deviation from σ_i at h_i^{n-1}. Because σ_i is one-stage unimprovable by assumption, it must be that

$$v_i(\sigma_i^{n-2}, h_i^{n-1}) = v_i(\sigma_i, h_i^{n-1}) \geq v_i(\sigma_i^{n-1}, h_i^{n-1}).$$

Because σ_i^{n-2} and σ_i^{n-1} share the same information sets h_i^1 through h_i^n along every path generated by the two, it must be that

$$v_i(\sigma_i^{n-2}, h_i^1) \geq v_i(\sigma_i^{n-1}, h_i^1). \tag{9.2}$$

Combining (9.1) and (9.2) we conclude that

$$v_i(\sigma_i^{n-2}, h_i^1) > v_i(\sigma_i, h_i^1).$$

Using an induction argument we must conclude that

$$v_i(\sigma_i^1, h_i^1) > v_i(\sigma_i, h_i^1), \tag{9.3}$$

but σ_i^1 is a one-stage deviation from σ_i, implying that (9.3) contradicts the assertion that σ_i is one-stage unimprovable. Therefore if σ_i is one-stage unimprovable then it must be optimal. ■

Though the formal proof is a bit tedious, the intuition is not. If the strategy σ_i were not a best response to σ_{-i} (not "optimal" in the proof just given), then given that the game is finite, there must be a finite number of deviations to improve upon σ_i. We can look for the "last" such deviation in the sequence, and if we restrict attention to the subgame that starts at that information set (or at the closest possible root that precedes it, if it is not a singleton) then in that subgame player i has a one-stage deviation that makes him better off.

9.6 Summary

- Multistage games are defined by a series of normal-form stage-games that are played in sequence, in which players obtain payoffs after every stage-game and future payoffs are discounted.

- Any sequence of stage-game Nash equilibrium play can be supported as a subgame-perfect equilibrium in the multistage game, regardless of the discount factor.

- Players can use credible threats and promises in later stages to provide long-term incentives for short-term actions that may not be self-enforcing in the earlier-period stage-games.

- Because future payoffs are discounted, the effectiveness of long-term incentives will depend on how patient the players are.

- The set of outcomes that can be supported by a subgame-perfect equilibrium will often depend on the discount factor.

9.7 Exercises

9.1 **Multistage Practice:** Consider the following simultaneous-move game that is played twice (the players observe the first-period outcome prior to the second-period play):

		Player 2		
		L	*C*	*R*
	T	10, 10	2, 12	0, 13
Player 1	*M*	12, 2	5, 5	0, 0
	B	13, 0	0, 0	1, 1

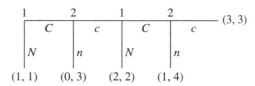

FIGURE 9.2 The Centipede Game.

a. Find all the pure-strategy subgame-perfect equilibria with no discounting ($\delta = 1$). Be precise in defining history-contingent strategies for both players.

b. For each of the equilibria you found in (a), find the smallest discount factor that supports it.

9.2 **Centipedes Revisited:** Two players are playing two consecutive games. First, they play the Centipede Game described in Figure 9.2. After the Centipede Game they play the following coordination game:

		Player 2	
		a	*b*
Player 1	A	1, 1	0, 0
	B	0, 0	3, 3

a. What are the Nash equilibria of each stage-game?

b. How many pure strategies does each player have in the multistage game?

c. Find all the pure-strategy subgame-perfect equilibria with extreme discounting ($\delta = 0$). Be precise in defining history-contingent strategies for both players.

d. Now let $\delta = 1$. Find a subgame-perfect equilibrium for the two-stage game in which the players receive the payoffs (2, 2) in the first stage-game.

e. What is the lowest value of δ for which the subgame-perfect equilibrium you found in (d) survives?

f. For δ greater than the value you found in (e), are there other outcomes of the first-stage Centipede Game that can be supported as part of a subgame-perfect equilibrium?

9.3 **Campaigning Revisited:** Two political candidates are scheduled to campaign in two states, in one in period $t = 1$ and in the other in $t = 2$. In each state each candidate can choose either a positive campaign that promotes his own agenda (P for player 1, p for player 2) or a negative one that attacks his opponent (N for player 1, n for player 2). Residents of the first-period state do not mind negative campaigns, which are generally effective, and payoffs in this state are given by the following matrix:

		Player 2	
		p	*n*
Player 1	P	2, 2	0, 5
	N	5, 0	3, 3

In the second-period state, residents dislike negative campaigns despite their effectiveness, and the payoffs are given by the following matrix:

Player 2

		p	n
Player 1	P	6, 6	1, 0
	N	0, 1	2, 2

a. What are the Nash equilibria of each stage-game? Find all the pure-strategy subgame-perfect equilibria with extreme discounting ($\delta = 0$). Be precise in defining history-contingent strategies for both players.

b. Now let $\delta = 1$. Find a subgame-perfect equilibrium for the two-stage game in which the players choose (P, p) in the first stage-game.

c. What is the lowest value of δ for which the subgame-perfect equilibrium you found in (b) survives?

d. Can you find a subgame-perfect equilibrium for this game in which the players play something other than (P, p) or (N, n) in the first stage?

9.4 **Online Gaming:** Consider a two-stage game between two firms that produce online games. In the first stage, they play a Cournot competition game (each chooses a quantity q_i) with demand function $p = 100 - q$ and zero marginal production costs ($c_i(q_i) = 0$ for $i = 1, 2$). In the second stage, after observing the pair (q_1, q_2) and after profits have been distributed, the players play a simultaneous-move "access" game in which each can open its platform to allow players on the other platform to play online with players on its platform (O for player 1, o for player 2), or choose to keep its platform noncompatible (N for player 1, n for player 2), in which case each platform's players can play only with others on the same platform. If they choose (N, n) then second-stage payoffs are $(0, 0)$. If only one firm chooses to open its platform, it bears a cost of -10 with no benefit, since the other firm did not allow open access. Finally if both firms choose (O, o) then each firm gets many more eyeballs for advertising, and payoffs for each are 2500. Both players use the same discount factor δ to discount future payoffs.

a. Find the unique Nash equilibrium in the first-stage Cournot game and all of the pure-strategy Nash equilibria of the second-stage access game. Find all the pure-strategy subgame-perfect equilibria with extreme discounting ($\delta = 0$). Be precise in defining history-contingent strategies for both players.

b. Now let $\delta = 1$. Find a subgame-perfect equilibrium for the two-stage game in which the players choose the monopoly (total profit–maximizing) quantities and split them equally (a symmetric equilibrium).

c. What is the lowest value of δ for which the subgame-perfect equilibrium you found in (b) survives?

d. For this part assume that if both firms choose open access, (O, o), then the payoffs to each firm is 250 instead of 2500. Now let $\delta = 0.4$. Can you support a subgame-perfect equilibrium for the two-stage game in which the players choose the monopoly quantities and split them equally? If not, what are the highest profits that the firms can make in a symmetric equilibrium?

9.5 **Campaign Spending:** Two political candidates are destined to play the following two-stage game. Assume throughout that there is no discounting

($\delta = 1$). First, they compete in the primaries of their respective parties. Each candidate i can spend $s_i \geq 0$ resources on ads that reach out to voters, which in turn increases the probability that candidate i wins the race. Given a pair of spending choices (s_1, s_2), the probability that candidate i wins is given by $\frac{s_i}{s_1+s_2}$. If neither spends any resources then each wins with probability $\frac{1}{2}$. Each candidate values winning at a payoff of $16 > 0$, and the cost of spending s_i is equal to s_i. After each player observes the resources spent by the other and a winner in the primaries is selected, they can choose how to interact. Each can choose to be pleasant (P for player 1 and p for player 2) or nasty (N and n, respectively). At this stage both players prefer that they be nice to each other rather than nasty, but if one player is nasty then the other prefers to be nasty too. The payoffs from this stage are given by the following matrix (where $w > 0$):

		Player 2	
		p	n
Player 1	P	w, w	$-1, 0$
	N	$0, -1$	$0, 0$

a. Find the unique Nash equilibrium of the first-stage game and the two pure-strategy Nash equilibria of the second-stage game.

b. What are the Pareto-optimal outcomes of each stage-game?

c. For which value of w can the players support the path of Pareto-optimal outcomes as a subgame-perfect equilibrium?

d. Assume that $w = 1$. What is the "best" symmetric subgame-perfect equilibrium that the players can support?

e. What happens to the best symmetric subgame-perfect equilibrium that the players can support as w changes? In what way is this related to the role played by a discount factor?

9.6 **Augmented Competition:** Consider two firms playing a two-stage game with discount factor δ. In the first stage they play a Cournot quantity-setting game in which each firm has costs $c_i(q_i) = 10q_i$ for $i \in \{1, 2\}$ and the demand is given by $p(q) = 100 - q$, where $q = q_1 + q_2$. In the second stage, after the results of the Cournot game are observed, the firms play the following standard-setting game:

		Player 2	
		a	b
Player 1	A	100, 100	0, 0
	B	0, 0	300, 300

a. Find the unique Nash equilibrium of the first-stage game and the two pure-strategy Nash equilibria of the second-stage game.

b. As far as the two firms are considered, what are the symmetric Pareto-optimal outcomes of each stage-game?

c. For which values of δ can the Pareto-optimal outcomes be supported as a subgame-perfect equilibrium?

d. Assume that $\delta = 0.5$. What is the "best" symmetric subgame-perfect equilibrium that the players can support?

e. What happens to the best symmetric subgame-perfect equilibrium that the players can support as δ drops toward zero?

10

Repeated Games

The previous chapter on multistage games demonstrated two important lessons. First, when players play a sequence of games over time, it will be to their benefit to use conditional strategies in later stage-games to support desirable behavior in early stage-games. More importantly the behavior that can be supported need not be a static best response in the early stage-game because the conditional strategies in later stage-games can act as a powerful incentive scheme to help players resist the short-run temptation of deviating from the proposed path of play. Second, and equally important, the future that the players face must be important enough to support these dynamic incentives as self-enforcing. Using so-called reward-and-punishment strategies to sustain static non-best-response behavior is possible only if the players do not discount the future too heavily.

A special case of multistage games has received considerable attention over the years: the case of **repeated games.** A repeated game is simply a multistage game in which the *same stage-game* is being played at every stage. These games have been studied for two primary reasons. The first is that repeated games seem to capture many realistic settings. These include firms competing in the same market over long periods of time, politicians engaging in pork-barrel negotiations in session after session of a legislature, and workers on a team production line who perform some joint task day after day. The second reason is that repeating the same game over time results in a very convenient mathematical structure that makes the analysis somewhat simple and elegant. As such, social scientists were able to refine this analytical tool and provide a wide range of models that can be applied to understand different social phenomena. This chapter provides a glimpse into the analysis of repeated games and shows some of the extreme limits to what we can do with reward-and-punishment strategies.

10.1 Finitely Repeated Games

A finitely repeated game is, as its name suggests, a stage-game that is repeated a finite number of times. Given that we have defined a multistage game in the previous chapter, we can easily define a finitely repeated game as follows:

Definition 10.1 Given a stage-game G, $G(T, \delta)$ denotes the **finitely repeated game** in which the stage-game G is played T consecutive times, and δ is the **common discount factor.**

As an example, consider the two-stage repeated game in which the following game is repeated twice with a discount factor of δ:

		Player 2		
		m	f	r
	M	4, 4	$\overline{-1, 5}$	0, 0
Player 1	F	5, -1	$\overline{1, 1}$	0, 0
	R	0, 0	0, 0	$\underline{3, 3}$

As the matrix shows, there are two pure-strategy Nash equilibria, and they are *Pareto ranked,* that is, (R, r) is better than (F, f) for both players. Thus, as we have seen before in Chapter 9 on multistage games, we have a "carrot" and a "stick" to use in an attempt to discipline first-period behavior. This implies that for a high enough discount factor we may be able to find subgame-perfect equilibria that support behavior in the first stage that is not a Nash equilibrium of the one-shot stage-game.

You should be able to convince yourself that for a discount factor $\delta \geq \frac{1}{2}$ the following strategies[1] constitute a subgame-perfect equilibrium of this two-stage game:

Player 1: Play M in stage 1. In stage 2 play R if (M, m) was played in stage 1, and play F if anything but (M, m) was played in stage 1.

Player 2: Play m in stage 1. In stage 2 play r if (M, m) was played in stage 1, and play f if anything but (M, m) was played in stage 1.

In fact, the players' ability to provide reward-and-punishment strategies in this game is very similar to that in the Prisoner-Revenge Game we analyzed in the previous chapter. In the two-stage repeated game, if any one of the players deviates from the proposed path of play (M, m) then both players will incur a loss of -2 in the second-period stage-game because the players will switch from the "good" second-stage equilibrium, (R, r), to the "bad" one, (F, f). The temptation to deviate from (M, m) for either player is the added payoff of $+1$ in the first period that a player gets by deviating from M (or m) to F (or f).[2] The difference between this example and the Prisoner-Revenge Game of the previous chapter is that in this example the *same game* is repeated twice. It is the multiplicity of equilibria in the stage-game that is giving the players the leverage to use conditional second-stage strategies of the reward-and-punishment kind.

---

1. Because this stage-game has nine outcomes, it is a bit cumbersome to write down the formal representation of these strategies. A simple way to write down this strategy for player 1 would be $s_1^* = (s_1^1, s_1^2(h_1))$, where $s_1^1 = M$ and

$$s_1^2(h_1) = \begin{cases} R & \text{if } h_1 = (M, m) \\ F & \text{if } h_1 \neq (M, m). \end{cases}$$

2. In case you had trouble convincing yourself that these strategies form a subgame-perfect equilibrium, here is the argument: In the second stage the players are clearly playing a Nash equilibrium regardless of the history of play. To check that this is a subgame-perfect equilibrium we need to check that players would not want to deviate from M in the first stage of the game. Consider player 1 and observe that $v_1(M, s_2) = 4 + 3 \times \delta$ and $v_1(F, s_2) = 5 + 1 \times \delta$, which implies that M is a best response if and only if $-1 \geq -2\delta$, or $\delta \geq \frac{1}{2}$.

Because finitely repeated games are a special case of multistage games, the conclusion must be that we need *multiple continuation Nash equilibria* to be able to support behavior that is not a Nash equilibrium of the stage-game in earlier stages. This immediately implies the following result, which is a special case of proposition 9.3 that was stated for multistage games:

Proposition 10.1 *If the stage-game of a finitely repeated game has a* unique *Nash equilibrium,* then the finitely repeated game has a **unique** subgame-perfect equilibrium.

The proof is identical to that of proposition 9.3. What are the consequences of this proposition? It immediately applies to a finitely repeated game like the Prisoner's Dilemma or any of the simultaneous-move market games (Cournot and Bertrand competitions) that we analyzed in Section 5.2. If such a stage-game is played T times in a row as a finitely repeated game, then there will be a *unique subgame-perfect equilibrium* that is the repetition of the static noncooperative Nash equilibrium play in every stage.

This result may seem rather disturbing, but it follows from the essence of credibility and sequential rationality. If two players play the Prisoner's Dilemma, say, 500 times in a row, then intuition suggests that they can use the future to discipline behavior and try to cooperate by playing mum in early periods, with deviations from mum being punished by a later move to fink. However, the "unraveling" process that proves proposition 10.1 will apply to this finitely repeated Prisoner's Dilemma precisely because it is common knowledge that the game will end after 500 periods. In the last stage-game at $T = 500$, the players must each play fink because it is the unique Nash equilibrium. (In fact it is the unique dominant strategy!) As a consequence, in the stage-game at $T = 499$ the players cannot provide reward-and-punishment incentives, so in this stage they must again play fink. This argument will continue, thus proving that it is impossible to support a subgame-perfect equilibrium in which the players can cooperate and play mum in early stages of the game.

For many people this argument is at first counterintuitive. Because the players have a long future ahead of them, this should seem enough to construct equilibrium strategies that provide them with incentives to cooperate. The problem is that in order to provide such incentives, the players must be able to construct reward-and-punishment continuation equilibrium strategies. More importantly, these continuation strategies themselves must be equilibrium strategies and hence must rely on multiple equilibria in the continuation of the repeated game.

As we will see in the next section, however, putting a somewhat different interpretation on the idea of a long future will restore our intuition—albeit in a rather surprising way.

10.2 Infinitely Repeated Games

The problem identified with the finitely repeated Prisoner's Dilemma is a consequence of the common knowledge among the players that *the game has a fixed and finite number of periods*. If a repeated game is finite, and if its stage-game has a unique Nash equilibrium, then starting in the last period T the "Nash unraveling" of the static Nash equilibrium follows from sequential rationality.

What would happen if we eliminated this problem by assuming that the game does not have a final period? That is, what if, regardless of the stage at which the players find themselves and whatever they have played, there is always a "long" future ahead of them? As we will see, this slight but critical modification will give the players the ability to support play that is not a static Nash equilibrium of the stage-game, even when the stage-game has a unique Nash equilibrium. In fact we will see that once the stage-game is repeated infinitely often, the players will have the freedom to support a wide range of behaviors that are not consistent with a static Nash equilibrium in the stage-game. First we need to define payoffs for an infinitely repeated game.

10.2.1 Payoffs

Let G be the stage-game, and denote by $G(\delta)$ the infinitely repeated game with discount factor δ. The natural extension of the present-value concept that we used in the previous chapter is given by the following:

Definition 10.2 Given the discount factor $0 < \delta < 1$, the **present value** of an *infinite sequence of payoffs* $\{v_i^t\}_{t=1}^{\infty}$ for player i is

$$v_i = v_i^1 + \delta v_i^2 + \delta^2 v_i^3 + \cdots + \delta^{t-1} v_i^t + \cdots = \sum_{t=1}^{\infty} \delta^{t-1} v_i^t.$$

Notice that unlike the case of a finitely repeated game (or finite multistage game) we *must have* a discount factor that is less than 1 for this expression to be well defined. This follows because for $\delta = 1$ this sum may not be well defined if it approaches infinity. If the payoffs v_i^t are *bounded,* that is, all stage payoffs v_i^t are less than some (potentially large) number, then with a discount factor $\delta < 1$ this sum is a well-defined bounded number.[3]

As we argued in Section 2.4.2, discounting future payoffs is a natural economic assumption for a sequence of payoffs that are received over time. The question is whether it is reasonable to assume that players will engage in an infinite sequence of play. Intuitively a sensible answer is that they won't. We can, however, motivate this idea by interpreting discounting as a response to an uncertain future.

Imagine that the players are playing a given stage-game today and with some probability $\delta < 1$ they will play the game again tomorrow. With probability $1 - \delta$, however, their relationship ends for some exogenous reason and the players will no longer interact with each other. If the relationship continues for one more day then again, with probability $\delta < 1$, they will continue for another period, and so on.

---

3. To see this, assume that there exists some bound $b > 0$ such that $v_i^t < b$ for all t. First observe that $v_i = \sum_{t=1}^{\infty} \delta^{t-1} v_i^t < \sum_{t=1}^{\infty} \delta^{t-1} b$. We are left to show that $B = \sum_{t=1}^{\infty} \delta^{t-1} b$ is finite. Notice that

$$B(1 - \delta) = \sum_{t=1}^{\infty} \delta^{t-1} b - \delta \sum_{t=1}^{\infty} \delta^{t-1} b = \sum_{t=1}^{\infty} \delta^{t-1} b - \sum_{t=1}^{\infty} \delta^t b$$

$$= b + \sum_{t=1}^{\infty} \delta^t b - \sum_{t=1}^{\infty} \delta^t b = b,$$

which in turn implies that $B = \frac{b}{1-\delta} < \infty$ for any $0 < \delta < 1$.

Now imagine that in the event these players continue playing the game, the payoffs of player i are given by the sequence $\{v_t^i\}_{t=1}^{\infty}$. If there is no additional discounting,[4] we can think of the *expected payoff* from this sequence as

$$Ev = v_i^1 + \delta v_i^2 + (1 - \delta) \times 0 + \delta^2 v_i^3 + \delta(1 - \delta) \times 0 + \delta^3 v_i^4 + \delta^2(1 - \delta) \times 0 + \cdots$$

$$= \sum_{t=1}^{\infty} \delta^{t-1} v_i^t.$$

That is, we can think of the present value from the sequence of payoffs as the expected value from playing this stochastic sequence of stage-games. There is something appealing about the interpretation of $\delta < 1$ as the probability that the players continue their ongoing relationship for one more period. In particular we can ask a simple question: what is the probability that the players will play this game infinitely many times? The answer is clearly $\delta^{\infty} = 0$! What makes this interesting is that the game will end in finite time with probability 1, but the *potential* future is always present at any given period.

To drive the point home, consider a *very high* probability of continuation, say $\delta = 0.97$. The probability of playing for at least 100 periods is $0.97^{100} = 0.0475$, and the probability of playing for at least 500 periods is $0.97^{500} = 2.4315 \times 10^{-7}$, which is less than one in four million! Nonetheless, as we will soon see, even though the likelihood that players play for so long is infinitesimally small, this structure will give the players a lot of leverage. In summary, we need not literally imagine that these players will play infinitely many games, but merely that they will have a chance to continue the play after every stage is over.

Before continuing with our analysis of equilibrium behavior in infinitely repeated games, it is useful to introduce the following definition:

Definition 10.3 Given $\delta < 1$, the **average payoff** of an *infinite sequence of payoffs* $\{v_i^t\}_{t=1}^{\infty}$ is

$$\overline{v}_i = (1 - \delta) \sum_{t=1}^{\infty} \delta^{t-1} v_i^t.$$

That is, the average payoff from a sequence is a *normalization* of the net present value: we are scaling down the net present value by a factor of $1 - \delta$. This is convenient because of the following mathematical property. Consider a player who receives a finite sequence of fixed payments v in each period $t \in \{1, 2, \ldots, T\}$. His net present value is then the sum

$$s(v, \delta, T) = v + \delta v + \delta^2 v + \cdots + \delta^{T-2} v + \delta^{T-1} v. \tag{10.1}$$

---

4. If there *were* discounting in addition to an uncertain future, then let $\delta < 1$ be the probability of continuation, and let $\rho < 1$ be the discount factor. Then the *effective discount factor* would just be $\delta\rho$, and the expected discounted payoff would be given by

$$Ev = v_i^1 + \delta\rho v_i^2 + (1 - \delta) \cdot 0 + (\delta\rho)^2 v_i^3 + \delta(1 - \delta) \cdot 0 + (\delta\rho)^3 v_i^4 + \delta^2(1 - \delta) \cdot 0 + \cdots$$

$$= \sum_{t=1}^{\infty} (\delta\rho)^{t-1} v_i^t.$$

Consider this sum multiplied by δ, that is,

$$\delta s(v, \delta, T) = \delta v + \delta^2 v + \delta^3 v + \cdots + \delta^{T-1} v + \delta^T v, \qquad (10.2)$$

and notice that from (10.1) and (10.2) we obtain that $s(v, \delta) - \delta s(v, \delta) = v - \delta^T v$, implying that

$$s(v, \delta, T) = \frac{v - \delta^T v}{1 - \delta}.$$

Now if we take the limit of this expression when T goes to infinity (the game is infinitely repeated) then we get

$$s(v, \delta) = \lim_{T \to \infty} s(v, \delta, T) = \frac{v}{1 - \delta}.$$

How does this relate to the average payoff? If we look at the net present value of a stream of fixed payoffs v, then by definition the average payoff is

$$\bar{v} = (1 - \delta)s(v, \delta) = (1 - \delta)\frac{v}{1 - \delta} = v.$$

That is, the average payoff of an infinite fixed sequence of some value v is itself equal to v. As we will soon see, using average payoffs will help simplify the analysis of subgame-perfect equilibria in infinitely repeated games. This approach will also help us identify the set of payoffs that can be supported by strategies that form a subgame-perfect equilibrium of a repeated game.

10.2.2 Strategies

Consider the extensive-form representation of an infinitely repeated game. In particular imagine a tree with a root, from which it continues to expand both in "length"—because of the added stages—and in "width"—because more and more information sets are created after each period. Because the stage-game is repeated infinitely many times, there will be an *infinite number* of information sets! Because a player's strategy is a *complete contingent plan* that specifies what the player will do in each information set, it may seem that defining strategies for players in an infinitely repeated game will be quite cumbersome, if not outright impossible.

It turns out that there is a convenient and rather natural way to describe strategies in this setup. Notice that every information set of each player is identified by a *unique path of play* or *history* that was played in the previous sequences. We already hinted at this fact when we discussed multistage games in the previous chapter. For example, if we consider the infinitely repeated Prisoner's Dilemma, then in the fourth stage each player has 64 information sets.[5] Each of these 64 information sets corresponds to a unique path of play, or history, in the first three stages. For example, the players playing (M, m) in each of the three previous stages is one such history. Any other combination of three consecutive plays will be identified with a different and unique history.

---

5. Because the stage-game has four distinct outcomes, after three repetitions of the stage-game there will be $4^3 = 64$ distinct outcomes, or histories.

This observation implies that there is a one-to-one relationship between information sets and *histories of play*. Because of this relationship, from now on we use the word "history" to describe a particular sequence of action profiles that the players have chosen up until the stage that is under consideration (for which past play is indeed the history). To make things precise, we define histories, and history-contingent strategies, as follows:

Definition 10.4 Consider an infinitely repeated game. Let H_t denote the set of all possible **histories** of length t, $h_t \in H_t$, and let $H = \cup_{t=1}^{\infty} H_t$ be the set of all possible histories (the union over t of all the sets H_t). A **pure strategy** for player i is a mapping $s_i : H \to S_i$ that maps histories into actions of the stage-game. Similarly a **behavioral strategy** of player i, $\sigma_i : H \to \Delta S_i$ maps histories into stochastic choices of actions in each stage.

The interpretation of a strategy is basically the same as what we had used for multistage games, which was demonstrated by the Prisoner-Revenge Game that we analyzed in the last chapter (see Figure 9.1). Because every information set is identified with a unique history, this is a complete definition of a strategy that is fairly intuitive to perceive.

10.3   Subgame-Perfect Equilibria

Now that we have identified a precise and concise way of describing strategies for the players in an infinitely repeated game, it is straightforward to define a subgame-perfect equilibrium:

Definition 10.5   A profile of pure strategies $(s_1^*(\cdot), s_2^*(\cdot), \ldots, s_n^*(\cdot)), s_i : H \to S_i$ for all $i \in N$, is a **subgame-perfect equilibrium** if the restriction of $(s_1^*(\cdot), s_2^*(\cdot), \ldots, s_n^*(\cdot))$ is a Nash equilibrium in every subgame. That is, for any history of the game h_t, the continuation play dictated by $(s_1^*(\cdot), s_2^*(\cdot), \ldots, s_n^*(\cdot))$ is a Nash equilibrium. (Similarly for behavioral strategies $(\sigma_1^*(\cdot), \sigma_2^*(\cdot), \ldots, \sigma_n^*(\cdot))$.)

As with the case of strategies, this may at first seem like an impossible concept to implement. How could we check that a profile of strategies is a Nash equilibrium for *any history,* especially because each strategy is a mapping that works on any one of the infinitely many histories? As it turns out, one familiar result will be a useful benchmark:

Proposition 10.2   *Let $G(\delta)$ be an infinitely repeated game, and let $(\sigma_1^*, \sigma_2^*, \ldots, \sigma_n^*)$ be a (static) Nash equilibrium strategy profile of the stage-game G. Define the repeated-game strategy for each player i to be the history-independent Nash strategy, $\sigma_i^*(h) = \sigma_i^*$ for all $h \in H$. Then $(\sigma_1^*(h), \ldots, \sigma_n^*(h))$ is a subgame-perfect equilibrium in the repeated game for any $\delta < 1$.*

The proof is quite easy and basically mimics the ideas of the proof of proposition 9.1 for finite multistage games. If player i believes that his opponents' behavior is independent of the history of the game then there can be no role for considering how current play affects future play. As a consequence, if he believes that his opponents' current play coincides with their play in the static Nash equilibrium of the stage-game, then by the definition of a Nash equilibrium his best response must be to choose his part of the Nash equilibrium.

This proposition demonstrates that for the infinitely repeated Prisoner's Dilemma playing fink unconditionally in every period by each player is a subgame-perfect equilibrium. The more interesting question that remains is whether or not we can support other types of behavior as part of a subgame-perfect equilibrium. In particular can the players choose mum for a significant period of time in equilibrium? The answer is yes, and in many ways the intuition rests on arguments that are similar to those underlying the reward-and-punishment strategies we saw in the Prisoner-Revenge Game analyzed in the previous chapter, but much more subtle.

Before discussing some general results, we start with an example of the infinitely repeated Prisoner's Dilemma with discount factor $\delta < 1$, in which the following stage-game is infinitely repeated:

		Player 2	
		m	*f*
	M	4, 4	−1, 5
Player 1	*F*	5, −1	1, 1

Consider the path of play in which the two players choose (M, m) in *every period.* Following this path the players' *average payoffs* as defined in definition 10.3 are $(\overline{v}_1, \overline{v}_2) = (4, 4)$. We now ask whether this path of play can be supported as a subgame-perfect equilibrium. Clearly the history-independent strategies of "play M (or m) regardless of the history" cannot be a subgame-perfect equilibrium for a simple reason: following a deviation of any player from mum to fink at any stage, the other player's strategy suggests that he is committed to continue to play mum, making the deviation profitable. This implies that if we can support (M, m) in each and every period then we need to find some way to "punish" deviations.

The Prisoner-Revenge Game that we analyzed in the previous chapter hints at what we need to support any "good" behavior as a subgame-perfect equilibrium. First we need a "carrot" continuation equilibrium to reward good behavior and a "stick" continuation equilibrium to punish bad behavior. Second we need a high enough discount factor for the reward-and-punishment strategies to be effective.

An obvious candidate for a "stick" continuation equilibrium is the unconditional repeated play of the (F, f) static Nash equilibrium, which results in the players receiving an average payoff $(\overline{v}_1, \overline{v}_2) = (1, 1)$. What then can be the "carrot" continuation equilibrium? This is where things get interesting, and in a very ingenious way. Recall that we are trying to support the play (M, m) in every period as a subgame-perfect equilibrium. If we can indeed do this, it means that playing (M, m) forever *would be a consequence of some subgame-perfect equilibrium* with appropriately defined strategies. But if this is true then we have our answer: the subgame-perfect equilibrium that implies the play of (M, m) forever will be the "carrot" continuation equilibrium that supports itself.

Now this may at first come across as a circular argument: the reason that playing (M, m) forever can be a subgame-perfect equilibrium is that it can be supported as a subgame-perfect equilibrium! Indeed this *is* a circular argument, but it is not inconsistent. It is for this reason that an equilibrium of this nature is often called a *bootstrap equilibrium,* in that we are using the proposed equilibrium to support itself. This argument is consistent precisely because any infinitely repeated game has the *same set of possible future plays at any stage of the game.* It is this observation that makes the analysis manageable and offers us the ability to use this circular logic in a consistent way.

To see how this works consider the Prisoner's Dilemma previously described. Assuming that playing (M, m) in every future period can be the reward, or "carrot" continuation equilibrium, we need to determine whether playing (F, f) unconditionally forever can be a sufficient "stick." To do this consider the following strategies for each player:

> **Player 1:** In the first stage play $s_1^1 = M$. For any stage $t > 1$, play $s_1^t(h_{t-1}) = M$ if and only if the history h_{t-1} is a sequence that consists only of (M, m), that is, $h_{t-1} = \{(M, m), (M, m), \ldots, (M, m)\}$. Otherwise, if some player ever played fink and $h_{t-1} \neq \{(M, m), (M, m), \ldots, (M, m)\}$ then play $s_1^t(h_{t-1}) = F$.

> **Player 2:** Play the mirror strategy to player 1: $s_2^1 = m$. For any stage $t > 1$, play $s_2^t(h_{t-1}) = m$ if and only if $h_{t-1} = \{(M, m), (M, m), \ldots, (M, m)\}$, while if $h_{t-1} \neq \{(M, m), (M, m), \ldots, (M, m)\}$ then play $s_2^t(h_{t-1}) = f$.

Before checking if this pair of strategies is a subgame-perfect equilibrium, it is worth making sure that they are well defined. We have to define the players' actions for *any* history, and we have done this as follows: in the first stage they both intend to be "good" and play mum. Then at any later stage they look back at history and ask: was there *ever* a deviation in *any previous stage* from always playing mum? If the answer is no, meaning that both players always cooperated and played mum, then in the next stage each player chooses mum. If, however, there was *any deviation* from cooperation in the past, be it once or more than once, be it by one player or both, the players will revert to playing fink, and by the definition of the strategies they will stick to fink thereafter.

These strategies are commonly referred to as **grim-trigger strategies** because they include a natural trigger: once someone deviates from mum, this is the trigger that causes the players to revert their behavior to fink forever, resulting in a very grim future. More generally the idea behind grim-trigger strategies is as follows. If the stage-game has a Nash equilibrium, and we are trying to support subgame-perfect equilibrium behavior that results in outcomes that are *better* than any Nash equilibrium, we use the established subgame-perfect equilibrium of playing the grim static Nash outcome forever to support the more desirable outcomes that are not supported by static best responses being played. Hence we use the grim trigger to provide incentives for the players to stick to behavior for which short-run temptations to deviate are present.

To verify that the grim-trigger strategy pair is a subgame-perfect equilibrium we need to check that there is no profitable deviation in any subgame. This may seem like an impossible mission because we have an infinite number of subgames. As with multistage games, however, it is not that tedious thanks to the power of the one-stage deviation principle introduced in Section 9.5. Indeed the following powerful proposition is extremely useful:

Proposition 10.3 *In an infinitely repeated game $G(\delta)$ a profile of strategies $\sigma^* = (\sigma_1^*, \ldots, \sigma_n^*)$ is a subgame-perfect equilibrium if and only if there is no player i and no single history h_{t-1} for which player i would gain from deviating from $s_i(h_{t-1})$.*

The idea of the proof, which is omitted, closely follows that of theorem 9.1.[6] The one-stage deviation principle implies that in order to confirm that a profile of strategies is a subgame-perfect equilibrium, we need only to check that no player has a single history from which he would want to deviate unilaterally. Applying this to our repeated Prisoner's Dilemma may not seem helpful because there are still an infinite number of histories. Nonetheless the task is not a hard one because even though there are an infinite number of histories, they fall into *one of two relevant categories:* either there was no deviation and the players intend to play (M, m) or there was some previous deviation and the players intend to play (F, f). These two categories of histories are common to any application of grim-trigger strategies in repeated games, which makes them relatively easy to check as part of a subgame-perfect equilibrium. That is, whenever we are trying to support one kind of behavior forever with the threat of resorting to another kind of behavior, we have two "states" in which the players can be, and we need only to check that they would not want to deviate from each of these two states.

Consider first the category of histories that *are not* consecutive sequences of only (M, m), implying that play ought to be the grim Nash equilibrium of the stage-game. Notice that these histories are *off the equilibrium path* because following the proposed strategies would result in (M, m) being played forever. Thus *in any subgame that is off the equilibrium path,* the proposed strategies recommend that the players play (F, f) in this period, and in any subsequent period, because a deviation from (M, m) has occurred sometime in the past. Clearly if this is the case then no player would want to choose mum instead of fink because, given his belief that his opponent will play fink, such a deviation from fink to mum will cause him a loss of -2 at this stage (getting -1 instead of 1), together with no gains in subsequent stages because the previous deviation will keep the players on the grim-trigger Nash equilibrium path. Thus in any subgame that is off the equilibrium path no player would ever want to deviate unilaterally from fink to mum.

Now consider the other category of histories that are consecutive sequences of (M, m), which are all the histories that are *on the equilibrium path.* If a player chooses to play mum, his current payoff is 4, and his continuation payoff from the pair of strategies is an infinite sequence of 4s starting in the next period. Therefore his payoff from following the strategy and not deviating will be

$$v_i^* = 4 + \delta 4 + \delta^2 4 + \cdots = 4 + \delta \frac{4}{1 - \delta}.$$

The division of the player's payoff into today's payoff of 4 and the continuation payoff from following his strategy of $\delta \frac{4}{1-\delta}$ is useful. It recognizes the idea that a player's actions have two effects: The first is the direct effect on his immediate payoffs today;

---

6. Here we cannot assert that there is a finite number of deviations that improves upon σ_i. Instead the discounting of payoffs implies a useful notion of continuity of the payoffs from paths of play that are "close" in that they coincide over many information sets. In particular if some deviation from σ_i to σ_i' at h_i^t improves upon σ_i, then there is a strategy σ_i'' that is identical to σ_i' for $n < \infty$ stages after h_i and identical to σ_i everywhere else, and σ_i'' must improve upon σ_i. Then, since $n < \infty$, we can use an induction argument similar to that for theorem 9.1. For a proof see Fudenberg and Tirole (1991), theorem 4.2.

the second is the indirect effect on his *continuation equilibrium payoff,* which results from the way in which the equilibrium continues to unfold.

If the player deviates from mum and chooses fink instead, then he gets 5 instead of 4 in the immediate stage of deviation, followed by his continuation payoff, which is an infinite sequence of 1s. Indeed we had already established that in the continuation game following a deviation off the equilibrium path, every player will stick to playing fink because his opponent is expected to do so forever. Therefore his payoff from deviating from the proposed strategy will be

$$v_i' = 5 + \delta 1 + \delta^2 1 + \cdots = 5 + \delta \frac{1}{1 - \delta}.$$

Thus we can easily see that the trade-off between sticking to the proposed strategy and deviating boils down to a simple comparison. A deviation will yield an immediate gain of 1 because the player gets 5 instead of 4 in the deviation stage, which needs to be weighed against the loss in continuation payoffs, which will drop from $\delta \frac{4}{1-\delta}$ to $\delta \frac{1}{1-\delta}$. We conclude that the player will not want to deviate if $v_i^* \geq v_i'$, or

$$4 + \delta \frac{4}{1 - \delta} \geq 5 + \delta \frac{1}{1 - \delta}, \tag{10.3}$$

which is equivalent to

$$\delta \geq \tfrac{1}{4}.$$

Once again, as with multistage games, we see the value of *patience.* If the players are sufficiently patient, so that the future carries a fair amount of weight in their preferences, then there is a reward-and-punishment strategy that will allow them to cooperate *forever.* The loss in continuation payoffs will more than offset the gains from immediate defection, and this will keep the players on the cooperative path of play.

Using the probabilistic interpretation of δ, these strategies will credibly allow the players to cooperate as long as the game is likely enough to be played again. It is reminiscent of the saying "what goes around comes around": to the extent that the players are likely to meet again, good behavior can be rewarded and bad behavior can be punished. If the future does not offer that possibility, there can be no credible commitment to cooperation.

To relate the current analysis to the analysis of the two-period multistage Prisoner-Revenge Game in Chapter 9, think of the repeated game as having two stages: "today," which is the current stage, and "tomorrow," which is the infinite sequence of play after this stage is over. In a two-period multistage game we determined that there are two necessary conditions that are required to support non-Nash behavior in the first stage: First, we need to have multiple equilibria in the second stage, and second, we need the players to be sufficiently patient. Clearly the inequality $\delta \geq \tfrac{1}{4}$ is the "sufficient patience" condition. The question is, where are the multiple equilibria coming from? This is where the infinite repetition creates "magic" through bootstrapping: from the unique equilibrium of the stage-game we get multiple equilibria of the repeated game. This so-called magic is so strong that many types of behavior can be supported as subgame-perfect equilibria, as Section 10.6 demonstrates.

10.4   Application: Tacit Collusion

One of the most celebrated applications of repeated-game equilibria with reward-and-punishment strategies has been to the study of tacit collusion among firms. Most developed countries forbid firms from entering into explicit contracts to restrict competition. In the United States section 1 of the Sherman Antitrust Act of 1890 proclaims as illegal any "conspiracies in restraint of trade," including such contracts to fix prices. However, even if there have been no actual meetings or discussions between competitors, can we be certain that they are not setting prices through some *implicit* understanding or agreement?

The use of repeated-game strategies can come in handy for firms, which through their beliefs about collusion-supporting strategies can support anticompetitive behavior without ever discussing it or reaching explicit agreements. Instead they employ implicit or *tacit collusion*. To demonstrate this point consider the Cournot duopoly problem in which each of the two firms has costs of production equal to $c_i(q_i) = 10q_i$ and demand is given by $p = 100 - q$, and imagine that this market game is played repeatedly in periods $t = 1, 2, \ldots$ with some discount factor $\delta \in (0, 1)$.

In Section 5.2.3 we solved for the Nash-Cournot equilibrium of the stage-game in which $q_1^* = q_2^* = 30$, the resulting equilibrium marker price was $p = 40$, and profits for each firm were $v_i^* = 900$. We also concluded that the sum of the Cournot profits is less than that of the monopoly profits. Indeed if the firms each produced $q_1 = q_2 = 22.5$ then each firm would earn profits equal to $v_i = (100 - 45) \times 22.5 - 10 \times 22.5 = 1012.5$, which is half the monopoly profits. If the firms could write a *binding contract* that restricted production for the stage-game to $q_i = 22.5$ then they would do so in order to improve profits. We now proceed to see how reward-and-punishment strategies similar to those previously proposed for the repeated Prisoner's Dilemma will allow our firms to coordinate on monopoly profits in a self-enforcing subgame-perfect equilibrium, without the need to use binding contracts that violate such regulations as the Sherman Act.

First, we have to decide how the firms will cooperate to split the monopoly profits, which will be determined by a pair of quantities q_1^c and q_2^c. We know that to achieve monopoly profits we must have $q_1^c + q_2^c = 45$, resulting in a price of $p = 55$. To make this example more interesting, consider a split given by the quantities $q_1^c = 22$ and $q_2^c = 23$, which is of course not symmetric. The implied stage-game profits are $v_1^c = (55 - 10) \times 22 = 990$ and $v_2^c = (55 - 10) \times 23 = 1035$.

Second, following the logic of the infinitely repeated Prisoner's Dilemma example, we know what the general form of the firms' strategies ought to be. Each firm i will play q_i^c in the first period. Then in every period $t > 1$ each firm will play q_i^c if the sequence of previous plays included only the pairs (q_1^c, q_2^c) and will instead play the grim-trigger static Nash equilibrium $q_i^* = 30$ if any other history occurred.

Third, we need to check that no firm wants to deviate from the proposed strategies. Again it is useful to recognize that there are two classes of histories: those that are on the equilibrium path, following sequences of (q_1^c, q_2^c), and those that are off the equilibrium path, which follow any other sequence of quantities. Thus, using the one-stage deviation principle, it suffices to check that, both on and off the equilibrium path, no player wants to deviate from his proposed subgame-perfect equilibrium strategy.

It is easy to see that no player would want to deviate from the proposed behavior off the equilibrium path. The reason is that at any stage off the equilibrium path player i expects player j to play $q_j^* = 30$, and $q_i^* = 30$ is a best response to this choice. As a consequence deviating in any off-the-equilibrium path history at any stage will cause an immediate loss, with no scope for future gain because once the grim-trigger strategy is triggered, it is history independent. Thus for any value of δ there is no temptation for deviations off the equilibrium path.

Now we check to see that no firm wants to deviate on the equilibrium path. Recall that on the equilibrium path each firm chooses q_i^c and gets a per-period (and average) payoff of v_i^c. Notice that in each stage no one of the two firms is playing a static best response. To see this, recall the best-response function derived in Section 5.2.3,

$$BR_i(q_j) = \frac{90 - q_j}{2}. \tag{10.4}$$

From (10.4) we can see that firm 1's best response to $q_2^c = 23$ is $BR_1(23) = 33.5$ and firm 2's best response to $q_1^c = 22$ is $BR_2(22) = 34$. Hence each firm i can gain in any given period by deviating from q_i^c, at the cost of losing the future stream of cooperative outcomes and resorting to the Nash-Cournot equilibrium forever after.

Furthermore we know that the most tempting deviation for each firm i in the stage-game is indeed to choose its static best response, which we denote as $q_i^d = BR_i(q_j^c)$. From the previous calculations, for firm 1 we have $q_1^d = 33.5$ and $q_2^d = 34$. This gives the deviating firm a one-stage deviation payoff of

$$v_i^d = (100 - q_i^d - q_j^c)q_i^d - 10q_i^d,$$

which for firm 1 is $v_1^d = (100 - 33.5 - 23) \times 33.5 - 10 \times 33.5 = 1122.25$ and for firm 2 is $v_2^d = (100 - 23 - 34) \times 34 - 10 \times 34 = 1122$.

We can now check the condition for a firm to prefer not to deviate from the proposed equilibrium path derived from the subgame-perfect equilibrium strategies. The stage-game payoff on the equilibrium path was denoted by v_i^c, the payoff from the best deviation is v_i^d, and the payoff from the Nash-Cournot outcome is v_i^*. Hence, similar to inequality (10.3) for the repeated Prisoner's Dilemma example, firm i will not want to deviate if the following inequality is satisfied:

$$\underbrace{v_i^c}_{\substack{\text{Current payoff} \\ \text{from cooperate}}} + \underbrace{\frac{\delta v_i^c}{1 - \delta}}_{\substack{\text{Future payoff} \\ \text{if cooperate}}} \geq \underbrace{v_i^d}_{\substack{\text{Current payoff} \\ \text{from best defect}}} + \underbrace{\frac{\delta v_i^*}{1 - \delta}}_{\substack{\text{Future payoff} \\ \text{if defect}}},$$

which, after some algebra, is equivalent to

$$\delta \geq \frac{v_i^d - v_i^c}{v_i^d - v_i^*}. \tag{10.5}$$

We proceed by calculating a "cutoff" discount factor for each firm by plugging the stage-game payoffs just calculated into (10.5) for each of the two firms. This calculation will result in a value of δ_i for each firm that is the lowest possible discount factor for which the firm will not be tempted to deviate from the equilibrium path. In particular,

$$\delta_1 \geq \frac{1122.25 - 990}{1122.25 - 900} = 0.595 \quad \text{and} \quad \delta_2 \geq \frac{1122 - 1035}{1122 - 900} = 0.392,$$

which nicely reflects the temptations to deviate. Firm 1 gains more from deviating in any period than does firm 2, and firm 1 has less to lose from the drop in the continuation payoffs than does firm 2. For this reason firm 1 has to value the future more than firm 2, and hence has to have a higher discount factor for this pair of strategies to be a self-enforcing subgame-perfect equilibrium.

Two points are worth making before we conclude this section. First, if the two firms somehow manage to share these beliefs, and never talk about any agreement, then they can circumvent the anticompetitive rules and regulations. One may be skeptical that such beliefs can converge so nicely without any communication between the firms. When rules against collusion and anticompetitive behavior do not apply, however, as in the case of OPEC, then in lieu of an enforceable contract the firms will indeed be able to resort to these kinds of reward-and-punishment strategies.

Second, and perhaps more intriguing, we know from history that even when such collusive agreements are in place there are occasional price wars between the firms, in which collusive behavior fails for some amount of time and is later restored.[7] Our analysis suggests that price wars are instances of off-the-equilibrium-path behavior and hence are puzzling when they do occur. That is, if the collusive behavior is sustainable as a subgame-perfect equilibrium then no firm would want to deviate at any stage on the equilibrium path. Why then would the firms engage in a price war? One explanation may be that they cannot easily detect deviations because, for example, actual produced quantities are not readily observable.

To demonstrate the idea behind this argument, consider the case of OPEC, a cartel that really does manipulate prices through the quantities that are produced (à la Cournot). Arguably it is prohibitively difficult to observe the quantities produced by each country (each "firm"), which in turn implies that using trigger strategies is problematic. Indeed if the players cannot observe the choices made by other players then how will they know if they should move from the "carrot" equilibrium to the "stick" equilibrium?

A clever observer will make the following argument: In the repeated Cournot model we do not *have* to observe quantities—as long as no firm deviates, the market price will remain fixed (at $p = 45$ in the example previously solved). If the price suddenly falls, it is immediately apparent that someone deviated from the proposed quantities, and the grim-trigger strategy results in a price war.

Such an observation is clever indeed, but it fails in the potential instance of random prices. In the market for oil, for example, the price is determined by supply (quantity produced) and demand. In the real world, however, demand is not constant as in our simplistic example. In some weeks there is a greater need for energy (say, in colder weeks when homes need heating) and in others the need is less. As a consequence prices will no longer be "clean" indicators of deviations by firms. Instead they can act only as "noisy" signals that imply *either* a reduction in demand *or* a deviation in quantities by some player. This in turn implies that players will be tempted to deviate and offer the excuse "I didn't deviate; demand must have been low!" So it turns out that occasional price wars are an inherent reality of trying to collude when detection is not perfect.

This interesting idea was first put forth in a seminal paper by Green and Porter (1984), which led to a large literature aimed at analyzing repeated games with *imperfect monitoring*. In such games players cannot perfectly monitor the behavior of other players, and as a consequence they must rely on signals that are not perfect indicators

---

7. In the repeated Cournot model this is a "quantity war" that causes a reduction in prices. You are asked to perform a similar analysis for the repeated Bertrand game in exercise 10.8 at the end of this chapter.

of the behavior of those players. This literature is extremely technical and is beyond the scope of this textbook. The interested (and suitably trained) reader should consult Mailath and Samuelson (2006, parts 2 and 3) for a deep dive into these technically challenging topics.

10.5 Sequential Interaction and Reputation

In the infinitely repeated Prisoner's Dilemma or the infinitely repeated Cournot game, players were able to use their ongoing relationship to support cooperative behavior that was not self-enforcing in the short run because of the temptation to deviate from it. In both examples two critical elements were required: that there be no predetermined terminal period and that the discount factor not be too small. As we have seen, if there is a terminal period then it is impossible to provide incentives to cooperate at the end of the game and, predicting this, the players will resort to noncooperative self-enforcing behavior. Similarly if the discount factor is too small then the incentives provided by the game's ongoing continuation are not strong enough to overcome the temptation to deviate from the cooperative path of play.

10.5.1 Cooperation as Reputation

One interpretation that has been used to describe the ability to create long-term incentives to overcome short-term temptations is that players in a repeated relationship can create a reputation to cooperate with each other. As long as a player maintains his reputation of being cooperative, other players will trust him and respond in kind. If a player fails to be cooperative at any stage then he will lose his goodwill and the players will move to a noncooperative phase of their engagement (e.g., using grim-trigger strategies such as those described in Section 10.3). Such reputational concerns may be dampened—or disappear altogether—when there is no future or when the future is not important enough.

Let's examine another game in which reputational incentives can facilitate better outcomes than those possible in a one-shot or finitely repeated setting. Recall the "trust game" that we introduced in Section 7.1.1. Player 1 first chooses whether to ask for the services of player 2. He can trust player 2 (T) or not trust him (N). If player 1 plays T, then player 2 can choose to cooperate (C), which represents offering player 2 some fair level of service, or he can defect (D), which is better for player 2 at the expense of player 1. The players prefer engaging in a cooperative exchange over not interacting at all (no trust), but the temptations to deviate make that impossible. As mentioned earlier, this game captures many real-life exchanges in which one party must trust another in order to achieve some gains from trade, but the second party can abuse that trust. The game is depicted in Figure 10.1.

If this one-shot game is repeated infinitely often with a discount factor of δ then we can create incentives that support cooperative behavior in ways similar to those demonstrated earlier for the infinitely repeated Prisoner's Dilemma game.[8] For instance, we can prescribe grim-trigger strategies of the following kind: Player 1's

---

8. Notice that this one-shot game is not a simultaneous-move game, yet we defined repeated games as repetitions of a normal-form game. Of course every extensive-form game can be represented by a normal form, so this can be done without any modifications to what we have already defined.

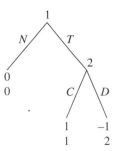

FIGURE 10.1 The one-shot trust game.

will be "In period 1 I will trust player 2, and as long as there were no deviations from the pair (T, C) in any period then I will continue to trust him. Once such a deviation occurs then I will not trust him forever after." Similarly player 2's strategy will be "In period 1 I will cooperate, and as long as there were no deviations from the pair (T, C) in any period then I will continue to do so. Once such a deviation occurs then I will deviate forever after."[9]

These two strategies will form a subgame-perfect equilibrium if no player wants to deviate at any stage. Using the one-stage deviation principle, it is first easy to see that no player wants to deviate off the equilibrium path: the strategies prescribe that player 2 will deviate so player 1 should indeed not trust, and they prescribe that player 1 will never trust again, so player 2, if found off the equilibrium path, should deviate. On the equilibrium path it is easy to see that player 1 is playing a best response because the stream of payoffs of 1 is better than the stream of payoffs of 0. Player 2 will prefer to stick to the equilibrium path if and only if the infinite stream of a payoff of 1 is better than deviating to get 2 today and 0 thereafter, that is,

$$\frac{1}{1-\delta} \geq 2$$

or

$$\delta \geq \tfrac{1}{2}.$$

This example illustrates again the power of continued relationships. If the two players play the game only once, we know that player 2 will not be able to commit to cooperate, and anticipating player 2's defection, player 1 will not trust player 2 at all.

10.5.2 Third-Party Institutions as Reputation Mechanisms

When we consider the trust game, or similar games, an important question arises: can we find a way to provide incentives to player 2 that extend beyond the terminal period of a one-shot interaction? One way to proceed is to use a third player who acts

---

9. Notice that I did not specify player 2's strategy in the following way: "As long as player 1 trusts me I will cooperate." The reason is that combining this strategy with player 1's proposed strategy is a Nash equilibrium, but it is not subgame-perfect. The problem is that some off-the-equilibrium-path prescriptions are not best responses. Showing this is left as exercise 10.4.

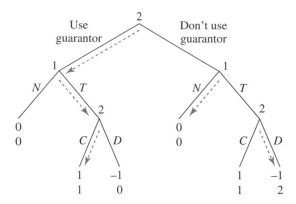

FIGURE 10.2   Trust game with a bond option.

as a guarantor, or enforcement institution. Imagine that player 2 deposits a payment equivalent to 2 in the hands of the guarantor, player 3, with the following contract: "If player 1 trusts me and I defect then you keep the payoff of 2 that I deposited in your hands. In any other circumstance you return it to me." Assume for the moment that player 3 must follow the contract and has no discretion of his own. This effectively modifies the game to a new game, described in Figure 10.2, and gives player 2 the option to post a "performance bond" in the hands of player 3 that guarantees the right incentives to cooperate for player 2. To see this, we can apply backward induction to this modified game (marked in the figure by the dotted arrow lines). In the unique subgame-perfect equilibrium, player 2 chooses to use the guarantor and player 1 then rationally anticipates that player 2 will cooperate, which indeed happens.

Of course two questions arise. First, why would such a committed guarantor exist in the first place? This modifies the nature of the trust game, and one might argue that, more often than not, such a guarantor does not exist. Second, if the guarantor is really to be thought of as a third player, then we should give him the option to choose whether or not to return the money to player 2 after player 2 cooperates. If we add the guarantor as player 3 instead of the terminal nodes at the left side of the tree in Figure 10.2, then it is obvious that player 3 has a dominant strategy: don't return the money to player 2!

Now consider the following situation. Imagine that player 3 is some local official such as a judge or arbitrator, who can provide his services in every period $t = 1, 2, \ldots$ and uses a discount factor $\delta < 1$ to evaluate future payoffs. In every period t a new pair of players P_1^t and P_2^t arrive in the jurisdiction of the judge, each pair playing a one-shot trust game with no future interaction. Player P_2^t can choose to post a bond equal to 2 with player 3, as described in Figure 10.2, but must pay him a fee of $w = 0.1$ for his services. Alternatively he can choose not to use player 3's services.

Can we support an equilibrium in which every pair of players P_1^t and P_2^t chooses to use player 3's services and successfully engages in cooperative trade in every period, despite not having their own mutual future trades on which to rely? Interestingly the answer is yes. Consider the following strategies for the players of this game:

> **Player P_1^t:** If player P_2^t posted a bond with player 3 then trust him, and otherwise don't.

Player P_2^t: Bond posting: In period $t = 1$ post a bond with player 3. In periods $t > 1$, if player 3 followed the bond contract in all periods before period t then post a bond with player 3, and otherwise don't.
Response to P_1^t: If a bond was posted in period t then choose to cooperate with P_1^t, and otherwise choose to defect.

Player 3: If player P_2^t posts a bond in period T and pays the wage then return the bond to player P_2^t if he cooperated, and otherwise don't return the bond. (This is the bond contract.)

To confirm that this is a subgame-perfect equilibrium, we need to check that players are playing a best response on and off the equilibrium path, regardless of the history. Starting with the equilibrium path, it is easy to see that players P_1^t and P_2^t are playing a best response. Effectively, given the prescribed strategy of player 3, then in every period players P_1^t and P_2^t are exactly playing the game depicted in Figure 10.2 (with the small change that instead of receiving a payoff of 1 from cooperating on the equilibrium path, player 2 receives 0.9 because of the wage he pays to player 3). Now consider player 3. If he follows his proposed strategy then he is guaranteed an infinite stream of payoffs of $w = 0.1$ per period. If he defects in any period t then he will get 2 in that period, and 0 forever after. Hence he will follow his proposed strategy if and only if

$$\frac{0.1}{1 - \delta} \geq 2 \qquad \text{or} \qquad \delta \geq 0.95.$$

Thus if the discount factor is large enough then player 3 will happily play the role of bondholder and follow the contract as described.[10]

The concept that third parties can play the role of a guarantor and enforce good trading behavior was introduced by Milgrom et al. (1990). The idea is that instead of the players themselves trying to maintain a reputation for good behavior, a third party—be it a judge or some other institutional player—will be able to sustain a reputation for supporting the good behavior of others. Milgrom et al. use this observation to explain the role of law merchants (player 3) in medieval Europe as institutional actors who helped support trade among players who had no ongoing relationships of their own. The key factor that enabled such institutions to add value is that future traders somehow learned about the law merchant, and about his past adherence to the prescribed norms of behavior. See Greif (2006) for an in-depth review of repeated-game approaches to the role of institutions in the development of trade.

10.5.3 Reputation Transfers without Third Parties

As the previous section showed, when instead of two infinitely long-lived players we have a sequence of short-lived (one-shot) pairs of players, then with the help of an infinitely lived third party we can restore reputational incentives to support gains

---

10. Note that the bond needs only to be greater than 1 for player 2 to prefer C over D because the gain to any P_2^t is exactly 1 from defecting. In addition, because the benefit to player 2 from following the equilibrium strategy is 1, compared to 0 if he does not post a bond, he should be willing to pay any fee up to 1 to player 3. In exercise 10.10 you are asked to use this fact and show that cooperative behavior can be sustained for any positive discount factor.

from trade. This perhaps suggests that some player, or some kind of institution, must continue to be active forever for there to be any kind of reputational concern that supports trade between players who themselves do not have the incentives to play cooperatively.

Kreps (1990b) offered an interesting observation on this matter. He suggested that reputation acquired under the name of a firm, or *entity,* may be separated from the *identity* of the player who is operating under the firm's name. Perhaps we don't need players to live for infinitely many periods; instead we need only an abstract entity that can be passed on from one player to the next. The entity will have a value in its own right because it carries with it the reputational concerns needed to maintain good behavior, and guarding its value will provide valuable incentives to its owner.

As before, imagine that in every period t a new pair of players P_1^t and P_2^t are matched to play a one-shot trust game as described in Figure 10.1, with no future interaction. Just before the first period starts, player P_2^1 can create a unique company or brand name—call it Trusted Associates. Assume that once he creates the brand name player P_2^1 has the sole rights to sell that name in the future. Now consider the following strategies:

Player P_1^1: If player P_2^1 created a unique brand name then trust him, and otherwise don't.

Player P_2^1: Choose a unique name, and cooperate with player P_1^1. Afterwards offer the brand name for sale to player P_2^2 at a price $p^* > 1$.

Player P_1^t (t > 1): If (1) player P_2^t bought the brand name from player P_2^{t-1} and (2) no abuse of trust (defection) ever occurred under that brand name then trust him. Otherwise don't.

Player P_2^t (t > 1): If (1) player P_2^{t-1} bought the brand name from player P_2^{t-2} at price p^* and (2) no abuse of trust (defection) ever occurred under that brand name then buy the brand name from player P_2^{t-1}, and then cooperate with player P_1^t. Afterwards offer the brand name for sale to player P_2^{t+1} at price p^*.

Showing that this set of strategies is a subgame-perfect equilibrium is not difficult and follows logic similar to that of the game in Figure 10.2. (You are asked to do this in exercise 10.11.) The idea behind the construction of these strategies is simple and appealing: the so-called brand name acts as a bearer of firm reputation that is passed on from one player 2 to the next. The investment made in the name before play begins in period t acts like a bond, and it will only be recovered if that period's player 2 will indeed cooperate.

The innovation laid out by Kreps (1990b) is the argument that we don't need a third player to intermediate the exchange of that bond, but instead can have a firm's good name be the bearer of its reputation. As a consequence short-lived agents become infinitely lived in the sense that the concept of a terminal period is removed altogether. The idea that an institution like a firm can sustain cooperative behavior among its employees, even if these come and go over time, was developed earlier by Cremer (1986), and the concept of using a firm's brand value as a bond for its performance dates back to Klein and Leffler (1981).

This repeated-game approach to a firm's reputation does, however, leave some questions unanswered. The most critical is that this approach does not account for how reputations arise and become valuable assets. There is no account of how a firm's reputation, represented by the value of its name, may increase (or decrease) in value, as is commonly observed in reality.[11] This is, of course, a consequence of the nature of the "bootstrapping" equilibria in repeated games, as discussed in Section 10.3. As the next section demonstrates, the force provided by using the future to provide incentives for present behavior in infinitely repeated games results in a rather disturbing outcome.

Remark *The Princess Bride* is a wonderfully funny film (based on a novel by William Goldman) that includes a lot of game theoretic insights, including an unforgettable scene on common knowledge (regarding a poisoned cup of wine). With respect to the value of a name, the following scene speaks directly to the insights provided by Kreps. In this scene Westley, who plays the role of the feared and ruthless Dread Pirate Roberts, tells his true love, Buttercup, that he really is not who people believe him to be:

> Well, Roberts had grown so rich, he wanted to retire. He took me to his cabin and he told me his secret. "I am not the Dread Pirate Roberts," he said. "My name is Ryan; I inherited the ship from the previous Dread Pirate Roberts, just as you will inherit it from me. The man I inherited it from is not the real Dread Pirate Roberts either. His name was Cummerbund. The real Roberts has been retired fifteen years and living like a king in Patagonia." Then he explained the name was the important thing for inspiring the necessary fear. You see, no one would surrender to the Dread Pirate Westley.

10.6 The Folk Theorem: Almost Anything Goes

In this section we describe, in a somewhat formal way, one of the key results from the theory of infinitely repeated games. This insight goes to the heart of the "magic" that we described in Section 10.3, explaining that when a stage-game with a unique Nash equilibrium is repeated infinitely often, many outcomes can be supported by a subgame-perfect equilibrium. We begin by introducing some necessary mathematical concepts.

Definition 10.6 Consider two vectors $v = (v_1, v_2, \ldots, v_n)$ and $v' = (v'_1, v'_2, \ldots, v'_n)$ in $\mathbb{R}^n$. We say that the vector $\widehat{v} = (\widehat{v}_1, \widehat{v}_2, \ldots, \widehat{v}_n)$ is a **convex combination** of v and v' if there exists some number $\alpha \in [0, 1]$ such that $\widehat{v} = \alpha v + (1 - \alpha)v'$, or $\widehat{v}_i = \alpha v_i + (1 - \alpha)v'_i$ for all $i \in \{1, \ldots, n\}$.

To imagine this concept intuitively, consider two payoff vectors in the Prisoner's Dilemma, say, $v = (4, 4)$ and $v' = (5, -1)$. As depicted in Figure 10.3, if a line segment is drawn to connect between these in the plane $\mathbb{R}^2$, then any point on this line will be a convex combination of these two payoff vectors. For example, if we take $\alpha = 0.6$ then the vector $\widehat{v} = 0.6 \times (4, 4) + 0.4 \times (5, -1) = (4.4, 2)$ is on that line segment. Another way to think of a convex combination is as a weighted average, in which each point gets a (potentially) different weight. In this example $(4.4, 2)$ puts a weight of 0.6 on the vector $(4, 4)$ and a weight of 0.4 on the vector $(5, -1)$.

---

11. An alternative approach to understanding a firm's name as a bearer of reputation, which accounts for the way reputations fluctuate in value, is given in Tadelis (1999). Personally I find the topic of reputations fascinating, and the interested reader can consult Bar-Isaac and Tadelis (2008) for a more recent survey.

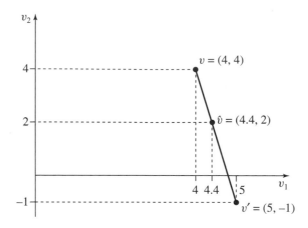

FIGURE 10.3 A convex combination of two vectors.

The definition of a convex combination can be generalized as follows. Consider a finite set of vectors $V = \{v^1, v^2, \ldots, v^k\}$ in $\mathbb{R}^n$. We say that $\widehat{v} = (\widehat{v}_1, \widehat{v}_2, \ldots, \widehat{v}_n)$ is a convex combination of the elements in V if there exist weights $(\alpha_1, \ldots, \alpha_k) \in \mathbb{R}^k_+$, $\sum_{j=1}^{k} \alpha_j = 1$, such that $\widehat{v} = \sum_{j=1}^{k} \alpha_j v^j$. Using this concept we offer the following definition:

Definition 10.7 Given a set of vectors $V = \{v^1, v^2, \ldots, v^k\}$ in $\mathbb{R}^n$, the **convex hull** of V is

$CoHull(V)$

$$
= \left\{ v \in \mathbb{R}^n : \exists (\alpha_1, \ldots, \alpha_k) \in \mathbb{R}^k_+, \sum_{j=1}^{k} \alpha_j = 1, \text{ such that } v = \sum_{j=1}^{k} \alpha_j v^j \right\}.
$$

That is, the convex hull of a set of k vectors includes all the vectors that can be generated by convex combinations of the k vectors. We will apply this definition to the set of average payoffs in a game. Recall from Section 10.2.1 that the average payoff of an infinite stream of some value v is itself v. This implies that if we take a profile of pure strategies s in a stage-game with n players and repeat it infinitely often then the players' average payoffs from the sequence of plays of s are equal to the stage-game payoffs from s. If the strategy sets are finite then the set of payoff vectors from pure strategies in the stage-game is finite and coincides with the set of average payoffs from playing any one of these pure strategies infinitely often. This will be a finite set of vectors in $\mathbb{R}^n$, playing the role of Z in the definition. For example, in the Prisoner's Dilemma given by the matrix

		Player 2	
		m	f
Player 1	M	4, 4	−1, 5
	F	5, −1	1, 1

the set of payoff vectors is $V = \{(4, 4), (5, -1), (-1, 5), (1, 1)\}$, and the convex hull is given by the gray shaded area in Figure 10.4.

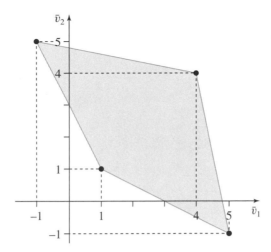

FIGURE 10.4 The convex hull of average payoffs for the Prisoner's Dilemma game.

We are ready to introduce one of the most celebrated results in the theory of repeated games. It turns out that for a high enough discount factor then *for almost any vector v in the convex hull* of the average payoffs in the repeated game it is possible to find a strategy profile that supports it as a subgame-perfect equilibrium. To state a version of this result formally, we call the convex hull of a game's payoffs the set of **feasible payoffs,** because each vector in the set can be achieved in the infinitely repeated game by combinations of strategies in the stage-game. We can now state the following result:

Theorem 10.1 (The Folk Theorem) *Let G be a finite, simultaneous-move game of complete information, let $(v_1^*, \ldots, v_n^*)$ denote the payoffs from a Nash equilibrium of G, and let $(v_1, \ldots, v_n)$ be a feasible payoff of G. If $v_i > v_i^*$, $\forall i \in N$, and if δ is sufficiently close to 1, then there exists a subgame-perfect equilibrium of the infinitely repeated game $G(\delta)$ that achieves an average payoff arbitrarily close to $(v_1, \ldots, v_n)$.*

Legend has it that the intuition for this result was floating around in the late 1960s and it is impossible to trace it to a single contributor; hence it is referred to as the *folk theorem.* However, one of the earliest publications of such a result was by James Friedman (1971). This result is not too hard to prove, but it is nonetheless omitted from this text. The interested reader can refer to Fudenberg and Tirole (1991, pp. 152–154) for a version of the folk theorem with an appropriate proof, and a more thorough analysis of a variety of versions of the folk theorem appears in Mailath and Samuelson (2006).

The idea behind the theorem is rather intuitive and is worth exploring. First, consider a payoff vector $(v_1, \ldots, v_n)$ that is in the convex hull of the stage-game's payoffs and offers every player more than he can get from a Nash equilibrium of the stage-game, which results in the payoffs $(v_1^*, \ldots, v_n^*)$. By the definition of a convex hull, $(v_1, \ldots, v_n)$ is a weighted average of some combination of payoffs that result from pure-strategy profiles of the stage-game. Second, it turns out that when the discount factor is close to 1, it is possible to construct a sequence of strategy profiles that will result in average payoffs equal to $(v_1, \ldots, v_n)$. The intuition is that the sequence of plays will choose different payoffs of the stage-game's pure strategies in a way that will imitate the weights required to achieve $(v_1, \ldots, v_n)$

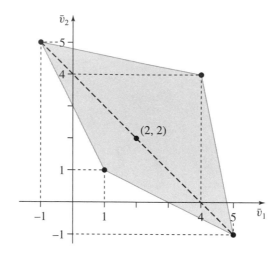

FIGURE 10.5 Constructing the folk theorem for the repeated Prisoner's Dilemma game.

as a convex combination. Third, because this feasible payoff vector $(v_1, \ldots, v_n)$ offers payoffs to each player that exceed his payoff from a Nash equilibrium of the stage-game, this provides us with the "wedge" that is needed to create reward-and-punishment strategies. Finally, with a high enough discount factor, we can make that wedge sufficiently large to deter short-run gains from deviation, no matter how large they are. This last point follows because even the biggest one-stage gain is dwarfed by an infinite series of even small losses when the discount factor is arbitrarily close to 1.

To see this through a simple example, take the Prisoner's Dilemma game described earlier and consider the average payoffs of $(2, 2)$. We will first show that for a discount factor δ that is close enough to 1 we can get average payoffs $(\bar{v}_1, \bar{v}_2)$ that are arbitrarily close to $(2, 2)$. As Figure 10.5 demonstrates, the payoff vector $(2, 2)$ is a convex combination of the payoffs $(5, -1)$ and $(-1, 5)$, each with an equal weight of 0.5. These two pairs of payoffs are achieved with the pure strategies (F, m) and (M, f), respectively. Hence, loosely speaking, it seems reasonable that if somehow the players can spend half their time playing each of these strategies then on average they should get the payoffs $(2, 2)$.

Consider the following strategies:

Player 1: In period 1 play F.
In even periods $t = 2, 4, \ldots$ play M *if and only if* the pattern of play was $(F, m), (M, f), (F, m), \ldots$, and otherwise play F.
In odd periods $t = 3, 5, \ldots$ play F.

Player 2: In period 1 play m.
In odd periods $t = 3, 5, \ldots$ play m *if and only if* the pattern of play was $(F, m), (M, f), (F, m), \ldots$, and otherwise play f.
In even periods $t = 2, 4, \ldots$ play f.

If the two players follow these proposed strategies then their path of play will result in the payoff realizations alternating between $(5, -1)$ and $(-1, 5)$, resulting in the following average payoffs:

$$\bar{v}_1 = (1-\delta) \sum_{t=1}^{\infty} v_t^1$$

$$= (1-\delta)[5 + \delta (-1) + \delta^2(5) + \delta^3(-1) + \cdots]$$

$$= (1-\delta)[5(1 + \delta^2 + \delta^4 + \cdots) + (-1)(\delta + \delta^3 + \delta^5 + \cdots)]$$

$$= (1-\delta)\left[5 \cdot \frac{1}{1-\delta^2} + (-1) \cdot \frac{\delta}{1-\delta^2}\right]$$

$$= \frac{5-\delta}{1+\delta},$$

and similarly for player 2,

$$\bar{v}_2 = (1-\delta)[(-1) + \delta (5) + \delta^2(-1) + \delta^3(5) + \cdots]$$

$$= \frac{-1+5\delta}{1+\delta}.$$

But notice that for both players

$$\lim_{\delta \to 1} \bar{v}_1 = \lim_{\delta \to 1} \bar{v}_2 = 2,$$

which shows that for both players the average payoffs approach (2, 2) as δ gets closer and closer to 1.

We now have to check to see if these strategies form a subgame-perfect equilibrium. As before, we need to check that no player would want to deviate unilaterally at any stage of the game. To do this we will use the one-stage deviation principle and the fact that every subgame is the same infinitely repeated game. It suffices therefore to check for three types of deviations:

1. **The punishment phase:** If there was a deviation from the proposed strategies at some period and the players enter the "punishment phase" (play (F, f) forever), no player will want to deviate because this guarantees a loss of -2 in the deviation period with no gains. (This is similar to the analysis in Section 10.3 that showed how to support the play of (M, m) forever as a subgame-perfect equilibrium.)

2. **A player's "good" period:** On the equilibrium path we have a form of symmetry: every odd period is "good" for player 1 (he gets 5) and "bad" for player 2 (he gets -1), while in every even period their roles are reversed. We first need to check if player 1 (player 2) wants to deviate in an odd (even) period. If he does not deviate then he obtains an average payoff of $\bar{v}_1$. If he deviates then he will get 4 instead of 5 in the deviating period and in subsequent periods the players will play (F, f) forever, giving a stream of 1s in each period instead of alternating between -1 and 5. It is easy to see that this cannot be a good idea: this deviation yields an average payoff of

$$v_1' = (1-\delta)[4 + \delta \cdot 1 + \delta^2 \cdot 1 + \cdots] = (1-\delta)4 + \delta.$$

It is easy to see that for high δ close to 1 the value of v_1' will be itself just barely above 1, while $\bar{v}_1$ is close to 2, implying that this deviation is not profitable. More precisely the player will not want to deviate if and only if $\frac{5-\delta}{1+\delta} \geq (1-\delta)4 + \delta$, which actually holds for all possible values of $\delta \in (0, 1)$.

3. **A player's "bad" period:** We are left to check if player 1 (player 2) wants to deviate in any even (odd) period. If he does not deviate then he obtains an average payoff of $\frac{-1+5\delta}{1+\delta}$. If he deviates then he will get 1 instead of -1 in the deviating period, but in subsequent periods the players will play (F, f) forever, giving a stream of 1s in each period instead of alternating between -1 and 5, so this deviation yields an average payoff of

$$v_1'' = (1-\delta)[1 + \delta \cdot 1 + \delta^2 \cdot 1 + \cdots] = 1.$$

Thus the player will not deviate if $\frac{-1+5\delta}{1+\delta} \geq 1$, which holds if and only if $\delta \geq \frac{1}{2}$.

We have concluded that the payoffs of $(2, 2)$ can (nearly) be achieved by the pair of strategies that we have specified when the discount factor is close enough to 1. The infinite repetition introduced the ability to include reward-and-punishment strategies, which as we have seen are intuitive incentive schemes used when people interact often, and believe that every interaction is likely to be followed by another one.

Remark It is actually possible to approach the payoffs of $(2, 2)$ in the repeated Prisoner's Dilemma even without δ approaching 1. For example, let $\delta = 0.9$. The strategies previously suggested will yield average payoffs of $\overline{v}_1 = \frac{5-0.9}{1+0.9} = 2.158$ and $\overline{v}_2 = \frac{-1+0.9\times5}{1+0.9} = 1.842$. Now imagine that we slightly alter the proposed path of play so that in the even period $t = 26$ instead of playing (F, m) the players will play (F, f). (And of course the strategies described earlier will be modified to account for this change.) In period $t = 26$ player 1 will receive a stage-game payoff of 1 instead of 5, while player 2 will receive a stage-game payoff of 1 instead of -1. It follows that the average payoff of player 1 drops by $\frac{(5-1)\times0.9^{25}}{1+0.9} = 0.151$ while the average payoff of player 2 increases by $\frac{(1-(-1))\times0.9^{25}}{1+0.9} = 0.076$. Hence the revised path gives our two players average payoffs of $\overline{v}_1 = 2.158 - 0.151 = 2.007$ and $\overline{v}_2 = 1.842 + 0.076 = 1.918$. We can continue to fine-tune the payoffs in this way. If in period $t = 36$ we require the players to play (M, m) instead of (M, f) then player 1 will receive a stage-game payoff of 4 instead of 5, while player 2 will receive a stage-game payoff of 4 instead of -1. The average payoff of player 1 drops further by $\frac{(5-4)\times0.9^{35}}{1+0.9} = 0.013$ while the average payoff of player 2 increases by $\frac{(4-(-1))\times0.9^{35}}{1+0.9} = 0.066$, resulting in $\overline{v}_1 = 2.007 - 0.013 = 1.994$ and $\overline{v}_2 = 1.918 + 0.066 = 1.984$. By manipulating payoffs that are further away in a similar fashion we can approach the average payoffs of $(2, 2)$ even when the discount factor is not approaching 1.

10.7 Summary

- If a stage-game that has a unique Nash equilibrium is repeated for finitely many periods then it will have a unique subgame-perfect equilibrium regardless of the discount factor's value.

- If a stage-game that has a unique Nash equilibrium is repeated infinitely often with a discount factor $0 < \delta < 1$, it will be possible to support behavior in each period that is not a Nash equilibrium of the one-shot stage-game.

- The "carrot" and "stick" incentives are created by bootstrapping the repetition of the stage-game's unique equilibrium, which becomes a more potent threat as the discount factor approaches 1.

- The folk theorem teaches us that as the discount factor approaches 1, the set of average payoffs that can be supported by a subgame-perfect equilibrium of the infinitely repeated game grows to a point that almost anything can happen.

- Repeated games are useful frameworks to understand how people cooperate over time, how firms manage to collude in markets, and how reputations for good behavior are sustained over time even when short-run temptations are present.

10.8   Exercises

10.1   **Medicare Drug Policy:** In early 2005 there was a discussion of a proposed U.S. government policy that supported the use of so-called discount cards which pharmaceutical firms could offer senior citizens for the purchase of medications. These cards would have a subscription fee and would in return offer discounts if prescription drugs were bought through the issuing companies. The federal government argued that any of the large pharmaceutical companies could enter this market for discount cards, and this in turn would promote competition. To ensure this the government maintained a web site with posted prices and posted discounts for each card. Some consumer advocates suggested that the companies would just hike up prices and offer a discount from those higher prices, resulting in less benefit for consumers. The government argued that this approach would not make too much sense because there was free entry into the program and resultant competition. Can you argue, using some formal ideas on tacit collusion, that given the way things are set up it is in fact possible—and maybe even easier—for the firms to squeeze more profits out at the expense of consumers?

10.2   **Grim Trigger:** Consider the infinitely repeated game with discount factor $\delta < 1$ of the following variant of the Prisoner's Dilemma:

		Player 2		
		L	C	R
	T	6, 6	−1, 7	−2, 8
Player 1	M	7, −1	4, 4	1, 5
	B	8, −2	5, −1	0, 0

a. For which values of the discount factor δ can the players support the pair of actions (M, C) played in every period?

b. For which values of the discount factor δ can the players support the pair of actions (T, L) played in every period? Why is your answer different from that for (a)?

10.3   **Not-So-Grim Trigger:** Consider the infinitely repeated Prisoner's Dilemma with discount factor $\delta < 1$ described by the following matrix:

Player 2

	m	f
M	4, 4	−1, 5
F	5, −1	1, 1

Player 1

Instead of using grim-trigger strategies to support a pair of actions (a_1, a_2) other than (F, f) as a subgame-perfect equilibrium, assume that the players wish to choose a less draconian punishment called a "length-T punishment" strategy. If there is a deviation from (a_1, a_2) then the players will play (F, f) for T periods and then resume playing (a_1, a_2). Let δ_T be the critical discount factor so that if $\delta > \delta_T$ then the adequately defined strategies will implement the desired path of play with length-T punishment as the threat.

    a.   Let $T = 1$. What is the critical value δ_1 to support the pair of actions (M, m) played in every period?

    b.   Let $T = 2$. What is the critical value δ_T to support the pair of actions (M, m) played in every period?[12]

    c.   Compare the two critical values in (a) and (b). How do they differ and what is the intuition for this?

10.4 **Trust Off the Equilibrium Path:** Recall the trust game depicted in Figure 10.1. We argued that for $\delta \geq \frac{1}{2}$ the following pair of strategies is a subgame-perfect equilibrium. For player 1: "In period 1 I will trust player 2, and as long as there were no deviations from the pair (T, C) in any period, then I will continue to trust him. Once such a deviation occurs then I will not trust him forever after." For player 2: "In period 1 I will cooperate, and as long as there were no deviations from the pair (T, C) in any period, then I will continue to do so. Once such a deviation occurs then I will deviate forever after." Show that if instead player 2 uses the strategy "as long as player 1 trusts me I will cooperate" then the path (T, C) played forever is a Nash equilibrium for $\delta \geq \frac{1}{2}$ but is not a subgame-perfect equilibrium for any value of δ.

10.5 **Negative Ad Campaigns Revisited:** Recall exercise 5.14 from Chapter 5, in which each one of two political parties can choose to buy time on commercial radio shows to broadcast negative ad campaigns against its rival. These choices are made simultaneously. Government regulations forbid a party from buying more than 2 hours of negative campaign time, so that each party cannot choose an amount of negative campaigning above 2 hours. Given a pair of choices (a_1, a_2), the payoff of party i is given by the following function: $v_i(a_1, a_2) = a_i - 2a_j + a_i a_j - (a_i)^2$.

    a.   Find the unique pure-strategy Nash equilibrium of the one-shot game.

    b.   If the parties could sign a binding agreement on how much to campaign, what levels would they choose?

    c.   Now assume that this game is repeated infinitely often, and the foregoing demonstrates the choices and payoffs per period. For which

12. Hint: You should encounter an equation of the form $a\delta^3 - (a + 1)\delta + 1 = 0$, for which it is easy to see that $\delta = 1$ is a root. In this case you know that the equation can be written in the form $(\delta - 1)(a\delta^2 + a\delta - 1) = 0$, and you can solve for the other relevant root of the cubic equation.

discount factors $\delta \in (0, 1)$ can the levels you found in (b) be supported as a subgame-perfect equilibrium of the infinitely repeated game?

d. Despite the parties' ability to coordinate, as you have demonstrated in your answer to (c), the government is concerned about the parties' ability to run up to 2 hours a day of negative campaigning, and it is considering limiting negative campaigning to $\frac{1}{2}$ hour a day, so that now $a_i \in [0, \frac{1}{2}]$. Is this a good policy to limit negative campaigns further? Justify your answer with the relevant calculations. What is the intuition for your conclusion?

10.6 **Regulating Medications:** Consider a firm (player 1) that produces a unique drug that is used by a consumer (player 2). This drug is regulated by the government so that its price is $p = 6$. This price is fixed, but the quality of the drug depends on the manufacturing procedure. The "good" (G) manufacturing procedure costs 4 to the firm and yields a value of 9 to the consumer. The "bad" (B) manufacturing procedure costs 0 to the firm and yields a value of 4 to the consumer. The consumer can choose whether or not to buy at the price p, and this decision must be made before the actual manufacturing procedure is revealed. However, after consumption, the true quality of the drug is revealed to the consumer. The choice of manufacturing procedure, and thus of the cost of production, is made before the firm knows whether the consumer will buy or not.

a. Draw the game tree and the matrix of this game, and find all the Nash equilibria.

b. Now assume that the game is repeated twice. (The consumer learns the quality of the product in each period only if he consumes.) Assume that each player tries to maximize the (nondiscounted) sum of his stage payoffs. Find *all* the subgame-perfect equilibria of this game.

c. Now assume that the game as repeated infinitely many times. Assume that each player tries to maximize the discounted sum of his or her stage payoffs, where the discount rate is $\delta \in (0, 1)$. What is the range of discount factors for which the good manufacturing procedure will be used as part of a subgame-perfect equilibrium?

d. Consumer advocates are pushing for a lower price for the drug, say 5. The firm wants to approach the Food and Drug Administration and argue that if the regulated price is decreased to 5 then this may have dire consequences for both consumers and the firm. Can you make a formal argument, using the given parameters, to support the firm? What about the consumers?

10.7 **Diluted Happiness:** Consider a relationship between a bartender and a customer. The bartender serves bourbon to the customer and chooses $x \in [0, 1]$, which is the proportion of bourbon in the drink served, while $1 - x$ is the proportion of water. The cost of supplying such a drink (standard 4-ounce glass) is cx, where $c > 0$. The customer, without knowing x, decides on whether or not to buy the drink at the market price p. If he buys the drink his payoff is $vx - p$, and the bartender's payoff is $p - cx$. Assume that $v > c$ and all payoffs are common knowledge. If the customer does not buy the drink he gets 0 and the bartender gets $-cx$. Because the customer has some experience, once the drink is bought and he tastes it, he learns the value of x, but this is only after he pays for the drink.

    a. Find all the Nash equilibria of this game.

    b. Now assume that the customer is visiting town for 10 days, and this "bar game" will be played on each of the 10 evenings that the customer is in town. Assume that each player tries to maximize the (non-discounted) sum of his stage payoffs. Find all the subgame-perfect equilibria of this game.

    c. Now assume that the customer is a local, and the players perceive the game as repeated infinitely many times. Assume that each player tries to maximize the discounted sum of his stage payoffs, where the discount rate is $\delta \in (0, 1)$. What is the range of prices p (expressed in the parameters of the problem) for which there exists a subgame-perfect equilibrium in which every day the bartender chooses $x = 1$ and the customer buys at the price p?

    d. For which values of δ (expressed in the parameters of the problem) can the price range that you found in (c) exist?

10.8 **Tacit Collusion:** Two firms, which have zero marginal cost and no fixed cost, produce some good, each producing $q_i \geq 0$, $i \in \{1, 2\}$. The demand for this good is given by $p = 200 - Q$, where $Q = q_1 + q_2$.

    a. First consider the case of Cournot competition, in which each form chooses q_i and this game is infinitely repeated with a discount factor $\delta < 1$. Solve for the static stage-game Cournot-Nash equilibrium.

    b. For which values of δ can you support the firms' *equally splitting* monopoly profits in each period as a subgame-perfect equilibrium that uses grim-trigger strategies (i.e., after one deviates from the proposed split, they resort to the static Cournot-Nash equilibrium thereafter)? (Note: Be careful in defining the strategies of the firms!)

    c. Now assume that the firms compete à la Bertrand, each choosing a price $p_i \geq 0$, where the lowest-price firm gets all the demand and in case of a tie they split the market. Solve for the static stage-game Bertrand-Nash equilibrium.

    d. For which values of δ can you support the firms' *equally splitting* monopoly profits in each period as a subgame-perfect equilibrium that uses grim-trigger strategies (i.e., after one deviates from the proposed split, they resort to the static Bertrand-Nash equilibrium thereafter)? (Note: Be careful in defining the strategies of the firms!)

    e. Now instead of using grim-trigger strategies, try to support the firms' equally splitting monopoly profits as a subgame-perfect equilibrium in which after a deviation firms would resort to the static Bertrand competition for *only two periods*. For which values of δ will this work? Why is this answer different than your answer in (d)?

10.9 **Negative Externalities:** Two firms are located adjacent to one another, and each imposes an external cost on the other: the detergent that firm 1 uses in its laundry business makes the fish that firm 2 catches in the lake taste funny, and the smoke that firm 2 uses to smoke its caught fish makes the clothes that firm 1 hangs out to dry smell funny. As a consequence each firm's profits are increasing in its own production and decreasing in the production of its neighbor. In particular if q_1 and q_2 are the firms' production levels then their per-period (stage-game) profits are given by $v_1(q_1, q_2) = (30 - q_2)q_1 - q_1^2$ and $v_2(q_1, q_2) = (30 - q_1)q_2 - q_2^2$.

a. Draw the firms' best-response functions and find the Nash equilibrium of the stage-game. How does this compare to the Pareto-optimal stage-game profit levels?

b. For which levels of discount factors can the firms support the Pareto-optimal level of quantities in an infinitely repeated game?

10.10 **Law Merchants Revisited:** Consider the three-person game described in Section 10.5.2. A subgame-perfect equilibrium was constructed with a bond equal to 2, a fee equal to $w = 0.1$ was paid by every player P_2^t to player 3, and it was shown that it is indeed an equilibrium for any discount factor $\delta \geq 0.95$. Show that a similar equilibrium exists in which players P_1^t trust players P_2^t who post bonds, players P_2^t post bonds and cooperate, and player 3 follows the contract in every period, for any discount factor $\frac{1}{2} < \delta < 1$.

10.11 **Trading Brand Names:** Show that the strategies proposed in Section 10.5.3 constitute a subgame-perfect equilibrium of the sequence of trust games.

10.12 **Folk Theorem Revisited:** Consider the infinitely repeated trust game described in Figure 10.1.

a. Draw the convex hull of average payoffs.

b. Are the average payoffs $(\overline{v}_1, \overline{v}_2) = (-0.4, 1.1)$ in the convex hull of average payoffs? Can they be supported by a pair of strategies that form a subgame-perfect equilibrium for a large enough discount factor δ?

c. Show that there is a pair of subgame-perfect equilibrium strategies for the two players that yields average payoffs that approach $(\overline{v}_1, \overline{v}_2) = \left(\frac{1}{3}, \frac{4}{3}\right)$ as δ approaches 1.

11

Strategic Bargaining

O f the many possible applications of strategic interaction among a small number of players, one ubiquitous situation surely comes to mind: bargaining. Examples are plentiful: a firm and a union bargaining over wages and benefits; a local municipality bargaining with a private provider over the terms of service; the head of a political party bargaining with other party members over campaign issues; and even the mundane bargaining that occurs between a buyer and a merchant at a street bazaar. In all of these examples, deals are struck by parties bargaining over terms, payments, and other aspects that all boil down to one basic fact: each party wants to get the best deal it can.

How can we model and analyze bargaining through the lens of game theory? As demonstrated by the examples just mentioned, the issue is often about a surplus that has to be split among the parties, and it involves the parties making proposals, responding to them, and trying to settle on an agreement. Following some early work by Ståhl (1972, 1977), Rubinstein (1982) offered a particular stylized model of bargaining that has become the standard in most applications. The simple framework considers two players that need to split a "pie" (representing the surplus from an agreement or the gains from trade). The pie is assumed to have a total value that is normalized to equal 1, and then the parties bargain on how to split the pie.

In order to analyze this process as a strategic game, the bargaining process itself needs to be structured to include actions, outcomes that result from actions, and preferences over the outcomes, as with any other game we can imagine. Ståhl suggested a prespecified procedure in which the game starts with player 1 making an offer to split the pie and player 2 either accepting or rejecting the offer. If the offer is accepted, the split proposed by player 1 is implemented, and the game ends. If player 2 rejects player 1's offer, the players then switch roles: player 2 makes the offer and player 1 responds by accepting or rejecting the offer. The game can either continue this way until some exogenous deadline arrives or simply go on forever.

Yet, as the saying goes, "time is money," and delay ought to impose some loss on the players. There are two ways to impose this loss. First, there may be a fixed cost of progressing from one round to another, so that a constant piece of the pie is removed every time a rejection occurs. Alternatively we can impose discounting in the same way we did for repeated games: after every period of rejection, the total size of the pie is only a fraction of what it was before. This second approach is the one we will be following here, and we can summarize the game as follows.

In the first round:

- Player 1 offers shares $(x, 1 - x)$, where player 1 receives x and player 2 receives $1 - x$.
- Player 2 then chooses to accept, causing the game to end with payoffs $v_1 = x$ and $v_2 = 1 - x$, or reject, causing the game to move to the second round.

In the second round:

- A share $1 - \delta$ of the pie is dissipated (the pie is discounted with discount factor $0 < \delta < 1$).
- Player 2 offers shares $(x, 1 - x)$, where player 1 receives x and player 2 receives $1 - x$.
- Player 1 chooses to accept, causing the game to end with payoffs $v_1 = \delta x$ and $v_2 = \delta(1 - x)$, or reject, causing the game to move to the third round.

In the third round and beyond:

- The game continues in the same way, where following rejection in an odd period player 2 gets to offer in the next even period, and vice versa.
- Each period has further discounting of the pie, so that in period t the total pie is worth δ^{t-1}.

From the specification of the bargaining procedure we see that once a player agrees to the offer made by his opponent, the game will end with an agreement. If instead the players keep rejecting offers then the game may continue either indefinitely or until some prespecified deadline is reached. Assume for now that the game ends in some prespecified final period T, which will be reached if the players did not reach an agreement in the $T - 1$ periods that preceded T. Furthermore assume that period T is a "hard deadline," so that both players receive a payoff of zero if they fail to reach an agreement in this last period. This, for example, can be the case with a fisherman bargaining with a restaurateur over a fresh catch. If they wait more than a few hours, the fish will go bad and there will no longer be gains from trade. The heuristic extensive form of this game with an odd number of periods T (so that player 1 is first and last) is depicted in Figure 11.1.

Notice that this game has an interesting structure that is different from what we have seen so far. On the one hand, it has some features of a finitely repeated game as follows: If we think of an "odd" round as one in which player 1 proposes and player 2 responds, and an "even" round as the converse, then we can treat each *pair of rounds* as *one stage* that repeats itself as long as an agreement is not reached. Furthermore as rounds proceed the value of the pie that these players need to split is shrinking according to the discount factor δ. On the other hand, there are two features that are not part of the repeated-game structure we developed in Chapter 10. First, the game can end during any round if the proposal is accepted, and second, payoffs are obtained only when the game actually ends and not as a flow of stage-game payoffs.

Before proceeding with the analysis, an obvious question is whether this stylized bargaining game is a reasonable caricature of reality. We need to convince ourselves that this game is, at least at some level, representative of what we believe bargaining is about. To a large extent, sequences of offers and counteroffers are natural components of any type of bargaining, and clearly once agreements are reached the bargaining stage is over. However, the fixed end date T, at which an all-or-nothing agreement

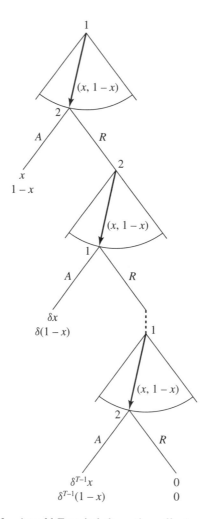

FIGURE 11.1 An odd T-period alternating-offer bargaining game.

is reached, is not extremely appealing. For simplicity, we will first analyze the game with a finite end date; later we will extend the analysis to eliminate this artificial termination stage.

11.1 One Round of Bargaining: The Ultimatum Game

We begin the analysis with an easy case in which there is only one round of bargaining, so that $T = 1$. This is often referred to in the literature as an **ultimatum game:** player 1 makes an ultimatum take-it-or-leave-it offer to player 2, and player 2 either accepts it, in which case the pie is split between the players, or rejects it, in which case both players receive a payoff of zero. Because this is a game of perfect information (player 2 sees the offer of player 1 before he needs to accept or reject it), we can apply backward induction to find which paths of play can be supported by a subgame-perfect equilibrium.

As a benchmark, we begin the analysis by finding which paths of play can be supported by a Nash equilibrium without requiring sequential rationality. The result is quite striking:

Proposition 11.1 *In the bargaining game if $T = 1$ then any division of surplus $x^* \in [0, 1]$, $(v_1, v_2) = (x^*, 1 - x^*)$, can be supported as a Nash equilibrium.*

Proof We will construct a pair of strategies that are mutual best responses and that lead to $(x^*, 1 - x^*)$ as the division of surplus. Let player 1's strategy be "I propose x^*," and let player 2's strategy be "I accept any offer $x \leq x^*$ and reject any offer $x > x^*$." It is easy to see that these two strategies are mutual best responses independent of the value of $x^* \in [0, 1]$. ∎

This proposition tells us that the concept of Nash equilibrium has no bite for this simple one-stage bargaining game because it can rationalize *any division of surplus*. In other words, we cannot predict in any meaningful way what the outcome of such a game will be if we require only that people choose strategies that are mutual best responses.

That said, a close observation of the strategies constructed to support an arbitrary division of surplus will immediately reveal that sequential rationality is almost always violated: Player 2 is saying "there is a minimum that I am willing to accept." But what if player 1 offers him less? In particular what if he offers him $(1 - x) = \varepsilon > 0$ with ε being a very small number? If player 2 rejects this offer he will get 0, while if he accepts it he will get a payoff of $\varepsilon > 0$, implying that his best response is to accept *any strictly positive payoff*. Anticipating this, player 1 should offer player 2 the smallest possible amount. This logic leads to the following result:

Proposition 11.2 *The bargaining game with $T = 1$ admits a unique subgame-perfect equilibrium in which player 1 offers $x = 1$ and player 2 accepts any offer $x \leq 1$.*

Proof We have established that player 2 must accept any positive share $(x < 1)$. Player 2 is indifferent between accepting or rejecting $x = 1$, so the proposed strategy is sequentially optimal, and the unique best response of player 1 to player 2's strategy is to offer $x = 1$. The only other sequentially rational strategy for player 2 is to accept any strictly positive share $(x < 1)$ and reject getting 0 $(x = 1)$. But player 1 does not have a best response to this strategy,[1] and therefore it cannot be part of a subgame-perfect equilibrium. ∎

This result is very stark, especially when compared with the analysis of Nash equilibrium. Requiring only mutual best responses yields no meaningful prediction, but sequential rationality predicts a unique and extreme outcome: player 2 should accept anything, and thus player 1 has an extreme form of take-it-or-leave-it first-mover advantage, giving him the whole pie of surplus.

Remark Many experiments have been performed by researchers to see what behavior actually prevails for the ultimatum game, in which player 1 can offer to split a sum of money with player 2. If player 2 accepts then the split is realized and both players receive the proposal of player 1. If player 2 rejects then the players receive

---

1. This is similar to what happens in the asymmetric-cost Bertrand competition model presented in Section 5.2.4, in which player 1's payoff function is discontinuous, and hence he does not have a well-defined best response. To see it in this context, imagine that player 2's strategy is to accept any offer $x < 1$ and reject $x = 1$. Clearly $x = 1$ is not a best response because any offer $x < 1$ is better. The problem is that no offer $x < 1$ is a best response either. If player 1 offers some $x' < 1$, then the offer $x = x' + \frac{1-x'}{2}$ is better than x'. The conclusion is that player 1's payoff function is discontinuous at $x = 1$, and he does not have a best response to player 2's strategy.

nothing. Contrary to the theory, and maybe not too surprisingly, the experiments show that those players in the role of player 1 offer significant shares to those in the role of player 2, typically between 25% and 50% of the total value. Furthermore, when player 2 is offered a small amount, typically less than 20–25% of the total value, then he will reject these offers. This suggests that subjects in the role of player 2 act with some "spite" when they are offered low shares because they prefer getting nothing to an outcome in which player 1 gets a large fraction of the total value. See Camerer (2003) for an excellent summary of the experimental results and some theories that predict this behavior.

11.2 Finitely Many Rounds of Bargaining

When we consider bargaining games that extend beyond one period but that will still end at some specified date $T < \infty$, the Nash equilibrium concept continues to have no predictive power. In particular we can easily construct the kind of strategies used in the proof of proposition 11.1 to *any* horizon, *including an infinite horizon*. To construct a Nash equilibrium that results in an arbitrary division of surplus $x^* \in [0, 1]$ we can use the following time-independent strategies: In every odd round in which player 1 proposes, player 1 will propose x^* and player 2 will accept proposals of $x \leq x^*$. In every even round in which player 2 proposes, player 2 will propose x^* and player 1 will accept proposals of $x \geq x^*$. It is easy to see that these strategies are mutual best responses, and an agreement is reached in the first round with the division of surplus $(x^*, 1 - x^*)$.

What is even more disappointing is that, using the Nash concept, we cannot even pin down when an agreement will be reached. For example, we can adopt the strategies that result in any $(x^*, 1 - x^*)$ with the following modification: in the first period player 1 offers $x = 1$ and player 2 rejects *any* offer. It is easy to see that these form a pair of best responses, resulting in a Nash equilibrium in which an agreement is reached in the second period. In exercise 11.1, you are asked to show that similar strategies can be constructed to support an agreement of *any* division of surplus in *any* period.

Imposing sequential rationality will, once again, offer a stark contrast. Indeed it should not be surprising that applying sequential rationality and subgame perfection to any finite-length bargaining game will result in a unique outcome, because backward induction will apply in a similar way that it did for the case of one round of bargaining. In the last round the proposer will offer to keep the entire pie to himself, the responder will agree, and the solution will proceed sequentially using backward induction.

In the one-round ultimatum game analyzed previously, player 1 got all the surplus in the unique subgame-perfect equilibrium because by construction the only round is the last round. Sequential rationality implies that in the last round the player making the offer has all the "bargaining power," which in turn implies that he will get the whole surplus. We can ask then, what will happen in a game with two rounds of bargaining?

As Figure 11.2 demonstrates, the second round is a one-round bargaining game, which implies that *if it is reached* then player 2 gets the whole pie in any subgame-perfect equilibrium. Continuing with backward induction, and assuming that $\delta \leq 1$, we can find the subgame-perfect equilibrium for the two-round game as follows. In the first round player 2 knows that in the next (and last) round he will receive the whole pie, which will be worth only δ because of discounting. Thus if he is offered less than δ for himself (player 1 offers $x > 1 - \delta$) then player 2 should reject the offer and get the whole pie in the next period. Hence the unique sequentially rational strategy for

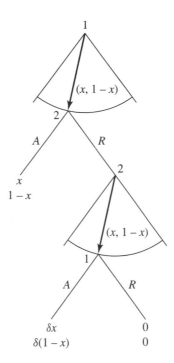

FIGURE 11.2 An alternating-offer bargaining game with two periods.

player 2 at the accept/reject stage of the first round is to accept any offer of $x \leq 1 - \delta$. Moving up the tree to the node at which player 1 makes his first offer, he realizes that any offer $x > 1 - \delta$ will be rejected, while any offer $x \leq 1 - \delta$ will be accepted. Thus player 1's best response is to offer $x = 1 - \delta$, leading to a division of surplus equal to $(v_1, v_2) = (1 - \delta, \delta)$.

Note that if $\delta = 1$ then player 2 gets the whole pie because he then has the ultimate form of a take-it-or-leave-it advantage, and with $\delta < 1$ player 1 has a bit of a first-mover advantage owing to the discounting. This observation is useful because it suggests an interesting pair of forces. Essentially the backward induction logic that we have used for the two-round bargaining model will generalize to any finite-round game. Therefore a tension will emerge between the ability to make a last take-it-or-leave-it offer and the first-mover advantage that is a result of discounting given $\delta < 1$. Whether player 1 or player 2 is the last mover will artificially depend on whether the game has an odd or even number of periods.

Consider the case with an odd number of rounds $T < \infty$, implying that player 1 has both the first-mover and last-mover advantages. The following backward induction argument applies:

- In period T player 2 accepts any offer, so player 1 offers $x = 1$ and payoffs are $v_1 = \delta^{T-1}$ and $v_2 = 0$.

- In period $T - 1$ (even period, player 2 offers), by backward induction player 1 should accept anything resulting in a payoff of $v_1 \geq \delta^{T-1}$. If player 1 is offered x in period $T - 1$ then $v_1 = \delta^{T-2}x$. This implies that in period $T - 1$ player 1 will accept any $x \geq \delta$ and by backward induction player 2 should offer $x = \delta$, which yields player 1 a payoff of $v_1 = \delta \times \delta^{T-2} = \delta^{T-1}$. Payoffs are then $v_1 = \delta^{T-1}$ and $v_2 = \delta^{T-2}(1 - \delta)$.

- In period $T - 2$ (odd period), conditional on the analysis for $T - 1$, player 2's best response is to accept any x that gives him $\delta^{T-3}(1 - x) \geq \delta^{T-2}(1 - \delta)$. Player 1's best response to this is to offer the largest x that satisfies this inequality, and solving it with equality yields player 1's best response: $x = 1 - \delta + \delta^2$. This offer followed by player 2's acceptance yields $v_1 = \delta^{T-3}x = \delta^{T-3} - \delta^{T-2} + \delta^{T-1}$ and $v_2 = \delta^{T-2} - \delta^{T-1}$.

- In period $T - 3$ (even period), using the same backward induction argument, player 1's best response is to accept any x that satisfies $\delta^{T-4}x \geq \delta^{T-3} - \delta^{T-2} + \delta^{T-1}$, and player 2 then makes an offer that satisfies this constraint with equality, $x = \delta - \delta^2 + \delta^3$, yielding $v_1 = \delta^{T-3} - \delta^{T-2} + \delta^{T-1}$ and $v_2 = \delta^{T-4}(1 - x) = \delta^{T-4} - \delta^{T-3} + \delta^{T-2} - \delta^{T-1}$.

We can continue with this tedious exercise only to realize that a simple pattern emerges. If we consider the solution for an odd period $T - s$ (s being even because T is assumed to be odd) then the backward induction argument leads to the sequentially rational offer

$$x_{T-s} = 1 - \delta + \delta^2 - \delta^3 + \cdots + \delta^s,$$

while for an even period $T - s$ (s being odd) then the backward induction argument leads to the sequentially rational offer

$$x_{T-s} = \delta - \delta^2 + \delta^3 - \delta^4 + \cdots + \delta^s.$$

We can use this pattern to solve for the subgame-perfect equilibrium offer in the first period, x_1, which by backward induction must be accepted by player 2 and it is equal to

$$
\begin{aligned}
x_1 &= 1 - \delta + \delta^2 - \delta^3 + \delta^4 + \cdots + \delta^{T-1} \\
&= (1 + \delta^2 + \delta^4 + \cdots + \delta^{T-1}) - (\delta + \delta^3 + \delta^5 + \cdots + \delta^{T-2}) \\
&= \frac{1 - \delta^{T+1}}{1 - \delta^2} - \frac{\delta - \delta^T}{1 - \delta^2} \\
&= \frac{1 + \delta^T}{1 + \delta},
\end{aligned}
$$

and this in turn implies that

$$v_1^* = x_1 = \frac{1 + \delta^T}{1 + \delta} \qquad \text{and} \qquad v_2^* = (1 - x_1) = \frac{\delta - \delta^T}{1 + \delta}. \tag{11.1}$$

We can now offer some insights into this solution, which has some very appealing properties. The first observation is important:

Proposition 11.3 *Any subgame-perfect equilibrium must have the players reach an agreement in the first round.*

The proof follows a simple logic. If an agreement is reached in a later round with payoffs (v_1', v_2') then discounting implies that part of the surplus is wasted and that $v_1' + v_2' < 1$. But then player 1 could deviate and offer $x = 1 - v_2' - \varepsilon$ for some small $\varepsilon > 0$, which guarantees player 2 the payoff $v_2' + \varepsilon$ immediately in the first round. Sequential rationality implies that player 2 should accept this immediately, and for ε small enough this gives player 1 a payoff greater than v_1'.

A second observation follows from the fact that player 1 has both the last-mover take-it-or-leave-it advantage and the first-mover discounting advantage. This implies that $v_1^* > v_2^*$ for any discount factor $\delta \in [0, 1]$, as can be seen from the equilibrium values in (11.1). Because T is fixed, however, it is hard to isolate these two effects, especially if we wish to measure the way in which patience, represented by δ, affects the equilibrium payoffs. For example, if players are very impatient and $\delta = 0$ then the first-mover advantage kicks in with full force. Regardless of T, this is basically equivalent to a one-round game in which player 1 gets the whole surplus. On the other hand, if we consider the limit of $\delta = 1$ then $v_1^* = 1$ and $v_2^* = 0$, because no discounting implies that moving last guarantees the whole surplus, regardless of the length of the game T.

This last observation is somewhat disturbing: if players are very patient then the last mover's take-it-or-leave-it advantage just flows up the game tree no matter how long the game is! This rather counterintuitive result is precisely due to the artificial stopping period T. If $\delta = 1$ then the only thing that matters is who has the last word—there is no loss from waiting, and hence the forward-looking players know that the last person to make an offer holds all the bargaining power.

To overcome this artificially dictated outcome, we will first address the source of the problem by fixing the discount factor δ and observing what happens to the equilibrium payoffs as the game gets longer and longer. More precisely,

$$\lim_{T \to \infty} v_1^* = \lim_{T \to \infty} \frac{1 + \delta^T}{1 + \delta} = \frac{1}{1 + \delta} \quad \text{and} \quad \lim_{T \to \infty} v_2^* = \lim_{T \to \infty} \frac{\delta - \delta^T}{1 + \delta} = \frac{\delta}{1 + \delta}.$$

$$(11.2)$$

As we can see from the limit payoffs in (11.2), the artificial end-period effect disappears. We can clearly observe the first-mover discounting advantage in that for any $0 < \delta < 1$ it must be that $\lim_{T \to \infty} v_1^* > \lim_{T \to \infty} v_2^*$. We can also use the limiting equilibrium payoffs to ask what happens as the patience of the players changes without the problem posed by an artificial end period T. For the extreme case of impatience with $\delta = 0$ we get the same result: the game is equivalent to a one-round ultimatum game. When the players become very patient, we now obtain a more sensible result,

$$\lim_{\delta \to 1} \lim_{T \to \infty} v_1^* = \lim_{\delta \to 1} \frac{1}{1 + \delta} = \frac{1}{2} \quad \text{and} \quad \lim_{\delta \to 1} \lim_{T \to \infty} v_2^* = \lim_{\delta \to 1} \frac{\delta}{1 + \delta} = \frac{1}{2}.$$

$$(11.3)$$

The limit payoffs in (11.3) suggest a rather appealing and intuitive result for very long horizons with very patient players: the very long horizon eliminates the last mover's take-it-or-leave-it advantage for any given discount factor δ, leaving only the first-mover advantage due to discounting. If we take the level of patience to its limit of $\delta = 1$ then this eliminates the first-mover advantage as well, and we have the two players splitting the pie equally in the unique subgame-perfect equilibrium.[2]

As exercise 11.3, you can solve for the subgame-perfect equilibrium when the number of periods is even, and see that at the limit we get the same results.

---

2. Note that to reach this appealing conclusion the order of limits matters. If we reverse the order of limits and take $\delta \to 1$ first, followed by $T \to \infty$ second, then $\lim v_1^* = 1$ and $\lim v_2^* = 0$, preserving the artificial take-it-or-leave-it advantage.

11.3 The Infinite-Horizon Game

We now analyze the infinite-horizon bargaining game in which, if an agreement is never reached, the players continue to alternate roles indefinitely. In this case the assumption that disagreement leads to zero payoffs is natural because of discounting: if the two players disagree forever then there is nothing left on which to agree. However, there is a crucial difference between analyzing the infinite-horizon game and considering the infinite limit of the finite-horizon game: the path of perpetual disagreement is of infinite length, and we cannot therefore apply a backward induction argument to find the subgame-perfect equilibrium.

There is, however, an interesting feature of the infinite-horizon bargaining game: the *stationary structure of the game* following a disagreement at any stage. Every odd period is the same, with player 1 making an offer, and the continuation game has a potentially infinite horizon. Similarly every even period is the same, with player 2 making the offer. This allows us to apply a rather appealing, and not too difficult, logic to solve for what turns out to be the unique subgame-perfect equilibrium of the infinite-horizon bargaining game.[3]

The first important observation is that, following the same logic we introduced earlier in proposition 11.3, sequential rationality implies that an agreement must be reached in the first period. The reason is that wasteful discounting will not be tolerated because it can be avoided in the first stage. Hence any subgame-perfect equilibrium must have agreement reached in the first period.

Now imagine that there is more than one subgame-perfect equilibrium. If this were the case then when player 1 makes an offer he must have a *best* subgame-perfect equilibrium, yielding him a value of $\bar{v}_1$, and a *worst* subgame-perfect equilibrium, yielding him a value of $\underline{v}_1$. Similarly when player 2 makes an offer after rejecting an offer made by player 1 then he too has a *best* subgame-perfect equilibrium, yielding him a value of $\bar{v}_2$, and a *worst* subgame-perfect equilibrium, yielding him a value of $\underline{v}_2$.

This is where the stationary structure of the game is useful, because it implies that $\bar{v}_1 = \bar{v}_2 = \bar{v}$ and $\underline{v}_1 = \underline{v}_2 = \underline{v}$. The reason is that in any odd period the game looks exactly the same as in any even period, just with the identities of the players switched. Hence if the structure of the game results in the existence of different subgame-perfect equilibria that support certain payoffs for player 1 starting in period 1, the same must be true for player 2 in period 2.

Now we take advantage of the fact that what one player gives up, the other player must receive (and recall that agreement must be reached in the first period), which results in a straightforward relationship between the best- and worst-payoff subgame-perfect equilibria of the game. What is worst for player 1 must be best for player 2 and vice versa. Hence in the subgame-perfect equilibrium that supports the payoff $\underline{v}$ for player 1 in period 1, it must be true that if player 2 rejects player 1's offer, he can secure himself a payoff of $\bar{v}$ when it is his turn to make the offer.

Immediate agreement implies that player 1 must offer player 2 a share of the pie that will deter player 2 from rejecting player 1's offer. If player 2 can get $\bar{v}$ after rejection, this implies that player 1 must offer player 2 a payoff of at least $\delta \bar{v}$ in the first period. Of course player 1 will offer the least he can to secure an agreement,

---

3. The method used here was developed by Shaked and Sutton (1984), but the infinite model was proposed, and the unique solution was identified, in the seminal paper by Rubinstein (1982).

which implies that he will offer player 2 a share exactly equal to $\delta\overline{v}$, which in turn implies that player 1's payoff from this subgame-perfect equilibrium must be

$$\underline{v} = 1 - \delta\overline{v}. \tag{11.4}$$

A symmetric argument applies for the subgame-perfect equilibrium that results in player 1 obtaining a payoff of $\overline{v}$ and player 2 obtaining a payoff of $\underline{v}$, which results in the equation

$$\overline{v} = 1 - \delta\underline{v}. \tag{11.5}$$

Taking (11.4) and (11.5) we obtain that

$$\overline{v} = \underline{v} = \frac{1}{1 + \delta}.$$

This result characterizes the unique subgame-perfect equilibrium of the infinite-horizon bargaining game. Notice that it coincides with the limit of the unique subgame-perfect equilibrium of the finite-horizon bargaining game as shown in Section 11.2. Furthermore because we know that agreement is reached in the first round we can use this payoff characterization to spell out the strategies that together support this outcome. In the unique subgame-perfect equilibrium the strategies are as follows: In each odd round player 1 offers to keep $x = \frac{1}{1+\delta}$ to himself and player 2 accepts the offer if $x \leq \frac{1}{1+\delta}$. In each even round player 2 offers to give player 1 $x = \frac{\delta}{1+\delta}$ and player 1 accepts any $x \geq \frac{\delta}{1+\delta}$.

Remark   As you might have imagined, bargaining games have received considerable attention in the literature, in both economics and political science, because bargaining situations are so ubiquitous. Many variants of the bargaining game described here have been analyzed, in which the relationship might break down, the players each have an "outside option" that is greater than zero, there are more than two players that can be matched randomly, . . . the list goes on and on. Nash (1950b) offered an analysis of the bargaining problem, but interestingly he did not follow the framework of noncooperative game theory that he himself had founded. Instead he offered an axiomatic approach to suggest a sensible solution to the two-person bargaining problem. The resulting "Nash bargaining solution" is widely used in economics. For a broad treatment of bargaining games and solutions see Osborne and Rubinstein (1990) and Muthoo (1999).

11.4   Application: Legislative Bargaining

The bargaining model analyzed here offers a reasonable depiction of many two-person bargaining situations. In particular there is the potential for give-and-take, and at the end an agreement has to be reached by both parties for the deal to be sealed. Another setting in which bargaining is ubiquitous is in legislatures, where lawmakers bargain over the allocation of surplus through either bills, budget agreements, or regulations. In most political settings the convention is that once a majority of players concur on an offer, the allocation is agreed upon despite the dismay of those who may still oppose it.

In an important contribution, Baron and Ferejohn (1989) build on the Rubinstein-Ståhl model of bargaining and offer a framework for analyzing multilateral bargaining with a simple majority rule. The two main departures from the two-player model are as follows: First, there is an odd number N of players, where $\frac{N+1}{2}$ votes are needed to secure an agreement. Second, instead of players alternating their roles in a prespecified order, the bargaining protocol follows a random-assignment rule: in every period each player has an equal probability of being the proposer.

Baron and Ferejohn consider two bargaining rules that to a large extent mimic common practices: the closed rule, under which proposals cannot be amended, and the open rule, under which they can. In what follows, we adhere to the approach used in Section 11.3 to analyze the outcomes and expected payoffs from a potentially infinite-horizon model. The technique is to use the payoffs that some players expect to get if they reject an offer to determine what the equilibrium payoffs must be to support the proposed equilibrium behavior.

11.4.1   Closed-Rule Bargaining

The closed-rule setting is very much an N-player version of the Rubinstein infinite-horizon model. When called upon to propose, the proposer makes an offer to the legislative body of the form $x = (x_1, x_2, \ldots, x_N)$, where x_i is the share of surplus offered to player i, with $\sum_{i=1}^{n} x_i \leq 1$. If a majority accepts the proposal then the game ends, and if not then the current "pie" is discounted according to the discount factor δ.

It turns out that when there are three or more players in an N-player version of the Rubinstein infinite-horizon bargaining game then there are many possible subgame-perfect equilibria.[4] What Baron and Ferejohn suggest is a simple way around this indeterminacy. They focus attention on *symmetric stationary equilibria* in which first, each player proposes the same split of the pie every time he is called to propose, regardless of the history of the game, and second, the respondents vote based only on the current proposal as it compares to their expectations about future proposals. What is clever about these assumptions is that together they imply that, like the two-player version of Rubinstein's model, the game effectively "starts over" every time a proposal is rejected, and because the proposer is chosen randomly at the beginning of every stage, the continuation value of each player is the expected utility of playing the game.[5]

Let v be the expected payoff for any player i to play the subgame-perfect equilibrium of this game at the beginning of any stage,[6] and consider a player i who is choosing to respond to a proposal that awards him x_i. If he rejects the proposal then

---

4. This was noted first by Herrero (1985). For a large number of players and a discount factor close to 1, any split of the pie can be supported by a subgame-perfect equilibrium. Showing this is beyond the scope of this text, but the interested reader can consult Baron and Ferejohn (1989) for a proof.

5. More precisely, Baron and Ferejohn (1989) define two subgames as *structurally equivalent* if (1) the agenda at the initial nodes of the subgames is identical; (2) the set of players who can be recognized is the same; and (3) the strategy sets of players are identical. This means that two subgames in which a previous offer was rejected and someone is chosen to propose a bill are structurally equivalent. An equilibrium is then defined to be *stationary* if the continuation values for each structurally equivalent subgame are the same. Symmetry then implies that each player behaves in the same way. As a consequence of these restrictions, before any subgame of proposal starts, all players have the same expected payoff from playing the game.

6. Given the assumption that players are playing a symmetric stationary equilibrium, and because each player is equally likely to be chosen to be a proposer at the beginning of the game and at any stage, then this value v must be the same for every player.

he expects to get v in the next period, and hence he will accept the proposal if and only if $x_i \geq \delta v$. As a consequence, any player who plays the role of a proposer anticipates that he must propose at least δv to at least $\frac{N+1}{2}$ other players.

Given this observation, the proposer's best response is to choose exactly $\frac{N-1}{2}$ other players and offer them exactly δv each, keeping the rest of the pie for himself. Notice that $\frac{N-1}{2}$ is half of the group of responders, which together with the proposer makes up a majority in favor of the proposed split. Hence the portion of the pie that the proposer keeps for himself in the symmetric subgame-perfect equilibrium is

$$k = 1 - \frac{N-1}{2}\delta v. \tag{11.6}$$

To maintain symmetry, each proposer chooses his $\frac{N-1}{2}$ coalition partners randomly from the remaining $N-1$ players, implying that each of the $N-1$ responders gets chosen to be part of the coalition with probability $\frac{1}{2}$.

The equilibrium analysis leads to the computation of v. At the beginning of the game (and of every stage after rejection), each player expects to play in one of three roles. He can either be the proposer and receive k, which occurs with probability $\frac{1}{N}$, or be a responder, which occurs with probability $\frac{N-1}{N}$. If he is a responder, then with probability $\frac{1}{2}$ he will be in the winning coalition and receive δv, or he can be outside the coalition and receive zero. This implies that

$$v = \frac{k}{N} + \frac{N-1}{2N}\delta v. \tag{11.7}$$

Combining (11.6) and (11.7) we obtain that $v = \frac{1}{N}$. This should not be surprising because, after all, we focused on a symmetric equilibrium. As a consequence, we force all the players to get an equal share in expectations. In addition, because any agreement that could be reached at a later period can be reached at the beginning of the game, there is no need to have a delay.

It is more interesting to focus attention on the value of k, the share obtained by the proposer, which is

$$k(N) = 1 - \delta\left(\frac{N-1}{2N}\right). \tag{11.8}$$

Two implications of (11.8) are noteworthy. The first, which is congruent with Rubinstein's two-player model, is that as the discount factor δ increases, the share obtained by the first-mover proposer decreases. The intuition is that the proposer has to pay more to the responders in order to discourage them from waiting for their turns because the future becomes more valuable, and in the future they have a chance of being proposers themselves.

The second is the relationship between the share that the proposer has to offer his coalition partners and the size of the coalition needed. On the one hand, as the number of players grows the per-partner offer goes down, because, as we have seen, $v = \frac{1}{N}$. On the other hand, the proposer needs to pay more partners to form a winning coalition. It turns out that the second force overcomes the first, and as N grows the share of the proposer decreases. More concretely if $N = 3$ then the share obtained by the proposer is $k(3) = 1 - \delta\frac{1}{3} > \frac{2}{3}$, while as N increases this share drops and at the limit is $\lim_{N \to \infty} k(N) = \frac{2-\delta}{2}$. It is interesting to try to interpret this result as bargaining power. As N increases, it is true that in *absolute* terms the share of the proposer drops. However, in *relative* terms, compared to the share of other players (that is, in terms of the ratio of the proposer's share to the share of a coalition member), it increases and approaches infinity. Intuitively, as there are more and more players, the proposer has more options to form a coalition, and hence has more relative to the others in his coalition. Nonetheless to form that coalition he has to give up more resources.

11.4.2   Open-Rule Bargaining

Now let's introduce the option of amending a proposal, as is allowed in the U.S. Congress. Of course there are many ways in which bills and proposals can be amended, and just as with the closed-rule model we have to make some assumptions on the protocol to make the model tractable. As such, assume that after the proposer makes his proposal $x = (x_1, x_2, \ldots, x_N)$, one of the remaining $N - 1$ players is selected to offer an amendment; we refer to him as the amender. The amender can second the proposer's original proposal and then bring it to a yes-or-no vote. In this case a simple majority of $\frac{N+1}{2}$ is needed to pass the proposal x.

Alternatively the amender can offer an amendment, $x' = (x'_1, x'_2, \ldots, x'_N)$, in which case the game proceeds in two steps: First, the players vote between the original proposal x and the amendment x'. The winner of this vote becomes the baseline proposal for the next period of play, in which a new amender is chosen randomly among the N players to either second the baseline proposal or offer a new amendment. The game proceeds in this way until a proposal both passes the stage at which an amender seconds it and is accepted by a simple majority. As you can imagine, this model is more complex than the closed-rule model, given its protocol of allowing for amendments. As such, we take the simplest case of $N = 3$.

In the open-rule model not only is the original proposer constrained to offer a proposal that might be accepted by a simple majority, but he must also offer enough to deter a random amender from proposing his own amendment that he hopes to pass instead of the original proposal. Hence, whatever the original proposer offers, it must be enough to discourage any potential amender from offering an alternative amendment.

We proceed therefore with two possible approaches. In the first, called "guaranteed success," the original proposal will offer the same share to each of the two potential amenders, and in equilibrium each amender will second the proposal, and it will pass. In the second approach, called "risky success," the original proposer will offer a share to only one of the two potential amenders, taking the risk that the other player will be selected as an amender and in turn will "punish" the original proposer by offering him nothing.

11.4.2.1 Guaranteed Success   We will construct a symmetric stationary subgame-perfect equilibrium in which the first proposal passes, regardless of who the amender is. Let the proposer keep k for himself and offer each of the other two players $\frac{1-k}{2}$, with the expectation that either of the two potential amenders will second the proposal and it will pass. As before, we will use the equilibrium conditions to compute the equilibrium level of k. In addition, symmetry implies that in expectations all the players get the same expected payoff from playing the game.

Just as we defined v in the closed-rule model to be the expected payoff for any player i to play the subgame-perfect equilibrium of this game at the beginning of any stage, let $v(k)$ be the equilibrium expected payoff of a player beginning an amendment stage with a previous proposal that offers k to one of the players, say i, and $\frac{1-k}{2}$ to the other two players.

Hence in an equilibrium in which the original proposal is adopted, a proposer must offer each of the other two players at least $\delta v(k)$ to induce each to second the proposal, implying that $\frac{1-k}{2} \geq \delta v(k)$. The proposer will seek to maximize his share given this constraint, so his best response is to choose $\frac{1-k}{2} = \delta v(k)$. But notice that owing to symmetry each of the potential amenders could follow the same strategy and get the equilibrium proposal adopted, with k for himself. As such, it must be the case that $v(k) = k$, and this observation leads to the equation that characterizes this symmetric subgame-perfect equilibrium:

$$\frac{1-k}{2} = \delta k \qquad \text{or} \qquad k = \frac{1}{1+2\delta}. \tag{11.9}$$

Comparing this outcome to that of the closed-rule model, there is one predictable similarity, together with a perhaps more surprising difference. As in the closed-rule case, and as in any dynamic bargaining model with discounting, the original proposer has a first-mover advantage, which is decreasing in the discount factor: the more patient the players are, the more the proposer has to offer the others.

What is more interesting is the effect of amendments. By comparing the solution in (11.9) with the solution of the closed-rule case (11.8), we can show that the original proposer is better off in the closed-rule case.[7] The reason is that in the closed-rule case the proposer can target a subset of the players and play them off against the others, thus gaining extra bargaining power. With an open rule, such targeting is impossible if the proposer wants to guarantee that his proposal will pass, because if he caters to some players at the expense of others then those players who were shortchanged will have an incentive to amend the proposal at the expense of the proposer. As (11.9) shows, when the discount factor approaches 1 then the proposer has no bargaining power at all and the pie is equally split among the players, unlike the closed-rule case, in which his share approaches $\frac{2}{3}$ of the pie.

11.4.2.2 Risky Success We concluded the previous analysis with the observation that the proposer can guarantee himself a share of the pie equal to $k = \frac{1}{1+2\delta}$. An interesting question is the following: can it be beneficial for the proposer to target only one of the other two lawmakers and offer him a larger share at the expense of the remaining one, while taking the risk that the proposal might fail?

To analyze this situation assume again that the proposer keeps k for himself, but instead of splitting the remaining surplus among the two other players he targets only one and offers him $1 - k$, while the remaining player is offered nothing. The equilibrium for which we are aiming is one in which the player who was offered a share of $1 - k$ moves to second the offer, and hence have it accepted by a majority of two players. However, the strategy is risky because the player who was offered nothing by the proposer will certainly propose an amendment. It is easy to see that any amendment he offers that would give the other player at least $1 - k$ will beat the original proposal.

---

7. Subtracting the proposer's guaranteed-success equilibrium payoff in the open-rule model from his equilibrium payoff in the closed-rule model with three players yields, for all $1 > \delta > 0$,

$$\left(1 - \frac{\delta}{3}\right) - \left(\frac{1}{1+2\delta}\right) = \frac{\delta(5 - 2\delta)}{3(1+2\delta)} > 0.$$

Of course we have to determine what the player who was offered nothing will himself propose as an amendment. To make things simpler, while focusing as before on a symmetric subgame-perfect equilibrium, we assume that players follow a simple *selective inclusion strategy*. To illustrate this strategy, imagine that the proposer, i, offers k to himself, $1 - k$ to player j, and 0 to player h. If j is the randomly chosen amender then he will second the proposal and have it accepted. If instead h is the randomly chosen amender then he will make an amendment in which he proposes k to himself, $1 - k$ to player j, and 0 to player i. Because player j is indifferent, we can assume that he will vote for the amendment and cause it to be accepted.

Now we can calculate the symmetric subgame-perfect equilibrium value of k given the equilibrium restrictions, which in this case are a bit trickier than in the guaranteed-success case because there is asymmetry between the two responders at the proposal stage. Define $v(k)$ to be the equilibrium expected payoff of a proposer who offers k to himself, $1 - k$ to his coalition partner, and 0 to the third player. Let $v(0)$ be the equilibrium expected payoff of the player who will be offered 0.

Given the selective inclusion strategies, the proposer's offer will be accepted with probability $\frac{1}{2}$, in which case he will get k, and will be rejected with probability $\frac{1}{2}$, in which case there will be a delay of one period and then the proposer will find himself in the shoes of the player who gets offered 0. Hence the equilibrium expected payoff of the proposer is

$$v(k) = \tfrac{1}{2}k + \tfrac{1}{2}\delta v(0). \tag{11.10}$$

Similarly consider the situation of the player who was offered 0. With probability $\frac{1}{2}$ he will not be selected as the amender, in which case he will receive the proposed 0 because the other responder will second the original proposal and it will pass. However, with probability $\frac{1}{2}$ he will be selected, in which case he will find himself in the position of the proposer after the pie is discounted. Hence the equilibrium expected payoff of the player who is offered 0 is

$$v(0) = \tfrac{1}{2}\delta v(k). \tag{11.11}$$

These two restrictions offer us two equations with three unknowns, $v(0)$, $v(k)$, and k. The final restriction is the equilibrium requirement that the player who is proposed the share of $1 - k$ will indeed accept it. Since he can put himself immediately in the shoes of the proposer if he is willing to delay the deal by one period, his equilibrium payoff must be at least as large as what he can get from becoming a proposer, hence $1 - k \geq \delta v(k)$, and because the proposer will offer as little as possible to get this player to respond, we get the third equilibrium restriction, which is

$$1 - k = \delta v(k). \tag{11.12}$$

Solving equations (11.10)–(11.12) yields the solution, which is

$$k = \frac{4 - \delta^2}{4 + 2\delta - \delta^2}$$

and

$$v(k) = \frac{2}{4 + 2\delta - \delta^2}. \tag{11.13}$$

Recall that because there is a probability of $\frac{1}{2}$ that the proposer's offer is rejected, his equilibrium expected utility is given by $v(k)$ in equation (11.13). By comparing the solution in (11.13) with the solution of the closed-rule case in (11.8), we can show that the original proposer is better off in the closed-rule case.[8] As before, the closed rule offers the proposer more bargaining power.

It is most interesting to compare the risky-success symmetric equilibrium payoff to the guaranteed-success symmetric equilibrium payoff. The risky-success outcome will yield a higher expected payoff than the guaranteed-success outcome if and only if the value in (11.13) is greater than the value in (11.9), that is,

$$\frac{2}{4 + 2\delta - \delta^2} = \frac{1}{1 + 2\delta}$$

or

$$\delta > \sqrt{3} - 1.$$

That is, when the discount factor is above the value $\sqrt{3} - 1$, the expected payoff from the risky-success equilibrium is greater than that obtained from the guaranteed-success equilibrium.

This result not only has interesting implications but also has some intuition behind it. When the discount factor is high then players are patient, and it becomes quite costly for the proposer to prevent amendments from both potential amenders. As a consequence the proposer would rather opt to cater to only one of the two potential amenders, at the risk of having an amendment on the table that hurts him.

As for the results implication, if we are willing to take seriously the idea that the equilibrium played by the players will be the one that yields them the highest expected payoff, then the result suggests two interesting features of open-rule legislative bargaining. The first is that sometimes proposals will pass with a large (in the guaranteed-success case, unanimous) coalition, while at other times there will be a minimum coalition. Second, in the event that the minimum coalition is expected, there is some probability that the proposal fails and amendments impose costly delay in equilibrium, because every time a proposal fails the surplus is discounted.

11.5   Summary

- Bargaining situations are common across many social settings. Stylized models of bargaining games can help shed light on the outcomes that are more likely to occur as a consequence of the bargaining protocol.

- A key determinant of how players will share the surplus from an agreement is the discount factor. The higher the discount factor the more patient the players, and the surplus will be shared more equally among them.

8. Subtracting the proposer's risky-success expected equilibrium payoff in (11.13) from his equilibrium payoff in the closed-rule model with three players yields, for all $1 > \delta > 0$,

$$\left(1 - \frac{\delta}{3}\right) - \left(\frac{2}{4 + 2\delta - \delta^2}\right) = \frac{6 + 2\delta - 5\delta^2 + \delta^3}{12 + 6\delta - 3\delta^2} > 0.$$

- The bargaining protocol and agreement rules will have a significant impact on how the surplus is shared. In finite games with prespecified alternating moves, the last player to move has a last-mover strategic advantage, while the first player has an advantage owing to discounting.

- When the game is not finite or when the role of proposer is determined randomly with equal probability, there is no last-mover advantage and only the first-mover advantage remains.

11.6   Exercises

11.1   **Disagreement:** Construct a pair of strategies for the ultimatum game ($T = 1$ bargaining game) that constitute a Nash equilibrium and together support the outcome that there is no agreement reached by the two players and the payoffs are zero to each. Show that this disagreement outcome can be supported by a Nash equilibrium regardless of the number of bargaining periods.

11.2   **Holdup:** Consider an ultimatum game ($T = 1$ bargaining game) in which before player 1 makes his offer to player 2, player 2 can invest in the size of the pie. If player 2 chooses a low level of investment (L) then the size of the pie is small, equal to v_L, while if player 2 chooses a high level of investment (H) then the size of the pie is large, equal to v_H. The cost to player 2 of choosing L is c_L, while the cost of choosing H is c_H. Assume that $v_H > v_L > 0$, $c_H > c_L > 0$, and $v_H - c_H > v_L - c_L$.

    a.   What is the unique subgame-perfect equilibrium of this game? Is it Pareto optimal?

    b.   Can you find a Nash equilibrium of the game that results in an outcome that is better for both players as compared to the unique subgame-perfect equilibrium?

11.3   **Even/Odd Symmetry:** In Section 11.2 we analyzed the alternating-offer bargaining game for a finite number of periods when T was odd. Repeat the analysis for T even.

11.4   **Constant Delay Cost:** Consider a two-player alternating-offer bargaining game in which instead of the pie shrinking by a discount factor $\delta < 1$, the players each pay a cost $c_i > 0$, $i \in \{1, 2\}$, to advance from one period to the next. So if player i receives a share of the pie that gives him a value of x_i in period t then his payoff is $v_i = x_i - (t - 1)c_i$. If the game has T periods then a sequence of rejections results in each player receiving $v_i = -(T - 1)c_i$.

    a.   Assume that $T = 2$. Find the subgame-perfect equilibrium of the game and show in which way it depends on the values of c_1 and c_2.

    b.   Are there Nash equilibria in the two-period game that are not subgame perfect?

    c.   Assume that $T = 3$. Find the subgame-perfect equilibrium of the game and show in which way it depends on the values of c_1 and c_2.

11.5   **Asymmetric Patience 1:** Consider a three-period sequential (alternating-offer) bargaining model in which two players have to split a pie worth 1 (starting with player 1 making the offer). Now the players have different discount factors, δ_1 and δ_2.

    a.   Compute the outcome of the unique subgame-perfect equilibrium.

b. Show that when $\delta_1 = \delta_2$ player 1 has an advantage.

c. What conditions on δ_1 and δ_2 give player 2 an advantage? Why?

11.6 **Asymmetric Patience 2:** Consider the analysis of the infinite-horizon bargaining model in Section 11.3 and assume that the players have different discount factors δ_1 and δ_2. Find the unique subgame-perfect equilibrium using the same techniques, and show that as δ_1 and δ_2 become closer in value, the solution you found converges to the solution derived in Section 11.3.

11.7 **Legislative Bargaining Revisited:** Consider a finite T-period version of the Baron and Ferejohn legislative bargaining game with an odd number N of players and with a closed rule as described in Section 11.4.1.

a. Find the unique subgame-perfect equilibrium for $T = 1$. Also find a Nash equilibrium that is not subgame perfect.

b. Find the unique subgame-perfect equilibrium for $T = 2$ with a discount factor $0 < \delta \leq 1$. Also find a Nash equilibrium that is not subgame perfect.

c. Compare what the first period's proposer receives in the subgame-perfect equilibrium you found in (b) to what a first-period proposer receives in the two-period, two-person Rubinstein-Ståhl bargaining game. What intuitively accounts for the difference?

d. Compare the subgame-perfect equilibrium you found in (b) to the solution of the infinite-horizon model in Section 11.4.1. What intuitively accounts for the similarity?

PART IV

STATIC GAMES OF INCOMPLETE INFORMATION

12

Bayesian Games

In all the examples and appropriate tools for analysis that we have encountered thus far, we have made an important assumption: that the game played is common knowledge. In particular we have assumed that the players are aware of who is playing, what the possible actions of each player are, and how outcomes translate into payoffs. Furthermore we have assumed that *this knowledge of the game is itself common knowledge*. These assumptions enabled us to lay the methodological foundation for such solution concepts as iterated elimination of dominated strategies, rationalizability, and most importantly Nash equilibrium and subgame-perfect equilibrium.

Little effort is needed to convince anyone that these idealized situations are rarely encountered in reality. For example, consider one of our early examples, the duopoly market game. We have analyzed both the Cournot and Bertrand models of duopolistic competition, and for each we have a clear and precise, easily understood outcome. One assumption of the model was that the payoffs of the firms, like their action spaces, are common knowledge. However, is it reasonable to assume that the production technologies are indeed common knowledge? And if they are, should we believe that the productivity of workers in each firm is known to the other firm? More generally, is it reasonable to assume that the cost function of each firm is precisely known to its opponent?

Perhaps it is more convincing to believe that firms have a reasonably good idea about their opponents' costs but do not know exactly what they are. Yet the analysis toolbox we have developed so far is not adequate to address such situations. How do we think of situations in which players have *some* idea about their opponents' characteristics but don't know for sure what these characteristics are? At some level this is not so different from the situation in a simultaneous-move game in which a player does not know what actions his opponents are taking, but instead knows what the set of actions *can be*. As we have seen earlier, a player must form a conjecture about the behavior of his opponents in order to choose his best response, and we identified this idea as the player's *belief* over the actions that his opponents will choose. We also required that these beliefs, and the appropriate best responses, be mutually *consistent and correct* for us to be able to use the concept of equilibrium as a method of analysis.

In the mid-1960s John Harsanyi realized the similarity between beliefs over a player's actions and beliefs over his other characteristics, such as costs and preferences. Harsanyi proceeded to develop an elegant and extremely operational way to capture the idea that beliefs over the characteristics of other players—their *types*—can be embedded naturally into the framework of game theory that we have already developed. This advancement set Harsanyi up to be the third Nobel Laureate to share the prestigious prize with John Nash and Reinhard Selten in 1994.

We call games that incorporate the possibility that players could be of different types (a concept soon to be well defined) *games of incomplete information.* As with games of complete information, we will develop a theory of equilibrium behavior that requires players to have beliefs about their opponents' characteristics and their actions, and furthermore requires that these beliefs be consistent or correct. It should be no surprise that this will require very strong assumptions about the cognition of the players: we assume that common knowledge reigns over the possible characteristics of players and over the likelihood that each type of player is indeed part of the game.

To develop this concept further, let's go back to the structure of a strategic-form game in which there is a set of players, $N = \{1, 2, \ldots, n\}$ and for each player $i \in N$, a set of actions A_i. Continue to assume that the set of players and the possible actions that each player has are common knowledge. The missing component is the set of payoff functions or preferences that the players have over outcomes. To capture the idea that a player's characteristics may be unknown to other players, we introduce *uncertainty over the preferences* of the players. That is, instead of having a unique payoff function for each player that maps profiles of actions into payoffs, games of incomplete information allow players to have one of possibly many payoff functions. We associate each of a player's possible payoff functions with the player's type, which captures the idea that a player's preferences, or type, may not be common knowledge.

To operationalize this idea and endow players with well-defined beliefs over the types of other players, Harsanyi (1967–68) suggested the following framework. Imagine that before the game is played Nature chooses the preferences, or type, of each player from his possible set of types.[1] Another way to think about this approach is that Nature is choosing a game from among a large set of games, in which each game has the same players with the same action sets, but with different payoff functions. If Nature is randomly choosing among many possible games, then there must be a well-defined probability distribution over the different games. It is this observation, together with the requirement that everything about a game must be common knowledge, that will make this setting amenable to equilibrium analysis.

At this stage an example is useful. Consider the following simple "entry game," depicted in Figure 12.1, in which an entrant firm, player 1, decides whether or not to enter a market. The incumbent firm in that market, player 2, decides how to respond to an entry decision of player 1 by either fighting or accommodating entry. The payoffs given in Figure 12.1 show that if player 1 enters, player 2's best response is to accommodate, which in turn implies that the unique subgame-perfect equilibrium is for player 1 to enter and for player 2 to accommodate entry. (Convince yourself that there is another pure-strategy Nash equilibrium that is not subgame perfect.)

Now imagine that there is one type of player 1 with payoffs as given in Figure 12.1, and there are two types of player 2. The first type, called "rational," has payoffs as shown in Figure 12.1. The second type, called "crazy," enjoys fighting and the payoff

---

1. As we will soon see, two different types of a player may not necessarily differ in that player's preferences, but they may differ in the *knowledge* that the player has about the types of other players, or about other characteristics of the game. Since this concept is a bit more subtle, we leave it for later, when we will be more comfortable with the notion of Bayesian games and incomplete information.

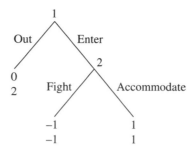

FIGURE 12.1   A simple entry game.

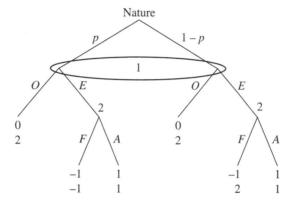

FIGURE 12.2   An incomplete-information entry game.

he gets from (Enter, Fight) is 2 instead of -1. The structure of the game is fixed by the set of players N and the action spaces A_i for each player $i \in N$, yet Nature chooses which type of player 2 is playing the game with player 1. To complete the structure of this game we need to state the likelihood or probability of each type being selected by Nature. Let p denote the probability that Nature chooses the rational type. We can now depict the extensive form of this incomplete-information entry game in Figure 12.2.

Before being able to analyze this game, we must address the issue of what players know when they play the game. Recall that our motivating examples were descriptions of situations in which players were uncertain about the preferences of other players. To discuss optimal behavior, we have to let the players maximize their payoffs given their beliefs about the situation they are in, just as we did with the analysis of single-person decisions and with games of complete information. Thus we must assume that players know their own preferences, which in turn will allow us to analyze a player's best response given his assumptions about the behavior of his opponents. This is the reason for the single information set in Figure 12.2, which shows that player 1 is uncertain about the preferences of player 2, but player 2 knows what his preferences are when he needs to make a decision.

A final issue still needs to be addressed. If players know their own preferences, but they do not know the preferences or types of their opponents, then what must players know in order for them to have a well-defined best response and in turn let us perform equilibrium analysis? The natural step, as Harsanyi realized, is to require players to form *correct beliefs* about the preferences and types of their opponents. This in turn makes it possible for players to form rational conjectures about the way in which their opponents will play the game.

For this reason we assume that, despite each player not necessarily knowing the *actual* preferences of his opponents, he does know the precise way in which Nature chooses these preferences. That is, each player knows the probability distribution over types, and this itself is common knowledge among the players of the game. In the entry game of Figure 12.2 this is given by specifying that p is common knowledge, and hence player 1 knows that the probability that he is at the left node in his information set is p. This is often called the *common prior assumption,* and it means that all the players agree on the way the world works, as described by the probabilities according to which Nature chooses the different types of players. This is of course a strong assumption, but without it, it is more challenging to explore equilibrium behavior.[2]

Now that we have completed the extensive-form representation of this game, we can turn to its normal form. Notice that in the normal form player 2 must have *four* pure strategies: in each of his information sets he has two actions from which to choose. Let's define a strategy of player 2 as $xy \in \{AA, AF, FA, FF\}$, where x describes what a rational player 2 does and y what a crazy one does. This is a preview of what we will soon see more generally: when we introduce incomplete information, a strategy of a player is now a prescription that tells each *type* of a player what he should do if this is the type that Nature chose for the game. The strategy set for player 1 is simply $\{E, O\}$.

As we have seen earlier in Section 7.3, each pair of pure strategies will result in a unique path of play that starts with Nature's choice and then follows with the actions of both players. In this example, if, say, player 2 plays AF (A if he is rational and F if he is crazy), and if player 1 plays E, then with probability p the outcome will yield payoffs of $(1, 1)$, and with probability $(1 - p)$ the payoffs will be $(-1, 2)$. Thus in expectations the pair of payoffs from the pair of strategies (AF, E) is

$$v_1 = p \times 1 + (1 - p) \times (-1) = 2p - 1$$
$$v_2 = p \times 1 + (1 - p) \times 2 = 2 - p.$$

In this way we can compute the expected payoffs of both players from each pair of strategies, which results in the following normal-form matrix game:

		Player 2 AA	Player 2 AF	Player 2 FA	Player 2 FF
Player 1	O	0, 2	0, 2	0, 2	0, 2
	E	1, 1	$2p - 1, 2 - p$	$1 - 2p, 1 - 2p$	$-1, 2 - 3p$

For concreteness, set $p = \frac{2}{3}$, which results in the following normal-form matrix game:

		Player 2 AA	Player 2 AF	Player 2 FA	Player 2 FF
Player 1	O	0, 2	0, 2	0, 2	0, 2
	E	1, 1	$\frac{1}{3}, \frac{4}{3}$	$-\frac{1}{3}, -\frac{1}{3}$	$-1, 0$

---

2. If the players do not share a common prior, that is, they have different beliefs about the world, but these beliefs themselves are common knowledge, then we can still use the tools developed here.

Now that we have the normal-form matrix of this game, we can be agnostic about its origins and the way we constructed it, and instead just treat it as we would any other normal-form game, using the standard tools to find the possible Nash equilibria. That is, we can treat the payoff entries in the matrix as if they came from some alternative game without types or Nature, in which player 1 has two pure strategies and player 2 has four. Considering the best responses of each player, this game has three pure-strategy Nash equilibria: (O, FA), (O, FF), and (E, AF). Also notice that of these three, only (E, AF) is a subgame-perfect equilibrium.

Though our analysis of this game seems complete (ignoring mixed strategies), it may still be confusing to appreciate fully the strides we took from the start of describing a world in which players were not sure about certain elements of the game. First, we modeled this situation as one in which players have uncertainty about the preferences of other players. Second, we assumed that players share the same beliefs about this uncertainty, which allowed us to create a new game for which the standard equilibrium analysis applied. In particular the matrix game has two players with *expected payoffs* derived from the probability distribution over the *different types* of each player (in this case only player 2 had types). This was John Harsanyi's ingenious solution: we cannot perform equilibrium analysis unless we assume that each player knows the distribution of his opponents' types (the common prior assumption). Then, with this requirement in place, once a player assumes some behavior of the different types of his opponents then he can calculate his expected payoff from his own different actions. In this way Harsanyi changed the complex and challenging concept of *incomplete information* into a well-known game of *imperfect information*, in which Nature chooses the players' types and we can then use our standard tools of analysis.

There remains one potentially confusing point that is worth clarifying before we go on to the more general formal definitions. We can see how each *type* of player i can calculate his expected payoffs given a belief over the actions of each type of his opponents. But by writing a single game of imperfect information as we did, we are averaging the payoffs of *all* the types for each player using the likelihood of each type as the weight. That is, we have produced a "meta-player" (e.g., player 2 in the example) who cares about this average payoff across his different types rather than looking at the game one type at a time.

The question is, why will this meta-player's optimal response coincide with the optimal responses of *each* of the different types of which he is composed? This is precisely where the definition of strategies in extensive-form games comes in. When player 2 in the matrix game chooses the strategy AA, for example, then conditional on player 1 choosing E, the rational type of player 2 is playing a best-response strategy A, while the crazy type is playing his dominated strategy, A. For this reason AA in the matrix game is not a best response to E, but instead AF is a best response to E. This follows because when player 1 plays E, AF gives the crazy type a higher payoff than AA, while the payoff to the rational type is unchanged. As a consequence AF gives the meta-player higher expected payoffs than AA when player 1 plays E—another elegant consequence of Harsanyi's solution.

There is, however, an important point worth emphasizing further. Harsanyi's solution to the problem of incomplete information is not something we get for free. We must take a big leap of faith by assuming that the distribution of types (in the example given by p) is common knowledge. Before introducing incomplete information, the Nash equilibrium concept required players to form conjectures, or beliefs, that in equilibrium have to match the choices of their opponents. Here we must ask for more: all the players agree on the way in which players' types differ from each other, and on the way in which Nature chooses among these profiles of types. Therefore when we apply our models and tools to realistic situations, we should be mindful of the strong informational assumptions that we must make, which should guide us in putting confidence in our predictions or prescriptions. This is where art meets science: the modeler must be conscientious about when it makes sense to apply models with such strong assumptions, and when this leap requires too much faith.

12.1 Strategic Representation of Bayesian Games

12.1.1 Players, Actions, Information, and Preferences

Recall from Section 3.1 that a normal-form game of complete information is represented by $\langle N, \{S_i\}_{i=1}^n, \{v_i(\cdot)\}_{i=1}^n \rangle$, where $N = \{1, 2, \ldots, n\}$ is the set of players, S_i is the strategy space of player i, and $v_i : S \to \mathbb{R}$ is the payoff function of player i, where $S \equiv S_1 \times S_2 \times \cdots \times S_n$. Also recall that in a simultaneous-move (or static) game, each player's strategy set S_i is associated with his action set A_i.

To model situations in which players know their own payoffs from outcomes (different profiles of actions), but do not know the payoffs of the other players, we introduce the concept of *incomplete information,* which is composed of three new components. First, a player's preferences are associated with his **type.** If a player can have several different preferences over outcomes, each of these will be associated with a different type. More generally information that the player has about his own payoffs, or information he might have about other relevant attributes of the game, is also part of what defines a player's type. Second, uncertainty over types is described by Nature choosing types for the different players. Thus we introduce **type spaces** for each player, which represent the sets from which Nature chooses the players' types. Last but not least, there is common knowledge about the way in which Nature chooses between profiles of types of players. This is represented by a **common prior,** which is the probability distribution over types that is common knowledge among the players. Because every agent learns his own type, he can use the common prior to form posterior beliefs over the types of other agents. Formally we offer the following definition:

Definition 12.1 The normal-form representation of an n-player **static Bayesian game of incomplete information** is

$$\langle N, \{A_i\}_{i=1}^n, \{\Theta_i\}_{i=1}^n, \{v_i(\cdot; \theta_i), \theta_i \in \Theta_i\}_{i=1}^n, \{\phi_i\}_{i=1}^n \rangle,$$

where $N = \{1, 2, \ldots, n\}$ is the set of players; A_i is the *action set* of player i; $\Theta_i = \{\theta_{i1}, \theta_{i2}, \ldots, \theta_{ik_i}\}$ is the *type space* of player i; $v_i : A \times \Theta_i \to \mathbb{R}$ is the type-dependent payoff function of player i, where $A \equiv A_1 \times A_2 \times \cdots \times A_n$; and ϕ_i describes the *belief* of player i with respect to the uncertainty over the other players' types, that is, $\phi_i(\theta_{-i}|\theta_i)$ is the (posterior) conditional distribution on θ_{-i} (all other types but i) given that i knows his type is θ_i.

As the definition states, aside from the three basic components of players, actions, and preferences, the addition of types, type-dependent preferences, and beliefs about the types of other players captures the ideas illustrated earlier. Notice that in the definition we assume that each player's type space is finite, so that there are a total of

k_i types for each player. It is not difficult to expand the definition to include infinite type spaces.[3]

It is convenient to think about a static Bayesian game as one that proceeds through the following steps:

1. Nature chooses a profile of types $(\theta_1, \theta_2, \ldots, \theta_n)$.

2. Each player i learns his own type, θ_i, which is his *private information,* and then uses his prior ϕ_i to form posterior beliefs over the other types of players.

3. Players simultaneously (hence this is a *static game*) choose actions $a_i \in A_i$, $i \in N$.

4. Given the players' choices $a = (a_1, a_2, \ldots, a_n)$, the payoffs $v_i(a; \theta_i)$ are realized for each player $i \in N$.

In the foregoing definition a player's payoff $v_i(a; \theta_i)$ depends on the actions of all players and only on i's type, but it does *not* depend on the types of the other players θ_{-i}. This particular assumption is known as the **private values** case because each type's payoff depends only on his private information. This case is not rich enough to capture all the interesting examples that we will analyze, and for this reason we will later introduce the case of **common values,** in which $v_i(a_1, a_2, \ldots, a_n; \theta_1, \theta_2, \ldots, \theta_n)$ is possible. For expositional clarity, we first explore the private values setting.

Remark We took a very simple approach by effectively assuming that a player's beliefs are about the preferences of other players. Of course life is more complex than that. To use the tools of strategic analysis rigorously, player i's beliefs need also be well defined over the beliefs of his opponents about the preferences of player i, about the beliefs of the other players about the beliefs of player i about the beliefs of the other players about i's preferences, and so on ad infinitum. As you can imagine, all this gets very messy very fast. Harsanyi's celebrated solution to this problem followed from his argument that rather than write down the entire hierarchy of beliefs, it is possible to consider a model in which each player's "type" describes his preferences and all of his beliefs. Years later Mertens and Zamir (1985) and Brandenburger and Dekel (1993) made Harsanyi's idea precise and showed that it is indeed possible to define a rich type space that would allow for all possible hierarchies of belief. (These two papers are very technical and mathematically sophisticated.)

12.1.2 Deriving Posteriors from a Common Prior: A Player's Beliefs

In the definition of a Bayesian game we introduced the concept of a common prior, that is, all the players share the same beliefs about the distribution over the choices made by Nature. In this section we explore the meaning of each player i using the common prior to derive a posterior belief about the distribution of the other types of players. This concept is a simple application of conditional probabilities. If this material is unfamiliar to you, you should consult Section 19.4.3.

---

3. For example, a type space can be an interval $\Theta_i = [\underline{\theta}_i, \overline{\theta}_i]$, and instead of probabilities p_i over the distribution of other players' types, there will be a well-defined density $f_i(\cdot)$. This point will be revisited, together with an example, in Section 12.2.2.

Conditional probabilities follow a mathematical rule that derives the way in which a player or decision maker should change a **prior** (initial) belief in the light of new evidence, resulting in a **posterior** (updated) belief. In our application the idea can be described as follows. First, before Nature chooses the actual type of each player, imagine that every player does not yet know what his type will be; but he does know the probability distribution that Nature uses to choose the types for all the players. Later, after Nature has chosen a type for each player, they all independently and privately learn their types. This new piece of information for each player, his type, may provide some new piece of evidence about how the other players' types may have been chosen. It is in this respect that a player may derive new beliefs about the other players once he learns his type.

As a concrete example, imagine that there are two of many possible **states of Nature** or **events** (the possible choices that Nature can make). One is that it will be sunny (S) and the other is that the waves will be high (H). These two states can occur exclusively or together according to some prior distribution $\phi(\cdot)$. That is, $\phi(\cdot)$ describes the probabilities assigned to any combination of these states being true. Let $\phi(S)$ be the prior probability that it will be sunny, $\phi(H)$ be the prior probability that the waves will be high, and $\phi(S \cap H)$ be the prior probability that it will be sunny and that the waves will be high. Let's imagine that you wake up and see that it is sunny; without seeing them, what can you infer about the probability that the waves are high? It is not necessarily true that it is $\phi(H)$, because you just learned that it is sunny and this new information may be relevant. This is where the conditional probability formula applies, because it precisely computes the probability of state H *given that you know* that state S happened. Formally we have

Definition 12.2 Conditional on event S being true, the **conditional probability** that event H is true is given by

$$\Pr\{H|S\} = \frac{\phi(S \cap H)}{\phi(S)}. \tag{12.1}$$

The intuition is simple. If we know that S is true then there are two possibilities: either H is true or it is not. We can therefore think of two possible "combined" events. The first event is that both S and H are true, which given the prior belief occurs with probability $\phi(S \cap H)$. The second event is that S is true but H is false, which happens with probability $\phi(S \cap [\text{not } H])$. Thinking of both S and H together, we can ask, what is the probability that S is true? It must be the sum $\phi(S) = \phi(S \cap H) + \phi(S \cap [\text{not } H])$ because S can be true either with H being true or with $[\text{not } H]$ being true, and clearly H and $[\text{not } H]$ are mutually exclusive events. As a consequence, if I know that S is true then *conditional on this knowledge,* the likelihood of H being true is the *relative likelihood* of both S and H being true, among all the states in which S can be true. Using our particular example,

$$\Pr\{H|S\} = \frac{\phi(S \cap H)}{\phi(S \cap H) + \phi(S \cap [\text{not } H])} = \frac{\phi(S \cap H)}{\phi(S)}$$

To put this into the context of a Bayesian game, consider the following example. Imagine that there are two players, 1 and 2, each having two possible types, $\theta_1 \in \{a, b\}$ and $\theta_2 \in \{c, d\}$. Nature chooses these types according to a prior over the four possible type combinations, where the following joint distribution matrix describes Nature's prior:

Player 2's type

	c	d
a	$\frac{1}{6}$	$\frac{1}{3}$
b	$\frac{1}{3}$	$\frac{1}{6}$

Player 1's type

That is, the prior probability that player 1 is type a and player 2 is type d is equal to the prior probability that player 1 is type b and player 2 is type c, and this probability is $\frac{1}{3}$. Similarly the prior probability that player 1 is type a and player 2 is type c is equal to the prior probability that player 1 is type b and player 2 is type d, and this probability is $\frac{1}{6}$. The common prior assumption implies that each of the two players takes as given that Nature chooses the types according to the matrix.

Now imagine that player 1 learns that his type is a. What must be his belief about player 2's type? Using the conditional probability formula in (12.1),

$$\phi_1(\theta_2 = c | \theta_1 = a) = \frac{\Pr\{\theta_1 = a \cap \theta_2 = c\}}{\Pr\{\theta_1 = a\}} = \frac{\frac{1}{6}}{\frac{1}{6} + \frac{1}{3}} = \frac{1}{3},$$

and similarly

$$\phi_1(\theta_2 = d | \theta_1 = a) = \frac{\Pr\{\theta_1 = a \cap \theta_2 = d|\}}{\Pr\{\theta_1 = a\}} = \frac{\frac{1}{3}}{\frac{1}{6} + \frac{1}{3}} = \frac{2}{3}.$$

12.1.3 Strategies and Bayesian Nash Equilibrium

Recall that in the static normal-form games of complete information described in Section 3.1 we did not make a distinction between actions and strategies because choices were made once and for all. For games of incomplete information, however, we need to be a bit more careful to specify strategies correctly. The representation of a Bayesian game described earlier has action sets, A_i, for each player $i \in N$. However, each player i can be one of several types $\theta_i \in \Theta_i$, and each type θ_i may choose a different action from the set A_i. Thus to define a strategy for player i we need to specify what each type $\theta_i \in \Theta_i$ of player i will choose when Nature calls upon this type to play the game. For this we define strategies as follows:

Definition 12.3 Consider a static Bayesian game

$$\left\langle N, \{A_i\}_{i=1}^n, \{\Theta_i\}_{i=1}^n, \{v_i(\cdot; \theta_i), \theta_i \in \Theta_i\}_{i=1}^n, \{\phi_i\}_{i=1}^n \right\rangle.$$

A **pure strategy** for player i is a function $s_i : \Theta_i \to A_i$ that specifies a pure action $s_i(\theta_i)$ that player i will choose when his type is θ_i. A **mixed strategy** is a probability distribution over a player's pure strategies.

This turns out to be a convenient way to specify strategies for Bayesian games. You can think of it as if each player chose his *type-contingent* strategy before he learned his type and then played according to that strategy. This should remind you of strategies for extensive-form games that are defined as mappings from information sets to actions, where a pure strategy is a rulebook of what to choose in each information set. In Bayesian games we can think of the types of players as being their information sets: when player i learns his type then it is as if he is in a unique information set that is identified with this type.

This observation is useful because it allows us to consider consistently the beliefs of players over the strategies of their opponents when their opponents can be of different types. To illustrate this point, consider the entry game used to demonstrate the central ideas developed here, shown in Figure 12.2 and in its normal form by the following matrix:

		Player 2			
		AA	AF	FA	FF
Player 1	O	0, 2	0, 2	0, 2	0, 2
	E	1, 1	$\frac{1}{3}, \frac{4}{3}$	$-\frac{1}{3}, -\frac{1}{3}$	$-1, 0$

As we can now see, this game is just a special case of the more general Bayesian game described in definition 12.1. The strategy of the incumbent player 2 depends on his type, and hence the matrix representation of this game includes four type-dependent pure strategies. Player 1 has correct beliefs about the distribution of types given by Nature's probability distribution, and together with a specific strategy of player 2, player 1 will have well-defined beliefs over the different continuation paths and outcomes of the game.

It is useful to stress the way in which beliefs over outcomes are determined in static Bayesian games of incomplete information, which allows players to evaluate their expected payoffs from their alternative choices. In the entry game in Figure 12.2, let $\theta_2 \in \{r, c\}$ denote player 2's type (for rational and crazy), with p being the common prior that player 2 is rational. Now consider the case in which player 1 believes that player 2 is using the pure strategy AF, which can be described as

$$s_2(\theta_2) = \begin{cases} A & \text{if } \theta_2 = r \\ F & \text{if } \theta_2 = c. \end{cases}$$

In this simple example there is only one type of player 1. From player 1's own perspective, his expected payoff from playing E will be

$$Ev_1\left(E, s_2\left(\theta_2\right)\right) = pv_1(E, s_2(r)) + (1 - p)v_1(E, s_2(c))$$

$$= p \times 1 + (1 - p) \times (-1),$$

which with $p = \frac{2}{3}$ yields $Ev_1\left(E, s_2\left(\theta_2\right)\right) = \frac{1}{3}$, corresponding to the payoff entry from the pure-strategy pair (E, AF) in the matrix that represents this game.[4]

There are two observations worth emphasizing at this point. First, notice that if player i is using a pure (type-dependent) strategy, while Nature chooses player i's types randomly, then as far as players $j \neq i$ are concerned they are facing a player i who is using a *mixed strategy*. Again, using the example of the entry game, if player 2 uses the strategy AF then as far as player 1 is concerned, it is as if player 2 is choosing A with probability p and F with probability $1 - p$.

---

4. Because player 1 has a single type, writing $Ev_1\left(E, s_2\left(\theta_2\right); \theta_1\right)$ and $p_1\left(\theta_2|\theta_1\right)$ has an obvious redundancy.

Second, as mentioned briefly earlier, we are effectively specifying player i's strategy for all his information sets—one for each type—just as in any extensive-form game. Thus, given a certain realization of Nature, we are specifying what any player i will do even for those types that have not been realized. The reason we must specify i's strategy completely is so that player i's opponents can form well-defined beliefs over i's behavior. Players $j \neq i$ need to combine their beliefs from their posterior over i's types together with their beliefs over what each type θ_i of player i plans to do. Without this complete specification, players $j \neq i$ cannot calculate their expected payoffs from their own actions.

Now that we have completely defined what a static Bayesian game is, and what the strategies for each player are, it is easy to define a solution concept that is derived from Nash equilibrium as follows:

Definition 12.4 In the Bayesian game

$$\left\langle N, \{A_i\}_{i=1}^n, \{\Theta_i\}_{i=1}^n, \{v_i(\cdot; \theta_i), \theta_i \in \Theta_i\}_{i=1}^n, \{\phi_i\}_{i=1}^n \right\rangle,$$

a strategy profile $s^* = (s_1^*(\cdot), s_2^*(\cdot), \ldots, s_n^*(\cdot))$ is a **pure-strategy Bayesian Nash equilibrium** if, for every player i, for each of player i's types $\theta_i \in \Theta_i$, and for every $a_i \in A$, $s_i^*(\cdot)$ solves

$$\sum_{\theta_{-i} \in \Theta_{-i}} \phi_i(\theta_{-i}|\theta_i) v_i(s_i^*(\theta_i), s_{-i}^*(\theta_{-i}); \theta_i) \geq \sum_{\theta_{-i} \in \Theta_{-i}} \phi_i(\theta_{-i}|\theta_i) v_i(a_i, s_{-i}^*(\theta_{-i}); \theta_i).$$

(12.2)

That is, regardless of the type realization, no player wants to change his strategy $s_i^*(\cdot)$.

To restate the definition, a Bayesian Nash equilibrium has each player choose a type-contingent strategy $s_i^*(\cdot)$ so that given any one of his types $\theta_i \in \Theta_i$, and his beliefs about the strategies of his opponents $s_{-i}^*(\cdot)$, his expected payoff from $s_i^*(\theta_i)$ is at least as large as that from any one of his actions $a_i \in A_i$. His expectations are derived from the strategies played by other players and the mixing that occurs owing to the *randomization of nature* that each player faces through his beliefs $\phi_i(\theta_{-i}|\theta_i)$.

Note that on the right side of the inequality (12.2) we can replace a_i with $s_i'(\theta_i)$ so that it reads

$$\sum_{\theta_{-i} \in \Theta_{-i}} \phi_i(\theta_{-i}|\theta_i) v_i(s_i^*(\theta_i), s_{-i}^*(\theta_{-i}); \theta_i) \geq \sum_{\theta_{-i} \in \Theta_{-i}} \phi_i(\theta_{-i}|\theta_i) v_i(s_i'(\theta_i), s_{-i}^*(\theta_{-i}); \theta_i),$$

which means that player i of type θ_i does not want to replace the strategy $s_i^*(\cdot)$ with any other $s_i'(\cdot) \in S_i$ because this inequality holds for every type $\theta_i \in \Theta_i$. It suffices to consider any deviation from $s_i^*(\cdot)$ to an action, a_i, instead of the more complex notion of deviating to a type-dependent strategy, $s_i'(\cdot)$. Also note that the definition is stated for pure-strategy Bayesian Nash equilibria. The definition of a mixed-strategy Bayesian Nash equilibrium is the straightforward extension in which, instead of $s_i(\cdot)$ that maps types into pure actions, we will include $\sigma_i(\cdot)$ that maps types into probability distributions over actions.

Remark In the definition of a Bayesian Nash equilibrium we can write the condition (12.2) of playing a best response more generally as follows:

$$E_{\theta_{-i}}[v_i(s_i^*(\theta_i), s_{-i}^*(\theta_{-i}); \theta_i)|\theta_i] \geq E_{\theta_{-i}}[v_i(a_i, s_{-i}^*(\theta_{-i}); \theta_i)|\theta_i] \quad \text{for all } a_i \in A_i.$$

Just as in (12.2) we take the expectations of player i over the realizations of types θ_{-i} when player i knows his own type (hence the conditional expectation takes the form $E_{\theta_{-i}}[\cdot|\theta_i]$, which is conditional on the realization of θ_i). This notation is more general in the sense that it is not restricted to a finite type space and instead can apply to *a continuum of types for each player.* For example, each player may have an infinite number of possible types drawn from an interval $\Theta_i = [\underline{\theta}_i, \overline{\theta}_i]$, with cumulative distribution $F_i(\theta_i)$ and density function $f_i(\theta_i) = F_i'(\theta_i)$. In this case the expected payoff of player i will be written as an integral (more precisely, $n - 1$ integrals) that forms the expectations over the realizations of the other players' types, and accordingly their actions as derived from their strategies.

12.2 Examples

This section offers two examples as illustrations of Bayesian games of incomplete information. In the first example each player has a finite number of actions and types, while in the second each player has a continuum of types. Sections 12.3–12.5 present some important applications of Bayesian games.

12.2.1 Teenagers and the Game of Chicken

It is often the case that a player who finds himself in some game of conflict will hold out and suffer in order to appear strong, instead of letting his rival get the better of him. Be it firms in the marketplace, politicians in government, countries at war, or even kids on the playground, the optimal behavior will depend on some combination of each player's tendency to be aggressive and his belief about his opponent's tendency to be aggressive.

To illustrate this idea, consider a simple game of aggression that is known to many teenagers: the game of "chicken." The 1955 film *Rebel Without a Cause* features James Dean as a juvenile delinquent and introduced to the silver screen one rather dangerous variant of the game. In the movie two teenagers simultaneously drive their cars toward the edge of a cliff. The first one to jump out is considered chicken and loses the contest. (In the movie one dies, and coincidentally Dean himself died in a car accident just before the film was released.) Many other films have featured variations on this game of chicken, a well-known example being *Grease* (1978), starring John Travolta. The following example is yet another variant of the game.

Two teenagers, players 1 and 2, have borrowed their parents' cars and decided to play the game of chicken as follows: They drive toward each other in the middle of a street, and just before impact they must simultaneously choose whether to be chicken and swerve to the right, or continue driving head on. If both are chicken then both gain no respect from their friends but suffer no losses; thus both get a payoff of 0. If i continues to drive while $j \neq i$ plays chicken then i gains all the respect, which is a payoff of R, and j gets no respect, which is worth 0. In this case both players suffer no additional losses. Finally if both continue to drive head on then they split the respect (because respect is considered to be relative), but an accident is bound to happen and they will at least be reprimanded by their parents, if not seriously injured. An accident imposes a personal loss of k on each player, so the payoff to each one is $\frac{R}{2} - k$.

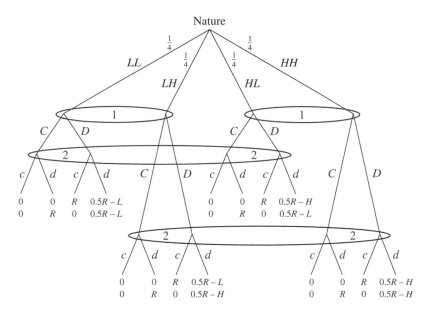

FIGURE 12.3 The game of chicken with incomplete information.

There is, however, a potential difference between these two youngsters: The punishment, k, depends on the type of parents they have. For each kid, parents can be either harsh (H) or lenient (L) with equal probability, and the draws from Nature on the types of parents are independently distributed. This means that the likelihood that player i's parents are harsh is equal to $\frac{1}{2}$ and is independent of the type of parents that player j has. If player i's parents are harsh then they will beat the living daylights out of their child, which imposes a high cost of an accident, denoted by $k = H$. If instead the parents are lenient then they will give their child a long lecture on why his behavior is unacceptable, which imposes a lower cost of an accident, denoted by $k = L < H$. Each kid knows the type of his parents but does not know the type of his opponent's parents. The distribution of types is common knowledge (this is the common prior assumption).

The extensive form of this game is depicted in Figure 12.3. There are four states of Nature, $\theta_1\theta_2 \in \{LL, LH, HL, HH\}$ that denote the types of player 1 and player 2, respectively, and each of these four states occurs with equal probability because of the probability distributions described earlier. The structure of the information sets follows from the knowledge of the players when they make their moves. Player 1, for example, cannot distinguish between the states of Nature LL and LH (his type is L), nor can he distinguish between HL and HH (his type is H). However, he knows what his own type is, and hence he can distinguish between these two pairs of states. A similar logic explains the information sets of player 2. The action sets of each player are $A_1 = \{C, D\}$ and $A_2 = \{c, d\}$, where C (or c) stands for "chicken" and D (or d) stands for "drive."

From the extensive form in Figure 12.3 we can derive the matrix form as follows. A strategy for player 1 is denoted by $xy \in S_1 = \{CC, CD, DC, DD\}$, where x is what he does if he is an L type and y is what he does if he is an H type. Similarly $S_2 = \{cc, cd, dc, dd\}$. The payoffs are calculated using the probabilities over the states of Nature together with the strategies of the players. For example, if player 1 chooses CD and player 2 chooses dd then the expected payoffs for player 1 are

$$Ev_1(CD, dd) = \frac{1}{4} \times v_1(C, d; L) + \frac{1}{4} \times v_1(C, d; L) + \frac{1}{4} \times v_1(D, d; H) + \frac{1}{4} \times v_1(D, d; H)$$

$$= \frac{1}{4} \times 0 + \frac{1}{4} \times 0 + \frac{1}{4} \times \left(\frac{R}{2} - H\right) + \frac{1}{4} \times \left(\frac{R}{2} - H\right) = \frac{R}{4} - \frac{H}{2}$$

and the expected payoffs for player 2 are

$$Ev_2(CD, dd) = \frac{1}{4} \times v_2(C, d; L) + \frac{1}{4} \times v_2(C, d; H) + \frac{1}{4} \times v_2(D, d; L) + \frac{1}{4} \times v_2(D, d; H)$$

$$= \frac{1}{4} \times R + \frac{1}{4} \times R + \frac{1}{4} \times \left(\frac{R}{2} - L\right) + \frac{1}{4} \times \left(\frac{R}{2} - H\right) = \frac{3R}{4} - \frac{L}{4} - \frac{H}{4}.$$

In a similar way we can calculate the payoffs from all the other combinations of pure strategies for each player (this would be useful practice for you if things still feel a bit shaky mathematically). This results in the following matrix-form Bayesian game:

		cc	cd	dc	dd
				Player 2	
	CC	$0, 0$	$0, \frac{R}{2}$	$0, \frac{R}{2}$	$0, R$
	CD	$\frac{R}{2}, 0$	$\frac{3R}{8} - \frac{H}{4}, \frac{3R}{8} - \frac{H}{4}$	$\frac{3R}{8} - \frac{H}{4}, \frac{3R}{8} - \frac{L}{4}$	$\frac{R}{4} - \frac{H}{2}, \frac{3R}{4} - \frac{L}{4} - \frac{H}{4}$
Player 1	DC	$\frac{R}{2}, 0$	$\frac{3R}{8} - \frac{L}{4}, \frac{3R}{8} - \frac{H}{4}$	$\frac{3R}{8} - \frac{L}{4}, \frac{3R}{8} - \frac{L}{4}$	$\frac{R}{4} - \frac{L}{2}, \frac{3R}{4} - \frac{L}{4} - \frac{H}{4}$
	DD	$R, 0$	$\frac{3R}{4} - \frac{L}{4} - \frac{H}{4}, \frac{R}{4} - \frac{H}{2}$	$\frac{3R}{4} - \frac{L}{4} - \frac{H}{4}, \frac{R}{4} - \frac{L}{2}$	$\frac{R}{2} - \frac{L}{2} - \frac{H}{2}, \frac{R}{2} - \frac{L}{2} - \frac{H}{2}$

To solve for the Bayesian Nash equilibria we need to have more information about the parameters R, H, and L. Assume that $R = 8$, $H = 16$, and $L = 0$, which results in the following matrix-form game:

		cc	cd	dc	dd
				Player 2	
	CC	$0, 0$	$0, 4$	$0, 4$	$\overline{0, 8}$
	CD	$4, 0$	$-1, -1$	$\overline{-1, 3}$	$-6, 2$
Player 1	DC	$4, 0$	$3, -1$	$\overline{3, 3}$	$\underline{2, 2}$
	DD	$\underline{8, 0}$	$2, -6$	$\overline{2, 2}$	$-4, -4$

A quick analysis reveals that the game has a unique pure-strategy Bayesian Nash equilibrium: (DC, dc). That is, the children of lenient parents will continue driving head on, while those of harsh parents will swerve to avoid the costly consequences. If the payoffs we assumed are indeed representative of this situation then the view that harsh upbringing yields better outcomes can be supported. Perhaps, however, children of lenient parents learn somehow to respect their parents' property (the car), while children of harsh parents, despite anticipating the pain of a harsh punishment, do not respect their parents' property. In light of this different description, we might assume that $L = 16$ and $H = 0$, which would result in the opposite prediction. Hence, as in any game, we need to get the payoffs right in order to feel comfortable offering policy recommendations.

12.2.2 Study Groups

Some if not most of you have participated in study groups, in which each student takes upon himself some effort, the fruits of which are shared with the other group members. Let's consider such a situation with two students, players 1 and 2, who have to hand in a joint lab assignment. Each student i can either put in the effort ($e_i = 1$) or shirk ($e_i = 0$), where the cost of putting in the effort is the same for each student and is given by some $c < 1$, while shirking involves no cost. If either one or both of the students put in the effort then the lab assignment is a success, while if both shirk then it is a failure.

Students vary in how much they care about their educational success, which is described by the following specification. Each student has a type $\theta_i \in [0, 1]$, which is independently and uniformly distributed over this interval. That is, for each player i the distribution of his type is $F(\theta_i) = \theta_i$ for $\theta_i \in [0, 1]$, and the density is $f(\theta_i) = 1$ for $\theta_i \in [0, 1]$. Student i's personal value from a successful assignment is given by the square of his type, θ_i^2. Hence if student i chooses to put in the effort then his payoff is guaranteed to be $\theta_i^2 - c$. If he chooses to shirk, however, then his payoff depends on what his partner does. If his partner j puts in the effort then student i's payoff is θ_i^2, while if his partner j shirks as well then student i's payoff is 0. Each student knows only his own type before choosing his effort e. It is also common knowledge that the types are distributed independently and uniformly on $[0, 1]$ and that the cost of effort is c.

This is an example of a Bayesian game with continuous type spaces and discrete sets of actions. It is not too useful to try to draw the game tree that represents this game because of the continuous types, nor can we derive a matrix because, despite the two possible actions, $e \in \{0, 1\}$, the continuous types imply that there is a continuum of strategies: a pure strategy for player i is a function $s_i : [0, 1] \rightarrow \{0, 1\}$ that identifies a choice of $e_i \in \{0, 1\}$ for every type $\theta_i \in [0, 1]$.

Given a belief of player i about the strategy of player j, the only factor that affects i's payoff is the probability that player j chooses $e_j = 1$. That is, if we consider two strategies of player j, $s_j(\theta_j)$ and $\widehat{s}_j(\theta_j)$—which result in the same probability that he chooses $e_j = 1$, so that $\Pr\{s_j(\theta_j) = 1\} = \Pr\{\widehat{s}_j(\theta_j) = 1\}$—then player 1 gets the same expected payoff from not choosing to put in his own effort. (Recall that player i's payoff from putting in effort $e_i = 1$ is equal to $\theta_i^2 - c$ regardless of the choice of player j.)

The best response of player i will be to choose effort $e = 1$ if this is (weakly) better than choosing $e = 0$, which will be true if and only if

$$\theta_i^2 - c \geq \theta_i^2 \Pr\{s_j(\theta_j) = 1\},$$

which can be rewritten as

$$\theta_i \geq \sqrt{\frac{c}{1 - \Pr\{s_j(\theta_j) = 1\}}}. \tag{12.3}$$

This in turn implies the following:

Claim 12.1 *The best response of player i to any strategy $s_j(\theta_j)$ is a **threshold rule**: there exists some $\widehat{\theta}_i$ such that i's best response is to choose $e = 1$ if $\theta_i \geq \widehat{\theta}_i$ and to choose $e = 0$ if $\theta_i \leq \widehat{\theta}_i$.*

The claim follows directly from (12.3) because for any strategy of player j the right side of the inequality is a constant. The intuition behind this result is quite simple. Imagine that player i believes that $\Pr\{s_j(\theta_j) = 1\} = 0$, so that he expects his partner to shirk for sure. If player i cares enough about the grade, namely $\theta_i^2 - c > 0$, then he should put in the effort $e = 1$ in order to succeed. This will be true for any $\theta_i \geq \sqrt{c}$, so that the type $\theta_i = \sqrt{c}$ of player i is just indifferent between working and shirking. If instead player i believes that his partner will choose $e = 1$ with some positive probability then player i can "free ride" a bit, and now the type $\theta_i = \sqrt{c}$ is no longer indifferent and strictly prefers to shirk, as will some small interval of types just above $\theta_i = \sqrt{c}$. For any given strategy of player j this type of player i who doesn't care enough about the grade

$$\left(\theta_i < \sqrt{\frac{c}{1 - \Pr\{s_j(\theta_j) = 1\}}} \right)$$

will shirk. If the probability that player j puts in the effort $e = 1$ is really high then $\frac{c}{1 - \Pr\{s_j(\theta_j)=1\}} > 1$ and player i will shirk for sure.

This means that we are looking for a Bayesian Nash equilibrium in which each student has a threshold type $\widehat{\theta}_i \in [0, 1]$ such that[5]

$$s_i(\theta_i) = \begin{cases} 0 & \text{if } \theta_i < \widehat{\theta}_i \\ 1 & \text{if } \theta_i \geq \widehat{\theta}_i. \end{cases}$$

This observation lets us calculate what a player's best-response function is given his belief about the threshold strategy of his opponent. If player j is using a threshold $\widehat{\theta}_j$ so that $e = 1$ if and only if $\theta_j \geq \widehat{\theta}_j$ then it follows that $\Pr\{s_j(\theta_j) = 1\} = 1 - \widehat{\theta}_j$. This in turn means that player i will choose $e = 1$ if and only if

$$\theta_i \geq \sqrt{\frac{c}{\widehat{\theta}_j}}. \tag{12.4}$$

This in turn results in the best response of player i. If $\widehat{\theta}_j > c$ then the right side of (12.4) is less than 1, implying that player i's threshold is less than 1. If $\widehat{\theta}_j < c$ then the right side of (12.4) is greater than 1, implying that player i's threshold is equal to 1 (it cannot of course be greater than 1). Define the best response of player i, $BR_i(\widehat{\theta}_j)$, as his best-response threshold strategy $\widehat{\theta}_i$ given that player j is using the threshold $\widehat{\theta}_j$. We therefore have

---

5. Notice that we are breaking the tie at $\theta_i = \widehat{\theta}_i$ in favor of $e = 1$. This is inconsequential for expected payoffs since the uniform distribution of θ_i puts zero probability on the type $\widehat{\theta}_i$ being realized.

$$BR_i(\widehat{\theta}_j) = \begin{cases} \sqrt{\frac{c}{\widehat{\theta}_j}} & \text{if } \widehat{\theta}_j \geq c \\ 1 & \text{if } \widehat{\theta}_j < c. \end{cases} \tag{12.5}$$

Claim 12.2 *In the unique Bayesian Nash equilibrium each player chooses the same threshold level $\theta_i = \theta^*$, where $0 < \theta^* < 1$.*

This claim follows directly from the best-response function given in (12.5). First, consider the possibility of a Bayesian Nash equilibrium in which each player i is choosing his threshold on the part of the best-response function where $\widehat{\theta}_j \geq c$ (the top part of (12.5)). We then have two best-response equations with two unknowns,

$$\widehat{\theta}_1 = \sqrt{\frac{c}{\widehat{\theta}_2}} \qquad \text{and} \qquad \widehat{\theta}_2 = \sqrt{\frac{c}{\widehat{\theta}_1}}, \tag{12.6}$$

which together imply that $\widehat{\theta}_1^2 \widehat{\theta}_2 = c$ and $\widehat{\theta}_2^2 \widehat{\theta}_1 = c$. Because we assumed that $0 < c < 1$ the equality can be satisfied only if $\widehat{\theta}_1 = \widehat{\theta}_2 = \theta^*$, where $0 < \theta^* < 1$. Substituting $\widehat{\theta}_1 = \widehat{\theta}_2 = \theta^*$ into any of the two best-response functions in (12.6) implies that $\theta^* = c^{\frac{1}{3}} < 1$ (because $c < 1$ and $c^{\frac{1}{3}} > c$).

Next we need to make sure that there is no Bayesian Nash equilibrium in which for some player the threshold is $\widehat{\theta}_j < c$, implying that player i will choose the threshold $\widehat{\theta}_i = BR_i(\widehat{\theta}_j) = 1$ (never choose $e = 1$). But notice that if $\widehat{\theta}_i = 1$ then player j's best response is to choose a threshold equal to $BR_j(1) = \sqrt{c}$, which contradicts the premise that $\widehat{\theta}_j < c$ because $\sqrt{c} > c$ for $c < 1$. Hence the unique Bayesian Nash equilibrium is the symmetric threshold choices $\theta^* = c^{\frac{1}{3}}$, which are implemented by the following strategies for each player $i \in \{1, 2\}$:

$$s_i^*(\theta_i) = \begin{cases} 0 & \text{if } \theta_i < c^{\frac{1}{3}} \\ 1 & \text{if } \theta_i \geq c^{\frac{1}{3}}. \end{cases}$$

Indeed it is also possible to verify this as the unique equilibrium by considering the plot of the best-response functions depicted in Figure 12.4. For example, player

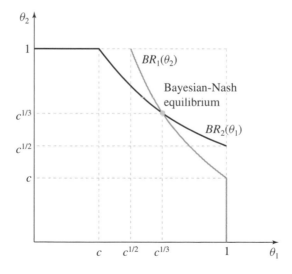

FIGURE 12.4 Best-response thresholds of the study group game.

2's best-response threshold is the lowest when player 1's threshold is $\widehat{\theta}_1 = 1$, where $BR_2(1) = c^{\frac{1}{2}}$, and it reaches a maximum when player 1's threshold is $\widehat{\theta}_1 = c^{\frac{1}{2}}$, where $BR_2(c^{\frac{1}{2}}) = 1$. Player 1's best-response function is symmetric, and the two cross at exactly the point where both equal $c^{\frac{1}{3}}$.

12.3 Inefficient Trade and Adverse Selection

One of the main conclusions of competitive market analysis in economics is that markets allocate goods to the people who value them the most. The simple intuition behind this conclusion works as follows: If a good is misallocated so that some people who have it value it less than people who do not, then so-called market pressures will cause the price of that good to increase to a level at which the current owners will prefer to sell it rather than hold on to it, and the people who value it more will be willing to pay that price. The determination of such a price is not clearly specified, but various mechanisms such as bargaining, auctions, or market intermediaries may help obtain it.

This powerful argument is based on some assumptions, one of which is that the value of the good is easily understood by all market participants, or in our terminology, there is *perfect information* about the value of the good. This is of course an assumption that applies to an ideal world, one that often departs from reality. Yet the assumption of perfect information not only implies the powerful result that markets will allocate goods to the agents who value them the most, but also allows us to obtain what is known as the Coase theorem. Its originator, Ronald Coase (1960), was awarded a Nobel Prize in part for this important contribution. It argues that in a world with perfect information and no market frictions (often referred to as transaction costs), the allocation of property rights will not affect economic efficiency. That is, even if for some reason goods are allocated to the people who do not value them the most, then with perfect information and well-functioning mechanisms to exchange goods, these goods will end up in the hands of those who value them the most.

It is important to understand the extent to which these arguments stand or fall in the face of incomplete information, when some people are better informed about the value of goods than others. To address this question we will develop a simple example that follows in the spirit of the important contribution made by George Akerlof (1970), a contribution that introduced the idea of *adverse selection* into economics and earned its author a Nobel Prize.

Imagine a scenario in which player 1 owns an orange grove. The yield of fruit depends on the quality of the soil and other local conditions, and we assume that through his experience only player 1 knows the quality of the land. Local geological surveys conclude that the quality of the land may be low, mediocre, or high, each with equal probability $\frac{1}{3}$. Thus we can think of the knowledge of player 1 about the quality of the land as his *type*, $\Theta_1 = \{L, M, H\}$, where each type has a different value for the land. Assume that the value for player 1 of owning land of type θ_1 is given by the following monetary-equivalent values:

$$v_1(\theta_1) = \begin{cases} 10 & \text{if } \theta_1 = L \\ 20 & \text{if } \theta_1 = M \\ 30 & \text{if } \theta_1 = H. \end{cases}$$

Now imagine that player 2 is a soybean grower who is considering the purchase of this land for production. Player 2's family expertise in growing soybeans has been

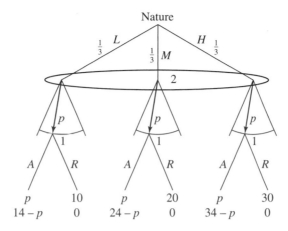

FIGURE 12.5 A trading game with incomplete information.

very profitable, but that pursuit also depends on the quality of the land. In particular player 2's monetary-equivalent values are given by

$$
v_2(\theta_1) = \begin{cases} 14 & \text{if } \theta_1 = L \\ 24 & \text{if } \theta_1 = M \\ 34 & \text{if } \theta_1 = H. \end{cases}
$$

The problem, however, is that player 2 knows only that the quality is distributed equally among the three options (the geological survey's results); he does not know which of the three it is.

Consider the following game: player 2 makes a take-it-or-leave-it price offer to player 1, after which player 1 can accept (A) or reject (R) the offer, and the game ends with either a transfer of land for the suggested price or no transfer. A strategy for player 2 is therefore a single price offer, $p \geq 0$, and a pure strategy for player 1 is a mapping from the offered price and his type space Θ_1 to a response, $s_1 : [0, \infty) \times \Theta_1 \to \{A, R\}$. The game is depicted in Figure 12.5.

The assumptions on the payoffs of the two players imply that from an efficiency point of view player 2 should own the land. Indeed if the quality of the land were common knowledge then there would be many prices at which both player 1 and player 2 would be happy to trade the property. For example, if the quality were known to be low then, as in the one-period bargaining model in Section 11.1 (the ultimatum game), the unique subgame-perfect equilibrium would have player 2 offering player 1 a price of 10 and player 1 accepting. Similarly any price between 10 and 14 would be supported by some Nash equilibrium.[6] We will now see which trades are possible in equilibrium when there is incomplete information as previously described.

Let's consider the value that player 2 places on the land. He knows that with equal probability it is worth one of the values $v_2 \in \{14, 24, 34\}$, so on average it is worth 24. He also knows that on average it is worth 20 to player 1. It would seem that the natural equilibrium candidate would be to offer the lowest price at which player 2 thinks that player 1 will accept, and such an offer would be $p = 20$.

---

6. For any $p^* \in [10, 14]$ a strategy for player 1 of accepting any offer $p \geq p^*$ and a strategy for player 2 of offering p^* would be a Nash equilibrium when the quality is low.

But what then would be player 1's best response? Recall that player 1 *knows* the quality of the land, in which case he would accept the offer only if his type is L or M. This implies that player 2 would get a parcel of land that is of low or mediocre quality, each with equal probability, and receive an expected value of $\frac{1}{2} \times 14 + \frac{1}{2} \times 24 = 19$. This implies that player 2's expected payoff from trading is $Ev_2 = 19 - 20 < 0$, implying that he would be better off not buying the land.[7]

Notice that for any offer $p \in [20, 30)$ player 1 would accept the offer only if his type is L or M, and player 2's expected value is still 19, making such a trade impossible in equilibrium. Thus if player 2 is to take into account the best response of player 1, he knows that he will get all types of player 1 to agree to sell *only if* player 2 offers 30, but in this case he would get only an expected value of 24, which is not profitable. Therefore we conclude that no trade can occur at a price greater than 20.

This implies that if trade is to occur it will occur at a price less than 20, which in turn implies that player 1 will agree to such a trade only if his type is L. Taking this into account, player 2 should offer a price no greater than $p = 14$, commensurate with player 1 trading if his type is indeed L. This logic yields the following result:

Proposition 12.1 *Trade can occur in a Bayesian Nash equilibrium only if it involves the lowest type of player 1 trading. Furthermore any price* $p^* \in [10, 14]$ *can be supported as a Bayesian Nash equilibrium.*

The reason that only the lowest type can trade in equilibrium follows from our analysis. To see that any price $p^* \in [10, 14]$ can be supported as a Bayesian Nash equilibrium, consider the following strategies: player 2 offers a price p^*, and the strategy for player 1 is

$$s_1(\theta_1) = \begin{cases} A \text{ if and only if } p \geq p^* \text{ and } R \text{ otherwise} & \text{when } \theta_1 = L \\ A \text{ if and only if } p \geq 20 \text{ and } R \text{ otherwise} & \text{when } \theta_1 = M \\ A \text{ if and only if } p \geq 30 \text{ and } R \text{ otherwise} & \text{when } \theta_1 = H. \end{cases}$$

In this case the strategies of the two players are mutual best responses, and therefore they constitute a Bayesian Nash equilibrium.

The conclusion of this result is that trade will occur only if the quality of the land is the lowest. This happens because of what is called **adverse selection.** When the buyer is willing to pay a price equal to his average value, then the type of seller who is willing to sell at this price is *below average,* because the best types select not to sell for an average price, hence the adverse selection of lower-than-average sellers. In the example this unraveling causes traded quality to drop to its lowest level, preventing the market from implementing efficient trade outcomes.

It is also worth mentioning that this scenario falls into the category of games with **common values.** In this category of games the type of one player affects the payoffs of another player. In this example the type of player 1 affects the payoffs of both player 1 and player 2. It is precisely this feature of the game that causes the adverse effects of equilibrium.

Remark Notice that in this setup the players would benefit if somehow player 1 could reveal to player 2 what the true quality of the land was. However, announcements of the type "my soil is of high quality" are useless: if player 2 were to believe such an

---

7. This follows from Bayes' rule: if only L or M types sell then *conditional on a sale,* each of these types occurs with probability $\frac{1}{2}$.

announcement then player 1 would always say "My soil is of high quality." If somehow player 1 could *credibly* reveal his type then the outcome would be different—a topic that is discussed in Chapter 16.

12.4 Committee Voting

Many decisions are made by committees. Examples include legislatures, firms, membership clubs, and juries. Each member of the committee will have different information, or different ways of interpreting information, and the goals of the committee members may be congruent or diverse. Because each committee member has private information about his own preferences or about the value of different decisions to other players, committee votes can be modeled as Bayesian games.

As an example, consider a jury made up of two players (jurors) who must collectively decide whether to acquit (A) or to convict (C) a defendant. The process calls for each player to cast a sealed vote, and the defendant is convicted only if both vote C. The problem is that there is uncertainty about whether the defendant is guilty (G) or innocent (I). The prior probability that the defendant is guilty is given by $q > \frac{1}{2}$ and is common knowledge. Assume that each player cares about making the right decision, so that if the defendant is guilty then each player receives a payoff of 1 from a conviction and 0 from an acquittal. If instead the defendant is innocent then each player receives a payoff of 0 from a conviction and 1 from an acquittal.

If the only information available to the players is the probability q then the game can be described by the following matrix:

Player 2

		A	C
Player 1	A	$1-q, 1-q$	$1-q, 1-q$
	C	$1-q, 1-q$	q, q

As the matrix shows, if either of the players chooses A then the defendant is acquitted and each player receives an expected payoff of $1 - q$. If both choose C then the defendant will be convicted and each will receive a payoff of q. Clearly because $q > \frac{1}{2}$ each player has a weakly dominant strategy, which is to vote C and convict the defendant.

Now imagine that things are a bit more complex. Each player i has a different expertise and, when observing the evidence, gets a private signal $\theta_i \in \{\theta_G, \theta_I\}$ that contains valuable information. In particular when the defendant is guilty player i is more likely to receive the signal $\theta_i = \theta_G$ than when the defendant is innocent, and vice versa for $\theta_i = \theta_I$. Assume that the signals are independent: the probability of receiving the signal θ_G when the defendant is guilty is equal to the probability of receiving the signal θ_I when the defendant is innocent, so that

$$\Pr\{\theta_i = \theta_G | G\} = \Pr\{\theta_i = \theta_I | I\} = p > \tfrac{1}{2} \quad \text{for } i \in \{1, 2\}$$

and

$$\Pr\{\theta_i = \theta_G | I\} = \Pr\{\theta_i = \theta_I | G\} = 1 - p < \tfrac{1}{2} \quad \text{for } i \in \{1, 2\}.$$

Note that the private information that each player receives, or his type, is not about how he evaluates outcomes. Instead it is information about the state of the world, namely the likelihood that the defendant is guilty. However, because the state of the world affects each player's payoff from the outcome, this means that the types are actually relevant for payoffs.

With this information structure in place we have a Bayesian game of incomplete information in which each player i will choose a signal-dependent action to maximize the probability that a guilty defendant is convicted while an innocent one is acquitted. Because each player has two types, given by the signal he observes, each player will have four pure strategies, $s_i \in \{AA, AC, CA, CC\}$, where strategy xy means that he chooses x when his signal is θ_G and he chooses y when his signal is θ_I. Notice that, like the adverse selection game in Section 12.3, this is a common-values game because both players want the right judgment to be rendered.

A natural first question is the following: if each player would be able to convict or acquit the defendant by himself, how would his signal determine his choice? Consider the decision problem with just one player. Without receiving the signal, the player knows that the defendant is guilty with probability $q > \frac{1}{2}$, so that he would choose to convict the defendant. After receiving the signal, however, the player will update his beliefs about the defendant as follows. If the signal was θ_G then the updated belief is[8]

$$\Pr\{G|\theta_i = \theta_G\} = \frac{\Pr\{G \text{ and } \theta_i = \theta_G\}}{\Pr\{\theta_i = \theta_G\}} = \frac{qp}{qp + (1-q)(1-p)} > q,$$

which means that the player is even more convinced that the defendant is guilty, and hence will choose to convict him.

If, however, the signal was θ_I then the updated belief is[9]

$$\Pr\{G|\theta_i = \theta_I\} = \frac{\Pr\{G \text{ and } \theta_I\}}{\Pr\{\theta_I\}} = \frac{q(1-p)}{q(1-p) + (1-q)p} < q.$$

This means that the player is less sure of the defendant's guilt than he was before the signal, and whether this is enough to persuade him to acquit the defendant depends on the value of p. In particular he will choose to acquit the defendant if and only if

$$\frac{q(1-p)}{q(1-p) + (1-q)p} < \frac{1}{2},$$

which reduces to $p > q$. Indeed the reason p has to be "high enough" is that the signal has to be informative enough about the defendant's actual condition. If, for example, $p = \frac{1}{2}$ the signal contains no information and the posterior $\Pr\{G|\theta_i = \theta_I\} = q$. As p increases above $\frac{1}{2}$, the signal becomes more informative and the posterior, $\Pr\{G|\theta_i = \theta_I\}$, decreases. Once p increases above the critical level of q then the signal is strong enough to reverse the player's prior conviction, and his posterior belief is that the defendant is more likely to be innocent than guilty.

The observation that if $p > q$ then each player would choose to vote according to his signal in the one-person decision problem leads to the obvious next question: is the pair of strategies in the game, $s_i = CA$, where each votes according to his signal, a Bayesian Nash equilibrium? Intuitively it seems like it should be because the signal

---

8. The inequality follows from the fact that the denominator is less than p because $qp + (1-q)(1-p) < qp + (1-q)p = p$.

9. The inequality follows from the fact that the denominator is greater than $1 - p$ because $q(1-p) + (1-q)p > q(1-p) + (1-q)(1-p) = 1 - p$.

that each player receives is informative about the actual condition of the defendant. To see whether this is indeed the case we need to verify whether CA is a best response to CA for each player.

Before proceeding to answer this question, it will be useful to calculate the probabilities of the different informational states of the two signals that the players can receive. There are four relevant informational states: either both receive the signal θ_G, both receive the signal θ_I, or each player receives a different signal (two states). For example, if the defendant is guilty then the probability that both receive a signal θ_G is equal to p^2, while if the defendant is innocent then the probability that both receive a signal θ_G is equal to $(1-p)^2$. Because the prior probability that the defendant is guilty equals q and the signals are independent conditional on the defendant's condition, the probability of both players receiving a signal θ_G is equal to $qp^2 + (1-q)(1-p)^2$. Similarly the probability that player 1 receives a signal θ_G and that player 2 receives a signal θ_I is equal to $qp(1-p) + (1-q)(1-p)p = p(1-p)$. Continuing in this way we can compute the probabilities of each of these states, given by the following table (which of course all add up to 1):

	$\theta_2 = \theta_G$	$\theta_2 = \theta_I$
$\theta_1 = \theta_G$	$qp^2 + (1-q)(1-p)^2$	$p(1-p)$
$\theta_1 = \theta_I$	$p(1-p)$	$q(1-p)^2 + (1-q)p^2$

To continue notice that given the rule that unanimity is needed to convict the defendant, a player is decisive, or "pivotal," only if the other player chooses C. The reason is that if player j chooses A then the defendant will be acquitted for sure, regardless of what player i chooses, while if player j chooses C then the decision of whether the defendant will be convicted or not depends on the choice of player i. For this reason, if player i believes that player j is playing according to the strategy CA, then he must also believe that his own vote matters only when player j observes a signal $\theta_j = \theta_G$. We therefore need to calculate the posterior belief that player i has about whether the defendant is guilty conditional on his own signal and on the belief that player j's signal is θ_G.

If player i's signal was $\theta_i = \theta_G$ then his updated belief is[10]

$$\Pr\{G|\theta_i = \theta_G \text{ and } \theta_j = \theta_G\} = \frac{\Pr\{G \text{ and } \theta_i = \theta_G \text{ and } \theta_j = \theta_G\}}{\Pr\{\theta_i = \theta_G \text{ and } \theta_j = \theta_G\}}$$

$$= \frac{qp^2}{qp^2 + (1-q)(1-p)^2} > q,$$

which means that the player is even more convinced that the defendant is guilty and hence will choose C to guarantee that the defendant is convicted when both have the signal θ_G.

If the signal was θ_I then the updated belief of player i conditional on the belief that player j's signal is θ_G is[11]

---

10. The inequality follows from the fact that the denominator is less than p^2 because $qp^2 + (1-q)(1-p)^2 < qp^2 + (1-q)p^2 = p^2$.

11. The inequality follows from the fact that the denominator is greater than $1-p$ because $q(1-p) + (1-q)p > q(1-p) + (1-q)(1-p) = 1-p$.

$$\Pr\{G|\theta_i = \theta_I \text{ and } \theta_j = \theta_G\} = \frac{\Pr\{G \text{ and } \theta_i = \theta_I \text{ and } \theta_j = \theta_G\}}{\Pr\{\theta_i = \theta_I \text{ and } \theta_j = \theta_G\}} = \frac{q(1-p)p}{p(1-p)} = q.$$

This means that when player i is conditioning the value of his signal on the event when his vote actually counts, which is when player j chooses C, then his signal $\theta_i = \theta_I$ becomes less convincing about the defendant's innocence. In fact, given the symmetric informational structure of the signals, if player i is in the situation in which he believes that he is pivotal then observing a signal of innocence is canceled out by his belief that player j saw a signal of guilt. That is, both signals cancel out, leaving the posterior equal to the prior q.

We conclude therefore that playing $s_i = CA$ for both players is not a Bayesian Nash equilibrium. In fact both players choosing CC, always convict regardless of the signal, is a Bayesian Nash equilibrium (showing this is left as part of exercise 12.8). This means that when both players receive a signal $\theta_i = \theta_I$, which occurs with probability $q(1-p)^2 + (1-q)p^2 > 0$, despite the fact that it is more likely that the defendant is innocent, he will still be convicted.

This example is a simple case of what Feddersen and Pesendorfer (1996) refer to as the "swing voter's curse" and is also closely related to independent work by Austen-Smith and Banks (1996). When a voter believes that his vote counts, he must condition this on the situation in which his vote counts. But then it means that he will interpret his information differently and not use it in an unbiased way. The jury game here is a simple two-player example of the more general analysis in Feddersen and Pesendorfer (1998), who show that with more than two players there will sometimes be conditions under which players will vote based on their information, but that the problem of the swing voter's curse is generally present.

This observation is quite important for scholars of voting in political science because it is in tension with the famous Condorcet jury theorem. That theorem states that under certain assumptions, the likelihood that a group of individuals will choose the correct alternative by majority voting exceeds the likelihood that any individual member of the group will choose that alternative. One of the assumptions is that individuals in the voting game behave exactly as they would if choosing alone—they will follow their information in voting. However, the insights from our example show that in strategic situations, individuals must form beliefs about when their vote counts and incorporate these beliefs into their own information.

12.5 Mixed Strategies Revisited: Harsanyi's Interpretation

Consider the static game of Matching Pennies introduced in Chapter 6, given by the following matrix:

		Player 2	
		H	T
Player 1	H	1, −1	−1, 1
	T	−1, 1	1, −1

and recall that the unique mixed-strategy Nash equilibrium has each player playing heads with probability $\frac{1}{2}$. One reason this solution may be somewhat unappealing is that players are indifferent between H and T, yet they are prescribed to randomize between these strategies in a unique, particular, and precise way for this to be a Nash equilibrium. Does it make sense to expect such precision when a player is indifferent?

This question has caused some discomfort with the notion of mixed-strategy equilibria. However, John Harsanyi (1973) offered a twist on the basic model of behavior to resolve this problem and alleviate, to some extent, the indifference problem. His idea works as follows. Imagine that each player may have some slight preference for choosing heads over tails or choosing tails over heads. This is done in such a way as to "break" the indifference of a player's best response if he believes that the probability of his opponent playing heads is exactly $\frac{1}{2}$.

In particular imagine that the payoffs are given by this "perturbed" Matching Pennies game:

Player 2

		H	T
Player 1	H	$1 + \varepsilon_1, -1 + \varepsilon_2$	$-1 + \varepsilon_1, 1$
	T	$-1, 1 + \varepsilon_2$	$1, -1$

and imagine that both ε_1 and ε_2 are independent and uniformly distributed on the interval $[-\varepsilon, \varepsilon]$ for some small $\varepsilon > 0$. This means that if $\varepsilon_i > 0$ is realized then player i has a strict preference for choosing H over T when he believes his opponent is choosing H with probability $\frac{1}{2}$, and similarly, if $\varepsilon_i < 0$ is realized, then player i has a strict preference for choosing T over H when he believes his opponent is choosing H with probability $\frac{1}{2}$.

Assume further that the value of ε_i is known only to player i but that the distribution of the values of ε_i is common knowledge. This perturbed Matching Pennies game is a Bayesian game of incomplete information with two actions for each player and a continuum of types, similar to the study group example solved in Section 12.2.2. Hence a pure strategy for each player is a mapping $s_i : [-\varepsilon, \varepsilon] \rightarrow \{H, T\}$ that assigns a choice to every type of player i.

Claim 12.3 *In the Bayesian perturbed Matching Pennies game, there is a unique pure-strategy Bayesian Nash equilibrium in which $s_i(\varepsilon_i) = H$ if and only if $\varepsilon_i \geq 0$, and $s_i(\varepsilon_i) = L$ if and only if $\varepsilon_i < 0$. This equilibrium converges in outcomes and payoffs to the Matching Pennies game when $\varepsilon \rightarrow 0$.*

It is quite easy to see that the proposed strategies are a Bayesian Nash equilibrium. If they are followed by player i, then because the distribution of ε_i is uniform over the interval $[-\varepsilon, \varepsilon]$, it follows that with probability $\frac{1}{2}$ player i is playing H, in which case the strategy of player j is a best response. To see that this is the unique Bayesian Nash equilibrium requires more work, but is not too hard.

This example is a simple special case of Harsanyi's purification theorem, following the idea that we can use incomplete information to "purify" any mixed-strategy equilibrium of a game of complete information. How should we interpret this result? It implies that if people are somewhat heterogeneous in the way monetary payoffs and actions are related, then we can have uncertainty over the types of players who are playing *pure strategies,* but the distribution of types makes a player have beliefs *as if* he were facing a player who is playing a mixed strategy. Harsanyi argues that using mixed-strategy equilibria in simple games of complete information can be thought of as a solution to the more complex games of incomplete information, in which players do not randomize but rather have strict best responses. The interested reader should refer to Govindan (2003) for a short and elegant presentation of Harsanyi's approach.

12.6 Summary

- In most real-world situations players will not know how much their opponents value different outcomes of the game, but they may have a good idea about the range of their valuations.

- It is possible to model uncertainty over other players' payoffs by introducing types that represent the different possible preferences of each player. Adding this together with Nature's distribution over the possible types defines a Bayesian game of incomplete information.

- Using the common prior assumption on the distribution of players' types, it is possible to adopt the Nash equilibrium concept to Bayesian games, renamed a Bayesian Nash equilibrium.

- Markets with asymmetric information can be modeled as games of incomplete information, resulting in Bayesian Nash equilibrium outcomes with inefficient trade outcomes.

- Harsanyi's purification theorem suggests that mixed-strategy equilibria in games of complete information can be thought of as representing pure-strategy Bayesian Nash equilibria of games with heterogeneous players.

12.7 Exercises

12.1 **Chicken Revisited:** Consider the game of chicken in Section 12.2.1 with the parameters $R = 8$, $H = 16$, and $L = 0$ as described there. A preacher, who knows some game theory, decides to use this model to claim that moving to a society in which all parents are lenient will have detrimental effects on the behavior of teenagers. Does equilibrium analysis support this claim? What if $R = 8$, $H = 0$, and $L = 16$?

12.2 **Cournot Revisited:** Consider the Cournot duopoly model in which two firms, 1 and 2, simultaneously choose the quantities they supply, q_1 and q_2. The price each will face is determined by the market demand function $p(q_1, q_2) = a - b(q_1 + q_2)$. Each firm has a probability μ of having a marginal unit cost of c_L and a probability $1 - \mu$ of having a marginal unit cost of c_H. These probabilities are common knowledge, but the true type is revealed only to each firm individually. Solve for the Bayesian Nash equilibrium.

12.3 **Armed Conflict:** Consider the following strategic situation: Two rival armies plan to seize a disputed territory. Each army's general can choose either to attack (A) or to not attack (N). In addition, each army is either strong (S) or weak (W) with equal probability, and the realizations for each army are *independent*. Furthermore the type of each army is known only to that army's general. An army can capture the territory if either (i) it attacks and its rival does not or (ii) it and its rival attack, but it is strong and the rival is weak. If both attack and are of equal strength then neither captures the territory. As for payoffs, the territory is worth m if captured and each army has a cost of fighting equal to s if it is strong and w if it is weak, where $s < w$. If an army attacks but its rival does not, no costs are borne by either side. Identify all the pure-strategy Bayesian Nash equilibria of this game for the following two cases, and briefly describe the intuition for your results:

   a. $m = 3$, $w = 2$, $s = 1$.
   b. $m = 3$, $w = 4$, $s = 2$.

12.4 **Grade Gambles:** Two students, 1 and 2, took a course with a professor who decided to allocate grades as follows: Two envelopes will each include a grade $g_i \in \{A, B, C, D, F\}$, where each of the five options is chosen with equal probability and the draws for each student $i \in \{1, 2\}$ are independent. The payoffs of each grade are 4, 3, 2, 1, and 0, respectively. Assume that the game is played as follows: Each student receives his envelope, opens it, and observes his grade. Then each student simultaneously decides if he wants to hold on to his grade (H) or exchange it with the other student (X). Exchange happens if and only if both choose to exchange. If an exchange does not happen then each student gets his assigned grade. If an exchange does happen then the grades are bumped up by one. That is, if student 1 had an initial grade of C and student 2 had an initial grade of D, then after the exchange student 1 will get a C (which was student 2's D) and student 2 will get a B (which was student 1's C). A grade of A is bumped up to an $A+$, which is worth 5.

   a. Assume that student 2 plays the following strategy: "I offer to exchange for every grade I get." What is the best response of student 1?

   b. Define a *weak exchange Bayesian Nash equilibrium* (WEBNE) as a Bayesian Nash equilibrium in which each student i chooses $s_i(g_i) = X$ whenever

$$E[v_i(X, s_{-i}(g_{-i}), g_i | g_i)] \geq E[v_i(H, s_{-i}(g_{-i}), g_i | g_i)].$$

That is, given his grade g_i and his (correct belief about his) opponent's strategy s_{-i}, choosing X is *as good as or better than H*. In particular a WEBNE is a pair of strategies (s_1, s_2) such that *given s_2* student 1 offers to exchange grades if exchange gives him at least as much as holding, and vice versa. Find all the symmetric (both students use the same strategy) WEBNE of this game. Are they Pareto ranked?

   c. Now assume that the professor suggests modifying the game: everything works as before, except that the students must decide if they want to exchange *before* opening their envelopes. Using equilibrium analysis, would the students prefer this game or the original one?

   d. From your conclusion in (c), what can you say about the statement "more information is always better"?

12.5 **Not All That Glitters:** A prospector owns a gold mine where he can dig to recover gold. His output depends on the amount of gold in the mine, denoted by x. The prospector knows the value of x, but the rest of the world knows only that the amount of gold is uniformly distributed on the interval $[0, 1]$. Before deciding to mine, the prospector can try to sell his mine to a large mining company, which is much more efficient in its extraction methods. The prospector can ask the company owner for any price $p \geq 0$, and the owner can reject (R) or accept (A) the offer. If the owner rejects the offer then the prospector is left to mine himself, and his payoff from self-mining is equal to $3x$. If the owner accepts the offer then the prospector's payoff is the price p, while the owner's payoff is given by the net value $4x - p$, and this is common knowledge.

    a.   Show that for a given price $p \geq 0$ there is a threshold type $x(p) \in [0, 1]$ of prospector, such that types below $x(p)$ will prefer to sell the mine, while types above $x(p)$ will prefer to self-mine.

    b.   Find the pure-strategy Bayesian Nash equilibrium of this game, and show that it is unique. What is the expected payoff of each type of prospector and of the company owner in the equilibrium you derived?

12.6   **Reap and Weep:** A farmer owns some land that he can farm to produce crops. Farming output depends on the talent of the farmer. The farmer knows his talent, but the rest of the world knows only that a farmer's talent is uniformly distributed: $\theta \in [0, 1]$. The farmer's payoff from farming his land is equal to his talent θ. Before setting up his farm, the farmer approaches the local manufacturing plant and offers to work on the production line. The farmer can ask the plant owner for any wage $w \geq 0$, and the owner can reject (R) or accept (A) the offer. If the owner rejects the offer then the farmer must return home and settle into his farming. If the owner accepts the offer then the farmer's payoff is the wage w, while the owner's payoff is given by the net value $\frac{3}{2}\theta - w$, and this is common knowledge.

    a.   Define the set of pure strategies for each player and find the pure-strategy Bayesian Nash equilibria of this game.

    b.   Averaging over the type of farmer, what are the possible levels of *social surplus* (sum of expected payoffs of the farmer and the owner in their potential relationship) from the equilibrium you derived in (a)?

    c.   A local policy maker who is advocating for the increase of social surplus is proposing to cut water subsidies to the farmers, which would imply that a farmer of type θ would get a payoff of $\frac{1}{2}\theta$ from farming his land. This policy has no effect on the productivity of manufacturing. Using the criterion of social surplus, can you advocate for this policy maker using equilibrium analysis?

12.7   **Trading Places:** Two players, 1 and 2, each own a house. Each player i values his own house at v_i. The value of player i's house to the other player, i.e., to player $j \neq i$, is $\frac{3}{2}v_i$. Each player i knows the value v_i of his own house to himself, but not the value of the other player's house. The values v_i are drawn independently from the interval $[0, 1]$ with uniform distribution.

    a.   Suppose players announce simultaneously whether they want to exchange their houses. If both players agree to an exchange, the exchange takes place. Otherwise no exchange takes place. Find a Bayesian Nash equilibrium of this game in pure strategies in which each player i accepts an exchange if and only if the value v_i does not exceed some threshold θ_i.

    b.   How would your answer to (a) change if player j's valuation of player i's house were $\frac{5}{2}v_i$?

    c.   Try to explain why any Bayesian Nash equilibrium of the game described in (a) must involve threshold strategies of the type postulated in (a).

12.8   **Jury Voting:** Consider the jury voting game in Section 12.4.

    a.   Show that both players choosing CC is a Bayesian Nash equilibrium of the two-player game.

    b.   Would it be better for the two players if only one of them decided on the fate of the defendant? Why or why not?

c. Consider a three-player game in which the defendant is convicted if at least two players vote to convict him, and acquitted otherwise (i.e., majority voting determines the defendant's fate). Assume that the informational structure is the same as in the two-player game. Are there conditions in which voting according to one's signal (C if $\theta_i = \theta_G$ and A if $\theta_i = \theta_I$) is a Bayesian Nash equilibrium?

12.9 **Grab the Dollar:** Each of two players has two possible actions: grab the dollar or don't grab the dollar. Player i's payoff is 1 if he is the only one to grab the dollar, and his payoff is 0 if he does not. Each player's payoff is -1 if both players grab the dollar.

a. Find the Nash equilibria of this game of complete information.
b. Consider the following perturbation of the game. The payoff structure is the same, with the following addition: When player i grabs the dollar his payoff is $1 + \theta_i$, where θ_i is player i's type and it is uniformly distributed over $[-\varepsilon, \varepsilon]$ with $\varepsilon > 1$. Each player knows only his own type, and the distribution of types is common knowledge. Find the pure-strategy Bayesian Nash equilibrium of this game of incomplete information.
c. Show that when ε converges to 0, the pure-strategy Bayesian Nash equilibrium converges to the mixed-strategy Nash equilibrium of the game with complete information.

12.10 **Purification:** Consider the perturbed Matching Pennies game described in Section 12.5 and show that the Bayesian Nash equilibrium identified is the unique Bayesian Nash equilibrium. (Hint: Some insights can be gained by looking at the solution to the study group example in Section 12.2.2.)

13

Auctions and Competitive Bidding

The use of auctions to sell goods has become commonplace thanks to the Internet auction platform eBay, which has become a popular shopping destination for over 100 million households across the globe. Before the age of the Internet, the thought of an auction raised visions of the sale of a Picasso or a Renoir in one of the prestigious auction houses, such as Sotheby's (founded in 1744) or Christie's (founded in 1766). In fact the use of auctions dates back much further. For a history of auctions see Cassidy (1967). Auctions are also used extensively by private- and public-sector entities to procure goods and services.[1]

The use of game theory to analyze both behavior in auctions and the design of auctions themselves, was introduced by the Nobel Laureate William Vickrey (1961), whose work spawned a large and still-expanding literature. The "big push" of game theoretical research on auctions happened after the successful use of game theory to advise both the U.S. government and the bidding firms when the Federal Communications Commission first decided to auction off portions of the electromagnetic spectrum for use by telecommunication companies in 1994. This auction was considered so successful that a reference to the work of many game theorists appears in an article in *The Economist* titled "Revenge of the Nerds" (July 23, 1994, page 70).

As we will soon see, auctions have many desirable properties, and these have made them a favorite choice of the U.S. Federal Acquisition Regulation as the legally preferred form of procurement in the public sector. They are very transparent, they have well-defined rules, they usually allocate the auctioned good to the party who values it the most, and, if well designed, they are not too easy to manipulate.

Generally speaking there are two common types of auctions. The first type, as we will refer to it, is the *open auction,* in which the bidders observe some dynamic price process that evolves until a winner emerges. There are two common forms of open auctions:

---

1. These are often referred to as "reverse auctions" because instead of selling at the highest price the auctioneer wishes to buy at the lowest price.

The English Auction: This is the classic auction we often see in movies (e.g., *The Red Violin*), in which the bidders are all in a room (or nowadays sitting by a computer or a phone) and the price of the good goes up as long as someone is willing to bid it higher. Once the last increase is no longer challenged, the last bidder to increase the price wins the auction and pays that price for the good. (The price may start at some minimum threshold, which would be the seller's *reserve* price.)

The Dutch Auction: This less familiar auction almost turns the English auction on its head. As with the English auction, the bidders observe the price changes in real time, but instead of starting low and rising by pressure from the bidders, the price starts at a prohibitively high value and the auctioneer gradually drops the price. Once a bidder shouts "buy," the auction ends and the bidder gets the good at the price at which he cried out. This auction was and still is popular in the flower markets of the Netherlands, hence its name.

The second common type of auction is the *sealed-bid auction,* in which participants write down their bids and submit them without knowing the bids of their opponents. The bids are collected, the highest bidder wins, and he then pays a price that depends on the auction rules. As with open auctions, there are two common forms of sealed-bid auctions:

The First-Price Sealed-Bid Auction: This very common auction form has each bidder write down his bid and place it in an envelope; the envelopes are opened simultaneously. The highest bidder wins and then pays a price equal to his own bid. A mirror image of this auction, sometimes referred to by practitioners as a *reverse auction,* is used by many governments and businesses to award procurement contracts. For example, if the government wants to build a new building or highway, it will present plans and specifications together with a request for bids. Each potential builder who chooses to participate will submit a sealed bid; the lowest bidder wins and receives the amount of its bid upon completion of the project (or possibly incremental amounts upon the completion of agreed-upon milestones).

The Second-Price Sealed-Bid Auction: As with the first-price sealed-bid auction, each bidder writes down his bid and places it in an envelope; the envelopes are opened simultaneously and the highest bidder wins the auction. The difference is that although the highest bidder wins, he *does not pay his bid* but instead pays a price equal to the second-highest bid or the *highest losing bid.* This auction may not seem common or familiar, but it turns out that it has very appealing properties and shares a strong connection to the very common English auction.

Regardless of the type of auction that is being administered, two things should be obvious. First, auctions are games in which the players are the bidders, the actions are the bids, and the payoffs depend on whether or not one receives the good and how much one pays for it (and possibly how much one pays for participating in the auction in the first place). In fact we have seen a simple two-player version of a different kind of auction, the all-pay auction, in Section 6.1.4. Second, it is hard to believe that bidders know exactly how much the good being sold is worth to the other bidders. Hence auctions have all the characteristics we have specified as appropriate for modeling as Bayesian games of incomplete information.

13.1 Independent Private Values

Buyers can have two reasons to purchase a good. They may wish to consume it, like some specialty food, thus using it for their own consumption benefit. Or they may wish to buy it as an investment, like gold or stocks, and possibly sell it at a later date. Motives can also be mixed, like those for buying a rare painting or a home, which can be both enjoyed by the buyer and considered an investment.

In the first case, buying a good for immediate consumption, the only valuation of the good that should matter to each potential buyer is how much it is worth from his own private perspective, with no consideration of how much others value the good. Consider a seller who offers a prime but extremely ripe piece of filet mignon that will go bad within the hour. With no time to turn around and resell it, the only thing one can do with the filet is to quickly grill it and eat it. The value one imputes to it, or one's willingness to pay for it, should depend only on how much one will enjoy this particular dish. This situation is referred to as one of **private values,** in which each person's willingness to pay depends only on his own type, and this in turn is private information. This differs from the **common-values** setting, in which the preferences of some players may depend on the types of other players (e.g., recall the adverse selection game in Section 12.3).

We start our analysis of auctions using the simpler private values setting and describe our auction setting as a game with the set of bidders $N = \{1, 2, \ldots, n\}$. We refrain from considering the seller as a player, because once the auction rules are set by him, he has no effect on bidding behavior or the outcomes that follow.

To capture the idea that players differ in their private willingness to pay, we resort to the familiar notion of types to describe the players' private valuations. In particular assume that the value for player i of obtaining the good is his type, θ_i, drawn from the interval $\theta_i \in [\underline{\theta}_i, \overline{\theta}_i]$ according to the cumulative distribution function $F_i(\cdot)$, where $F_i(\theta') = \Pr\{\theta_i \leq \theta'\}$ and $\underline{\theta}_i \geq 0$. We assume that the draws of θ_i from $[\underline{\theta}_i, \overline{\theta}_i]$ according to $F_i(\cdot)$ are independent and not correlated in any way. The payoff of a player who does not get the good is normalized to zero, and the payoff of a player who gets the good and pays a price $p \geq 0$ is $v_i = \theta_i - p$. Hence this setting is referred to as the **independent private values** or IPV case.

Following the methodology of Bayesian games, we assume that every player knows the distribution of all types of the other players and uses the $n - 1$ cumulative distribution functions $F_j(\cdot)$, $j \neq i$, to form beliefs about the types θ_{-i} of the other players.

13.1.1 Second-Price Sealed-Bid Auctions

We begin the analysis with second-price sealed-bid auctions, which are the easiest to analyze. The rules of the game are as follows. First, players learn their private valuations, but they know only the distribution of their opponents' valuations. Second, each player submits a bid $b_i \geq 0$, which is the action each player can choose. Finally the bids are collected, the highest bidder wins, and he pays a price equal to the second-highest bid. Thus we can write the payoff function of each player i as[2]

---

2. Notice that we resolve the possibility of two players being tied with the highest bid by assuming that in that case neither gets the good and both pay nothing. This can be easily changed to have

FIGURE 13.1   Weakly dominant strategies in second-price auctions.

$$v_i(b_i, b_{-i}; \theta_i) = \begin{cases} \theta_i - b_j^* & \text{if } b_i > b_j \text{ for all } j \neq i \text{ and } b_j^* \equiv \max_{j \neq i}\{b_j\} \\ 0 & \text{if } b_i \leq b_j \text{ for some } j \neq i. \end{cases}$$

Given that a player's action is his bid, a player's strategy assigns a bid to each of the player's types, in this case private valuations. Thus a strategy for player i is a function $s_i : [\underline{\theta}_i, \overline{\theta}_i] \to \mathbb{R}_+$ that assigns a nonnegative bid to each of his possible valuations.

Now we can write down the payoff of player i with valuation θ_i as a function of his own bid b_i and the strategies used by the other players $s_j(\cdot)$, $j \neq i$. First it is worth writing down a shorthand expression for player i's expected payoff as follows:

$$E_{\theta_{-i}}[v_i(b_i, s_{-i}(\theta_{-i}); \theta_i)|\theta_i] = \Pr\{i \text{ wins and pays } p\} \times (\theta_i - p) + \Pr\{i \text{ loses}\} \times 0.$$

In more specific terms, taking into account the rules of the second-price sealed-bid auction, the probability that player i wins is equal to the probability that all the other bids are below b_i, and in this event the price player i pays is the second-highest bid equal to p. The hard question is, what is the probability that all the other bids are below b_i? This of course will depend on the strategies $s_j(\theta_j)$ of the other players $j \neq i$. In particular the probability that i's bid is higher than j's bid is equal to the probability that j is a type that bids less than b_i. But without knowing exactly what j's strategy is we cannot write this probability down. This may look like quite a daunting task, but it turns out that the rules of this auction result in a rather surprising result:

Proposition 13.1   *In the second-price sealed-bid auction, each player has a weakly dominant strategy, which is to bid his true valuation. That is, $s_i(\theta_i) = \theta_i$ for all $i \in N$ is a Bayesian Nash equilibrium in weakly dominant strategies.*

If you are not already familiar with this famous result, first discovered by Vickrey (1961), then it is indeed quite remarkable. As we will see, it is also quite straight-forward once the analysis is laid out. We will prove this result by showing that for any valuation θ_i bidding θ_i weakly dominates both higher bids and lower bids. The analysis will argue the case for not bidding $b_i < \theta_i$, and the other argument is practically the mirror image.

Consider the case of a bidder who is bidding below his valuation, $b_i < \theta_i$, as described in Figure 13.1. There are three possible cases of interest with respect to the other $n - 1$ bids:

Case 1:   Player i is the highest bidder, in which case i wins and pays a price $p < b_i$. This corresponds to the situation in which all the other $n - 1$ bids are in region

---

one of them randomly be the winner and pay the second-highest price, which would be equal to his own bid (since the two tied as the highest bid). This case would add negligible complications and is remarked upon at the end of this section.

L in Figure 13.1, including the second-highest bid. If instead of bidding b_i player i would have bid θ_i then he would still win and pay the same price, so in case 1 bidding his valuation is as good as bidding b_i.

Case 2: The highest bidder j bids $b_j^* > \theta_i$, in which case i loses. This corresponds to the situation in which the winning bid is in region H in Figure 13.1. If instead of bidding b_i player i would have bid θ_i then he would still lose to b_j^*, so in case 2 bidding his valuation is as good as bidding b_i.

Case 3: The highest bidder j bids $b_i < b_j^* < \theta_i$, so that the highest bid is in region M and i does not win. If instead player i would have bid θ_i, he would have won the auction and received a payoff of

$$v_i = \theta_i - b_j^* > 0,$$

making this a profitable deviation, so in case 3 bidding his valuation is strictly better than bidding b_i.

Since cases 1–3 cover all the relevant situations, we conclude that bidding θ_i weakly dominates any lower bid because it is never worse and sometimes better. A similar argument shows that a bid $b_i > \theta_i$ will also be weakly dominated by bidding θ_i, and is left as exercise 13.1.

The fact that every player has a **weakly dominant strategy**, $s_i(\theta_i) = \theta_i$, implies that each player bidding his valuation is a Bayesian Nash equilibrium in weakly dominant strategies. This result is noteworthy not only because of its simple prescription—that players bid their valuations truthfully in a second-price sealed-bid auction—but also because it implies three other attractive attributes of this auction format.

First, in the IPV setting, bidders in a second-price sealed-bid auction *do not care* about the probability distribution over their opponents' types, and therefore the assumption of common knowledge of the distribution of types can be relaxed when such auctions are analyzed. In particular it means that we can apply this result even when we think that players have no idea about their opponents' valuations. This is a very nice feature of the second-price sealed-bid auction.

Second, in a second-price sealed-bid auction, even if types are correlated but values are private (a player's value depends only on his own type and not the types of other players), then it is a weakly dominant strategy to bid truthfully. This implies that if we are correct about our private values assumption, but we incorrectly assume that values are independent, it is still true that bidding your valuation is a weakly dominant strategy.

Last but not least, in the private values setting, the outcome of a second-price sealed-bid auction is *Pareto optimal* because the person who values the good most will be the one who gets it.

As we will now briefly see, the second-price sealed-bid auction is closely related to the very common English auction.

Remark We can change the game to be more appealing with respect to the treatment of tied highest bids. Assume that if $m \leq n$ bidders tie with the highest bid then they are each equally likely to win, and they pay the second-highest bid, which by definition is equal to their own bid in the case of a tie. Then the payoff function of each bidder can be written as

$$v_i(b_i, b_{-i}; \theta_i) = \begin{cases} \theta_i - b_j^* & \text{if } b_i > b_j \text{ for all } j \neq i \text{ and } b_j^* \equiv \max_{j \neq i}\{b_j\} \\ \frac{\theta_i - b_i}{\#\{\text{highest bidders}\}} & \text{if } b_i \geq b_j \text{ for all } j \neq i \text{ and } b_i = b_j \text{ for some } j \neq i \\ 0 & \text{if } b_i < b_j \text{ for some } j \neq i. \end{cases}$$

You should be able to convince yourself that the arguments leading to $b_i = \theta_i$ being a weakly dominant strategy are still valid, and thus that the foregoing analysis applies.

13.1.2 English Auctions

The typical English auction has an auctioneer with bidders in a room, and the auctioneer increases the standing bid until no bidder is willing to push the price higher. The process is observed by all the bidders as the price progresses toward the final price. This seems rather easy to describe as a game, but there is a slight problem: without prespecified discrete increments, a player i with a valuation greater than the current price ($p < \theta_i$) has no best response, because for any increment he can make, a smaller increment would be better. This problem is similar to the one we saw in the Bertrand duopoly model with different costs (see Section 5.2.4), and it was discussed further in Section 6.4.

There are two solutions to this problem. First, we can use a discrete action space such as dollars and cents. This is realistic, but if we would like to use calculus for the analysis then this would be impossible, implying a more cumbersome analytical approach. A second solution is to change the game slightly without losing the spirit of English auctions, which is precisely what Milgrom and Weber (1982) suggested in their influential paper. The "button-auction" game they proposed proceeds as follows.

As before, there are n players, each with valuation $\theta_i \in [\underline{\theta}_i, \bar{\theta}_i]$ drawn according to the cumulative distribution function $F_i(\cdot)$, which is independent across bidders. An added actor (not a strategic player) is an auctioneer who announces the current price, starting at $p = 0$, and raises it continuously over time. Each player has a button that is pressed at the beginning of the game when the starting price is $p = 0$. If the button of player i is continuously pressed and the current price is $p > 0$, this means that player i is willing to pay p if everyone else would drop out of the auction now. Once a player releases his button, he drops out of the auction and cannot reenter it. The winner is the last person to hold his button down, and the price is the posted price at which the second-to-last player let go of his button.

The similarity between the button-auction model and the English auction is apparent. As long as there is more than one player who is willing to push the current price up, the auctioneer will raise the price. Unlike the English auction, in which at a current price $\$p$ the auctioneer cries "Do I hear $\$(p + k)$?" in the button-auction model this happens automatically with an increment k that is infinitesimally small.

As in the second-price sealed-bid auction, we define strategies for each player as maps from their types, or valuations, to the price at which a player lets go of his button. Using these definitions of strategies, we can derive the expected payoff functions of every player, given his beliefs about the other players' strategies and types. This again may seem like a daunting task, but just as with the second-price sealed-bid auction, the following appealing result makes the analysis very simple:

Proposition 13.2 *In the button-auction model it is a weakly dominant strategy for each player to keep his button pressed as long as $p < \theta_i$ and to release it once $p = \theta_i$. This results in a Bayesian Nash equilibrium in weakly dominated strategies that is outcome-equivalent to the second-price sealed-bid auction.*

The proof of this proposition follows precisely from the following simple observations: First, if the current price is $p < \theta_i$ then player i (who is not the currently highest bidder) should continue holding his button, therefore causing the price to increase further. If he drops out, he may forgo the opportunity of winning the item at a price lower than his valuation, and if he continues with the proposed strategy he is guaranteed a nonnegative payment because he exits when $p = \theta_i$. Second, if $p = \theta_i$ then player i should drop out; otherwise he might win and pay more than his valuation.

The fact that the outcome is the same as in the second-price sealed-bid auction follows immediately: the player with the highest valuation wins, and he pays a price equal to the second-highest valuation because that was the price at which the second-to-last player dropped out. Hence the same three appealing features of the second-price sealed-bid auction are inherited by the English auction in the private values setting. First, each player has a best strategy that depends neither on the distributions of the other players' types nor on the strategies that they are using. Second, even if the distributions of the players' types are correlated, as long as each player's valuation depends only on his type (private values) then this is a weakly dominant strategy. Last but not least, the outcome is Pareto optimal.

Note that on the popular marketplace site eBay, the way the auctions work is very similar to this button-auction model: eBay uses a "proxy bidding" system that takes your bid and uses it as a proxy for your behavior. If you are the highest bidder then the current price is one increment above the second-highest bidder, in a similar fashion to the button-auction model with a positive increment. It is interesting to note that the instructions on eBay suggest that you bid truthfully: "When you place a bid, enter the maximum amount you are willing to pay for the item. eBay will bid on your behalf only if there is a competing bidder and only up to your maximum amount."[3]

13.1.3   First-Price Sealed-Bid and Dutch Auctions

Unlike the second-price sealed-bid auction, the first-price sealed-bid auction has each bidder pay his own bid. Obviously no bidder would want to bid more than his valuation. Not surprisingly, however, in the first-price sealed-bid auction we lose one of the appealing features of the second-price auction:

Claim 13.1   *In a first-price sealed-bid auction it is a dominated strategy for a player to bid his valuation.*

The reason this claim is true is straightforward. If a player bids his valuation then his payoff is zero: if he loses then he gets zero and if he wins then he pays his valuation, thus getting a payoff of zero. If instead the player bids slightly less than his valuation, then with some positive probability the other bids will be below his bid, in which case he will win the item and obtain a positive payoff. Hence his expected payoff is positive, which is more than he gets from bidding his valuation.

The fact that players will not want to bid their valuations implies a simple intuition. By lowering your bid incrementally you increase your net payoff margin when you win, which is the marginal benefit from reducing your bid. At the same time the probability of winning becomes lower, which is the marginal cost of reducing your

---

3. At the time of writing the quote appeared on http://pages.ebay.com/education/buyingtips/index.html#bid.

bid. However, this simple intuition is far from enough to help us find a Bayesian Nash equilibrium. It turns out that in the first-price sealed-bid auction the players do not have any weakly dominant strategy that maps types to bids because their best response depends on the belief over other players' bids, which depends on the other players' strategies.

To help make our analysis a bit more tractable, we will impose a reasonable assumption on the bidding behavior of players:

Assumption 13.1 The higher a player's valuation, the higher is his bid. That is, if $\theta'_j > \theta''_j$ then $s_j(\theta'_j) > s_j(\theta''_j)$.

This assumption implies that each player's bid function is *invertible*, meaning that for every bid b that a player j can make, there is a unique type of player j that makes this bid, and this type is given by the inverse of the player's strategy, $s_j^{-1}(b)$. The main benefit of imposing this assumption is that it results in a very simple expression for the probability that i's bid is higher than j's bid, which is

$$\Pr\{s_j(\theta_j) < b_i\} = \Pr\{\theta_j < s_j^{-1}(b_i)\} = F_j(s_j^{-1}(b_i)).$$

Recall that in the IPV setting we assumed that the different types are drawn independently, meaning that the valuations of the players are independent random variables. For this reason, the probability that i's bid is higher than all other bids is just the multiplication of the probabilities $F_j(s_j^{-1}(b_i))$, $j \neq i$. Thus the expected payoff of player i in a first-price sealed-bid auction is

$$E_{\theta_{-i}}[v_i(b_i, s_{-i}(\theta_{-i}); \theta_i)|\theta_i] = \prod_{j \neq i} \left[F_j(s_j^{-1}(b_i)) \right] \times (\theta_i - b_i). \qquad (13.1)$$

Therefore both the distributional assumptions on the players' types will matter, as well as the beliefs about their strategies. This is where the common knowledge assumption on a common prior has real bite, because without it, players cannot have consistent beliefs about the distribution of types. To complete the analysis of what a Bayesian Nash equilibrium for a first-price sealed-bid auction might be, we have to find a profile of strategies so that each player i's strategy maximizes his expected payoff given in (13.1), given the other $n - 1$ players' strategies.

To simplify our analysis further, consider the symmetric case in which for each player i the valuation is distributed on the same interval $[\underline{\theta}, \overline{\theta}]$ with $\underline{\theta} > 0$, according to the same distribution $F(\cdot)$ and density $f(\cdot)$. We will write down the conditions for a symmetric equilibrium (a restriction that makes the analysis much easier), where each of the n bidders uses the same strategy $s : [\underline{\theta}, \theta] \to \mathbb{R}_+$. Thus the objective of each player i when his type is θ_i, and when all other players are using the bidding strategy $s(\cdot)$, is derived from (13.1) and becomes

$$\max_{b \geq 0} E_{-i}[v_i(b, s_{-i}(\theta_{-i}); \theta_i)|\theta_i] = \left[F(s^{-1}(b)) \right]^{n-1} (\theta_i - b). \qquad (13.2)$$

Assuming an interior solution we can write the first-order condition of (13.2) as follows:

$$-\left[F(s^{-1}(b)) \right]^{n-1} + (n - 1) \left[F(s^{-1}(b)) \right]^{n-2} f(s^{-1}(b)) \frac{ds^{-1}(b)}{db} (\theta_i - b) = 0.$$

$$(13.3)$$

For the profile of bidding strategies $(s_1(\cdot), \ldots, s_n(\cdot)) = (s^*(\cdot), \ldots, s^*(\cdot))$ to be a Bayesian Nash equilibrium, $b = s^*(\theta)$ itself has to be the solution to (13.3) for any value of $\theta \in [\underline{\theta}, \overline{\theta}]$. If this is true then $s^{*-1}(b) = \theta$, and because $\frac{ds^{-1}(b)}{db} = \frac{1}{s'(s^{-1}(b))}$, we can derive the condition for a symmetric Bayesian Nash equilibrium from (13.3) to be

$$- [F(\theta)]^{n-1} + \frac{(n-1)[F(\theta)]^{n-2} f(\theta)(\theta - s(\theta))}{s'(\theta)} = 0. \qquad (13.4)$$

This may look a bit daunting, but it turns out to be a rather manageable differential equation.[4] We can rewrite (13.4) to obtain

$$[F(\theta)]^{n-1} s'(\theta) + (n-1)[F(\theta)]^{n-2} f(\theta)s(\theta) = (n-1)[F(\theta)]^{n-2} f(\theta)\theta. \qquad (13.5)$$

A careful look at the left side of (13.5) should convince you that it is the derivative of $[F(\theta)]^{n-1} s(\theta)$. It follows therefore that by integrating both sides of (13.5) we obtain

$$[F(\theta)]^{n-1} s(\theta) = \int_{\underline{\theta}}^{\theta} (n-1)[F(x)]^{n-2} f(x)x \, dx. \qquad (13.6)$$

The next step is to use integration by parts on the right side of (13.6), which yields the symmetric Bayesian Nash equilibrium bid (strategy) function[5]

$$s(\theta) = \theta - \frac{\int_{\underline{\theta}}^{\theta} [F(x)]^{n-1} dx}{[F(\theta)]^{n-1}}. \qquad (13.7)$$

As (13.7) demonstrates, in this Bayesian Nash equilibrium a player i with valuation θ will be *below* his valuation, as intuition suggests, and the amount by which he "shades" his bid will depend on the distribution of valuations $F(\cdot)$, the number of bidders n, and his type θ. Interestingly it turns out that in this symmetric example this is the unique symmetric Bayesian Nash equilibrium, but showing this result is beyond the scope of this text.

As a relatively simple illustration of such a Bayesian Nash equilibrium, consider the special case in which $F(\cdot)$ is uniform over $[0, 1]$. We can explicitly solve for (13.7) as follows:

---

4. A differential equation is an equation in which the derivatives of a function appear as variables. In this case we have a first-order differential equation since only the first derivative of $s(\cdot)$ appears together with $s(\cdot)$ in the equation. The solutions to these problems can be found in any standard textbook on the topic.

5. Given two functions $g(\theta)$ and $h(\theta)$, by the product rule of derivation we know that $(gh)' = g'h + gh'$, and integrating both sides of this equation implies that $g(\theta)h(\theta) = \int g'(\theta)h(\theta)d\theta + \int g(\theta)h'(\theta)d\theta$. Integration by parts uses this fact to express the following relationship: $\int g'(\theta)h(\theta)d\theta = g(\theta)h(\theta) - \int g(\theta)h'(\theta)d\theta$. Letting $g'(\theta) = (n-1)[F(\theta)]^{n-2} f(\theta)$ and letting $h(\theta) = \theta$, we have

$$\int_{\underline{\theta}}^{\theta} (n-1)[F(x)]^{n-2} f(x)x \, dx = [F(\theta)]^{n-1}\theta - \int_{\underline{\theta}}^{\theta} [F(x)]^{n-1} dx$$

because $F(\underline{\theta}) = 0$.

$$s(\theta) = \theta - \frac{\int_0^\theta x^{n-1}dx}{\theta^{n-1}}$$

$$= \theta - \frac{\theta^n}{n\theta^{n-1}}$$

$$= \theta\left(\frac{n-1}{n}\right). \tag{13.8}$$

In this case the equilibrium bid function or strategy of each player is his type multiplied by what we can call a "bid shading factor" equal to $0 < \frac{n-1}{n} < 1$. This of course implies that players bid below their valuation θ and that the amount of shading depends on the number of bidders. (The strategy already uses the uniform distribution to obtain this simple linear form.) As we can clearly see, the more bidders there are, the more fierce the competition is, and the closer is the bid to the player's true valuation θ.

It turns out that the Dutch and first-price sealed-bid auctions are closely related in the same way that the English and second-price sealed-bid auctions are. Recall that the Dutch auction starts at an obscenely high price and then drops continuously until the first bidder jumps in with the announcement "buy." Each bidder therefore must ask himself, at what price will I jump in and announce that I will buy? The longer a bidder waits, the more he obtains conditional on winning, but this comes at the risk of someone else jumping in, which reduces the probability of winning.

This simple observation suggests that the Dutch and first-price sealed-bid auctions are *strategically equivalent* in that these two games have the same normal form, and as a consequence have the same set of Bayesian Nash equilibria. In particular the solution to (13.7) will also be the solution to the equilibrium "buy" announcement of a bidder in the Dutch auction. The intuition is that the bid in the first-price sealed-bid auction, like the acceptance price in the Dutch auction, trades off the bidder's profit margin conditional on winning (which is the marginal benefit of shading the bid further below the valuation) with the reduced probability of winning (which is the marginal cost of shading the bid). If a player in a (symmetric) Dutch auction believes that all the other players are using the bid function given by (13.7) then by using the same bid function himself he will be playing a best response, because the Dutch and first-price sealed-bid auctions are strategically equivalent.

13.1.4 Revenue Equivalence

As we have seen, there are many auction formats that differ in their rules: sealed bids, open bids, first price, second price, . . . you can add your own imaginative twists. As a seller who plans to sell a good using an auction, the following question is natural: which of the many auction formats will yield the highest expected revenue? For the IPV setting, in which bidders' payoffs are given by $\theta - p$ if they win at price p and zero if they lose, the answer is quite surprising. It turns out that the expected revenue a seller obtains from any of the four auctions we considered is the same.

It should not be too surprising that the second-price sealed-bid auction and the English auction yield the same expected revenues because in both cases the players have a (weakly) dominant strategy, and it results in the same outcome. In addition, as argued earlier, the Dutch and first-price sealed-bid auctions result in similar revenues because in both the bid shading is a delicate play between the attempt to make some profit and the attempt to win the good, making these games strategically equivalent.

However, the fact that bid shading in the first-price auction and the seller settling for the second-highest valuation in the second-price auction both result in the same expected revenues is quite remarkable.

This result is called the **revenue equivalence theorem** and is considered one of the major findings of auction theory. A first proof of this result was given by William Vickrey (1961) and generalized by Myerson (1981) and independently by Riley and Samuelson (1981). Both Vickrey and Myerson were awarded the Nobel Prize in economics based partly on their contributions to auction theory, which have shaped the way economists and policy makers think about this commonly used allocation mechanism.

The revenue equivalence theorem states that any auction game that satisfies four conditions will yield the seller the same expected revenue, and will yield each type of bidder the same expected payoff. These conditions are as follows: (1) each bidder's type is drawn from a "well-behaved" distribution;[6] (2) bidders are risk neutral;[7] (3) the bidder with the highest type wins; and (4) the bidder with the lowest possible type ($\underline{\theta}$) has an expected payoff of zero.

Before turning to a simple illustration of this important result some background will be helpful. Imagine that we are drawing a series of realizations of a random variable using some distribution $F(\cdot)$. That is, we have some random variable $x \sim F(\cdot)$, and we are taking a series of draws, x_1, x_2, and so on until some x_n, all from the distribution $F(\cdot)$. Furthermore each of the draws x_i is drawn independently so that no two draws are correlated. Given the sample of draws $(x_1, x_2, \ldots, x_n)$, we can consider the ranking of the realized values and rank them from highest to lowest. The highest draw out of n, $x_n^{[1]} \equiv \max\{x_1, x_2, \ldots, x_n\}$, is called the **first-order statistic** of this sample. Similarly the second-highest draw, $x_n^{[2]}$, is called the **second-order statistic,** and the same for third-, fourth-, and kth-order statistics for any $k \in \{1, 2, \ldots, n\}$ for this sample in which we obtain n draws from the distribution $F(\cdot)$.

The relation of these statistical objects to bidding should now be rather apparent. Imagine n symmetric bidders with valuations that are independently drawn from some distribution $F(\cdot)$ over an interval $[\underline{\theta}, \overline{\theta}]$. In a second-price auction the seller will receive the second-highest valuation of the n bidders. The unique Bayesian Nash equilibrium in weakly dominant strategies has each player bid his valuation and the highest-bid winner pays the second-highest bid. This is just the second-order statistic of n draws from the distribution $F(\cdot)$. Hence the expected revenue of the seller is $E[\theta_n^{[2]}]$, which is the expected value of the second-order statistic.

In a first-price auction the seller will receive the highest bid of the n bidders. If the symmetric bidders all use the same bidding strategy $s(\theta)$ then this is just the first-order statistic of n draws from some distribution $G(\cdot)$, which is the distribution of bids, not valuations. Hence the expected revenue of the seller is $E[b_n^{[1]}]$, which is the expected value of the first-order statistic of the bid functions.

To see an illustration of the revenue equivalence theorem we will show that with n symmetric bidders whose types are drawn from the uniform distribution over the interval $[0, 1]$, the expected first-order statistic of the bid functions from a first-price sealed-bid auction is equal to the expected second-order statistic of the types (which is what the winner will pay in a second-price sealed-bid auction).

---

6. The distribution function $F_i(\cdot)$ from which each bidder's type is distributed must be strictly increasing and continuous.

7. You are asked to verify what happens when bidders are risk averse in exercise 13.7.

As we have shown in Section 13.1.3, the symmetric Bayesian Nash equilibrium with n bidders whose types are drawn uniformly over an interval $[\underline{\theta}, \overline{\theta}]$ is given by $s(\theta) = \theta\left(\frac{n-1}{n}\right)$, meaning that each bidder bids $\frac{n}{n-1}$ of his actual valuation θ. Hence the winning bid will be max $\left\{\frac{\theta_1(n-1)}{n}, \frac{\theta_2(n-1)}{n}, \ldots, \frac{\theta_n(n-1)}{n}\right\}$, which is equal to the first-order statistic of a series of n draws from the uniform distribution on the interval $\left[0, \frac{n-1}{n}\right]$. We can then compare this to the expected value from the second-price sealed-bid auction, which is the expected second-order statistic of a series of n draws from the uniform distribution over $[0, 1]$.

To compute these two expected revenues, we need to calculate the distribution of the first- and second-order statistics for the uniform distribution. The cumulative distribution of the first-order statistic $x_n^{[1]}$ of n draws of bids, which we denote by $F_n^{[1]}(\cdot)$, is not difficult to derive. In particular $F_n^{[1]}(x)$ is equal to the probability that all the n bids that were drawn are less than or equal to x, and hence

$$F_n^{[1]}(x) = \Pr\{\max\{b_1, b_2, \ldots, b_n\} \leq x\} = (F(x))^n,$$

where the last equality follows from the fact that the n values are drawn independently.[8] For the uniform distribution, the bid of each agent b_i is uniformly drawn from $[0, \frac{n-1}{n}]$, so that $F(b_i) = \frac{n}{n-1}b_i$, implying that the distribution of the first-order statistic is

$$F_n^{[1]}(x) = [F(x)]^n = \left(\frac{nx}{n-1}\right)^n \quad \text{for } 0 \leq x \leq \frac{n-1}{n}. \tag{13.9}$$

It follows from (13.9) that the density of the first-order statistic is given by

$$f_n^{[1]}(x) = \frac{dF_n^{[1]}(x)}{dx} = n\frac{n}{n-1}\left(\frac{nx}{n-1}\right)^{n-1} = \frac{n}{x}\left(\frac{nx}{n-1}\right)^n,$$

which in turn implies that the expected first-order statistic from Bayesian Nash equilibrium bids in a first-price auction is given by

$$E\left[b_n^{[1]}\right] = \int_0^{\frac{n-1}{n}} x f_n^{[1]}(x)dx = \int_0^{\frac{n-1}{n}} n\left(\frac{nx}{n-1}\right)^n dx$$

$$= \left[\frac{n}{n+1}\left(\frac{n}{n-1}\right)^n x^{n+1}\right]_0^{\frac{n-1}{n}} = \frac{n-1}{n+1}.$$

Thus the seller's expected revenue from the first-price sealed-bid auction with n bidders whose valuations are independently drawn from the uniform distribution over $[0, 1]$ is equal to $\frac{n-1}{n+1}$.

Now consider the distribution of a second-order statistic from n independent draws that is denoted by the cumulative distribution function $F_n^{[2]}(x) = \Pr\{x_n^{[2]} \leq x\}$. The event $\{x_n^{[2]} \leq x\}$ can occur in one of two *distinct* ways (i.e., from two mutually exclusive events). In the first event, all the draws x_i are below x, or $x_i \leq x$ for all $i = 1, 2, \ldots, n$. In the second event, for $n - 1$ of the draws, $x_i \leq x$, and for only one draw j, $x_j > x$. This second event can occur in n different ways: (1) $x_1 > x$ and for

---

8. That is, $\Pr\{b_i \leq x\} = F(x)$ for all $i = 1, 2, \ldots, n$, and since these draws are independent then $\Pr\{\max\{b_1, b_2, \ldots, b_n\} \leq x\} = F(x) \times F(x) \times \cdots \times F(x) = (F(x))^n$.

all $i \neq 1$, $x_i \leq x$; (2) $x_2 > x$ and for all $i \neq 2$, $x_i \leq x$; and so on up to (n) $x_n > x$ and for all $i \neq n$, $x_i \leq x$.

This description reveals that the cumulative distribution of the second-order statistic is as follows:

$$F_n^{[2]}(x) = \Pr\{\max\{b_1, b_2, \ldots, b_n\} \leq x\} + \sum_{i=1}^{n} \Pr\{x_i > x \text{ and for all } j \neq i, \ x_j \leq x\}$$

$$= [F(x)]^n + \sum_{i=1}^{n}(1 - F(x))[F(x)]^{n-1}$$

$$= [F(x)]^n + n(1 - F(x))[F(x)]^{n-1}$$

$$= n[F(x)]^{n-1} - (n-1)[F(x)]^n.$$

The bid of each agent b_i in the second-price sealed-bid auction is his valuation, which we assumed is uniformly drawn from $[0, 1]$, so that $F(b_i) = b_i$, and we obtain

$$F_n^{[2]}(x) = nx^{n-1} - (n-1)x^n. \tag{13.10}$$

It follows from (13.10) that the density of the second-order statistic is

$$f_n^{[2]}(x) = \frac{dF_n^{[2]}(x)}{dx} = n(n-1)x^{n-2} - n(n-1)x^{n-1},$$

and it follows that the expected second-order statistic is

$$E\left[\theta_2^{[2]}\right] = \int_0^1 x f_n^{[2]}(x)dx = \int_0^1 (n(n-1)x^{n-1} - n(n-1)x^n)dx$$

$$= \left[(n-1)x^n - \frac{n(n-1)x^{n+1}}{n+1}\right]\Big|_0^1 = (n-1) - \frac{n(n-1)}{n+1} = \frac{n-1}{n+1}.$$

We conclude therefore that the expected revenue of the seller is the same in both the first-price and second-price sealed-bid auctions. The revenue equivalence theorem is in fact much more general than the simplified version described here and is a very deep result that applies to an important class of games of incomplete information. The result is related to what is often called the envelope theorem. The interested reader is encouraged to consult the books by Krishna (2002) and Milgrom (2004) for an in-depth analysis of the revenue equivalence theorem and its implications for auction design.

13.2 Common Values and the Winner's Curse

Recall that the private values setting describes the situation in which each player's payoff depends on the profile of actions of all players and on his own type, but not on the types of other players. This setting is useful in describing such scenarios as how much different people value a hamburger or a bag of chips, but in many cases the payoff of one player will depend on the private information of other players.

Consider, for instance, a house that is on the market—how much would you be willing to pay for it? The answer will depend on two major components: first, your own private value of living in that house, and second, what you expect to get for the house if you choose to sell it at a later date. Other people may have investigated the house and hired inspectors who discovered different shortcomings that affect the value of the house. Hence the information of *other people* would enter into your willingness to pay for a house if you knew it, yet other people's information, or types, will generally be their own private information. This same argument will apply to a piece of art, a car, or even a movie—you may value a movie more if you think other people value it more, so you can later talk to them about it and all agree on how good (or bad) it was.

We refer to such scenarios as having a *common-values* component, similar to the adverse selection model of Section 12.3. To illustrate an extreme example of *pure common values* in which each player has the same value from winning the auction, imagine that two identical oil firms are considering the purchase of a new oil field. It is common knowledge that the amount of oil is either small, worth $10 million of net profits; medium, worth $20 million of net profits; or large, worth $30 million of net profits. Thus the oil field has one of three values, $v \in \{10, 20, 30\}$. Imagine that it is also common knowledge that these values are distributed so that it is equally likely that the amount is small or large, and twice as likely that it is medium, so that

$$\Pr\{v = 10\} = \Pr\{v = 30\} = \tfrac{1}{4}, \quad \text{and} \quad \Pr\{v = 20\} = \tfrac{1}{2}.$$

Now assume that the government, which currently owns the field, will auction it off in a second-price sealed-bid auction, and that before the auction each of the two firms will perform a (free) exploration that will provide some signal about the quantity of oil in the field. Specifically each firm $i \in \{1, 2\}$ receives a low or high signal, $\theta_i = \{L, H\}$, which is correlated with the amount of oil as follows:

1. If $v = 10$ then $\theta_1 = \theta_2 = L$.
2. If $v = 30$ then $\theta_1 = \theta_2 = H$.
3. If $v = 20$ then either $\theta_1 = L$ and $\theta_2 = H$, or $\theta_1 = H$ and $\theta_2 = L$, where each of these two events (conditional on $v = 20$) occurs with equal probability.

Thus the probabilities of each pair of signals materializing are given by the following probability distribution matrix:

$$
\begin{array}{cc|c|c|}
 & & \multicolumn{2}{c}{\theta_2} \\
 & & L & H \\
\hline
 & L & \frac{1}{4} & \frac{1}{4} \\
\hline
\theta_1 & H & \frac{1}{4} & \frac{1}{4} \\
\hline
\end{array}
$$

where the signal outcomes are *not independent*—they are correlated with the actual amount of oil.

We can associate each player's signal with his *type,* to the extent that given a signal, a player can form expectations about the signal of his opponent, and in turn about the quantity of oil. If player i observes a low signal L then he knows that the probability that player j's signal is low, $\Pr\{\theta_j = L | \theta_i = L\}$, is equal to the probability that player j's signal is high, $\Pr\{\theta_j = H | \theta_i = L\}$, and is equal to $\tfrac{1}{2}$. Similarly $\Pr\{\theta_j = H | \theta_i = H\} = \Pr\{\theta_j = L | \theta_i = H\} = \tfrac{1}{2}$.

Given the posterior updated beliefs that each type has, we can calculate the expected amount of oil in the field conditional on the signal that a player has. If player i observes a low signal, he knows that with probability $\tfrac{1}{2}$ the other signal is low and

$v = 10$, and with probability $\frac{1}{2}$ the other signal is high and $v = 20$. Therefore

$$E[v_i | \theta_i = L] = \frac{1}{2} \times 10 + \frac{1}{2} \times 20 = 15,$$

and similarly

$$E[v_i | \theta_i = H] = \frac{1}{2} \times 20 + \frac{1}{2} \times 30 = 25.$$

At this stage we have identified the way in which types, which are defined by the information a player has, map into expectations over the amount of oil and therefore into expected payoffs from owning the field. However, it is not true that one player's type alone determines the value of obtaining the oil field—there is valuable information in the type of the other player, which, as we will now see, makes the simple second-price sealed-bid auction somewhat less straightforward than in the IPV case.

To see this let's first consider the following question: in a second-price sealed-bid auction, is it a Bayesian Nash equilibrium for both players to submit truthful bids equal to their expected valuations? Formally, is $s_i(\theta_i) = E[v_i|\theta_i]$ for $i \in \{1, 2\}$ a Bayesian Nash equilibrium? To answer this question let's assume that player 2 is playing according to this prescription and then check whether following this strategy is a best response of player 1.

Consider the case in which $\theta_1 = L$ and player 1 bids 15 (his expected value), then with probability $\frac{1}{2}$ player 2 also bids 15, in which case they win with equal probability and pay the second-highest bid of 15. This happens only when both players have a low signal, in which case the value of the oil field is $v = 10$. With probability $\frac{1}{2}$ player 2 has a high signal and player 1 loses, giving him a payoff of zero. Therefore, player 1's expected payoff is

$$Ev_1 = \frac{1}{2} \times \left[\frac{1}{2} \times (10 - 15) \right] + \frac{1}{2} \times [0] = -1.25,$$

and player 1 would rather bid less than 15 and never win.

This perhaps surprising outcome is a direct consequence of the common-values setting in which the types are correlated. When player 1 wins the oil field it is because his opponent's bid was low (in this case they tie), a consequence of player 2 having a low signal. But if player 1 bids his average value given his signal, then he is not taking into account the fact that he wins only when player 2's signal is low. This is "bad news" because it implies that the quantity of oil is lower than the player thinks it is *on average*. This phenomenon, which occurs in common-values settings, is known as the **"winner's curse"**: a player wins when his signal is the most optimistic, which in the common-values setting means that he has overestimated the value of the good and is overpaying if he does not take this conditioning into account. In an equilibrium, therefore, players will have to choose their bids conditioning on the fact that when they win, this means that their signal is higher than everyone else's, which is likely to imply that the winner is too optimistic about the good's value. The reasoning is very similar to that behind the "swing voter's curse" described in Section 12.4. You are asked to solve for the Bayesian Nash equilibrium of this auction in exercise 13.8.

This phenomenon exists in both first- and second-price sealed-bid auctions, and it has important economic consequences as to which auctions may perform better in allocating a common-values good efficiently. For more on this topic see Krishna (2002) and Milgrom (2004).

13.3   Summary

- Auctions are commonly used games that allocate scarce resources among several potential bidders.

- There are two extreme settings of auction games. The first case is that of private values, in which the information of each player is enough for him to infer his value from winning the object. The second case is that of common values, in which the information of other players will determine how much the object is worth to any player.

- Auctions often differ in their rules, such as open or sealed bidding, or first or second price. Different rules will result in different equilibrium bidding behavior.

- In the private values setting, the second-price sealed-bid auction is strategically equivalent to the English auction. In both auctions each player has a simple weakly dominant strategy of bidding his true value for the object. The highest-value player wins and pays the second-highest value.

- In the private values setting, the first-price sealed-bid auction is strategically equivalent to the Dutch auction. In both auctions each player's best response depends on the strategies of other players, and calculating a Bayesian Nash equilibrium is not straightforward. In equilibrium each bidder shades his valuation when bidding in order to obtain a positive expected value from the auction.

- The revenue equivalence theorem identifies conditions under which each of the four kinds of auctions yields the seller the same expected revenues and results in the same outcomes for the participating bidders.

- If the auction is one of common values then players must take into account the downsides of the winner's curse and bid accordingly to avoid overpaying for the object.

13.4   Exercises

13.1 **Second-Price Auctions:** Show that in a second-price sealed-bid auction bidding your valuation weakly dominates bidding above your valuation.

13.2 **Complete Information:** Consider a set of N players participating in a sealed-bid second-price auction, but assume that there is complete information so that each player knows the valuation of every other player.

    a. Is it still true that each player bidding his valuation is a weakly dominant strategy?

    b. Are there other Nash equilibria of this game?

13.3 **Reserve Prices (Easier):** Consider a seller who must sell a single private value good. There are two potential buyers, each with a valuation that can take on one of three values, $\theta_i \in \{0, 1, 2\}$, each value occurring with an equal probability of $\frac{1}{3}$. The players' values are independently drawn. The seller will offer the good using a second-price sealed-bid auction, but he can set a "reserve price" of $r \geq 0$ that modifies the rules of the auction as follows. If both bids are below r then neither bidder obtains the good and it is destroyed. If both bids are at or above r then the regular auction rules prevail. If only one bid is at or above r then that bidder obtains the good and pays r to the seller.

    a.   Is it still a weakly dominant strategy for each player to bid his valuation when $r > 0$?

    b.   What is the expected revenue of the seller when $r = 0$ (no reserve price)?

    c.   What is the expected revenue of the seller when $r = 1$?

    d.   What explains the difference between your answers to (b) and (c)?

    e.   What is the optimal reserve price r for the seller, and what can you conclude about the value of reserve prices?

13.4   **Reserve Prices (Harder):** Consider a seller who must sell a single private value good. There are two potential buyers, each with a valuation that is drawn independently and uniformly from the interval [0, 1]. The seller will offer the good using a second-price sealed-bid auction, but he can set a "reserve price" of $r \geq 0$ that modifies the rules of the auction as follows. If both bids are below r then neither bidder obtains the good and it is destroyed. If both bids are at or above r then the regular auction rules prevail. If only one bid is at or above r then that bidder obtains the good and pays r to the seller.

    a.   Show that choosing $r = 0$ is not optimal for the seller. What is the intuition for this fact?

    b.   What is the optimal reserve price for the seller?

13.5   **Mixed Strategies:** Consider a seller who offers a single private value good using a first-price sealed-bid auction. There are two potential buyers, each with a valuation that can take on one of two values, $\theta_i \in \{4, 8\}$, each value occurring with an equal probability of $\frac{1}{2}$. The players' values are independently drawn. Call the type with $\theta_i = 4$ the "low" type and $\theta_i = 8$ the "high" type.

    a.   Show that in a Bayesian Nash equilibrium the low must be using a pure strategy bid $b_L = 4$. (Hint: First show that in a Bayesian Nash equilibrium the low type cannot be using a pure strategy bid $b_L < 4$. Next show that in a Bayesian Nash equilibrium the low type cannot be using a mixed strategy with two different bids.)

    b.   Show that given the fact described in (a), in any Bayesian Nash equilibrium the high type will never choose a bid $b_H > 6$.

    c.   Show that the low type choosing $b_L = 4$ and the high type choosing b_H uniformly from the interval [4, 6] together constitute a Bayesian Nash equilibrium.

    d.   Calculate the expected revenue to the seller given the Bayesian Nash equilibrium in (c). How does it compare to the expected revenue he would obtain from an English auction?

13.6   **Third-Price Sealed-Bid Auction:** Consider a third-price sealed-bid auction in which the highest bidder wins the auction and the winner pays the third-highest bid. Assume that n players' valuations are private and are drawn independently from the uniform distribution over the interval [0, 1].

    a.   Write down a bidder's expected payoff function for this auction.

    b.   Show that if each player chooses the following bid strategy then the players will be playing a Bayesian Nash equilibrium:

$$s_i(\theta_i) = \frac{n-1}{n-2}\theta_i.$$

      c.   Why do you think that bidders bid above their valuations?

      d.   Compute the seller's expected revenue from this auction and compare it to the expected revenue from the first- and second-price sealed bid auctions.

13.7 **Risk-Averse Bidders:** Imagine an auction for a single item with n bidders whose valuations θ_i are drawn uniformly from the interval $[0, 1]$. Assume that the players are risk averse and that the utility to player i of type θ_i from winning the item at a price p is given by $\sqrt{\theta_i - p}$ for $p \leq \theta_i$. (We will assume that no bidder ever bids or pays more than his valuation.)

      a.   Write down a bidder's expected payoff function for the first- and second-price sealed-bid auctions.

      b.   Show that in the first-price sealed-bid auction each bidder of type θ_i choosing the bidding strategy $s(\theta_i) = \theta_i \frac{m(n-1)}{m(n-1)+1}$ forms a symmetric Bayesian Nash equilibrium.

      c.   Show that in a second-price sealed-bid auction it is a (weakly) dominant strategy for each bidder to bid his valuation.

      d.   Compare the seller's expected revenue from the first- and second-price sealed-bid auctions. In what way does this example differ from the case of risk-neutral bidders?

13.8 **Common Values:** Consider the common-values oil field example of Section 13.2. Find a Bayesian Nash equilibrium for the second-price sealed-bid auction of the oil field.

13.9 **Private Value Linear Bidding:** Two players participate in a first-price sealed-bid auction in an IPV setting. The value of the good for each player is given by $0.5 + \theta_i$, where θ_i is observed only by player i and is uniformly distributed over the interval $[0, 1]$.

      a.   Show that there is a symmetric Bayesian Nash equilibrium in which each bidder i's bidding strategy is of the form $s_i^*(\theta_i) = a + b\theta_i$.

      b.   What is the expected payoff of a player if his type is θ_i?

13.10 **Common Values Linear Bidding:** Two players participate in a first-price sealed-bid auction in a common-values setting. The value of the good for each player is given by $\theta_1 + \theta_2$, where θ_i is a signal observed only by player i and is uniformly distributed over the interval $[0, 1]$. Hence if player i observes his signal θ_i then his expected valuation is equal to $0.5 + \theta_i$ (but unlike the previous exercise, this is a common-values setting).

      a.   Show that there is a symmetric Bayesian Nash equilibrium in which each bidder i's bidding strategy is of the form $s_i^*(\theta_i) = a + b\theta_i$.

      b.   What is the expected payoff of a player if his type is θ_i?

      c.   Show that for each value of θ_i the equilibrium bid of player i is less than what you found in (a) of the previous exercise.

14

Mechanism Design

There are many economic and political situations in which some central authority wishes to implement a decision that depends on the private information of a set of players. For example, a government agency may wish to choose the design of a public-works project based on the preferences of its citizens, who in turn have private information about how much they prefer one design over another. Alternatively, a monopolistic firm may wish to determine a set of consumers' willingness to pay for different products it can produce, with the goal of making as high a profit as possible.

In this section we provide a short introduction to the theory of **mechanism design,** which is the study of what kinds of mechanisms such a central authority can devise in order to reveal some or all of the private information that it is trying to extract from the group of players with which it is interacting. In essence the mechanism designer, our central authority, will design a game to be played by the players, and in equilibrium the mechanism designer wishes to both reveal the relevant information and act upon it. The study of mechanism design dates from the early work of Leonid Hurwicz (1972), and it has been an active area of theoretical research for the past four decades. In 2007 Hurwicz shared the Nobel Prize in economics with Eric Maskin and Roger Myerson for laying the foundations of this thriving research agenda.[1] Interestingly in the past decade there has been a growing interest among theoretical computer science scholars in mechanism design, which proves useful in the design of online systems, in particular those for online advertising and auctions.

14.1 Setup: Mechanisms as Bayesian Games

14.1.1 The Players

We start with a set of players, $N = \{1, 2, \ldots n\}$, and a set of public alternatives, X. The set X could represent many possible kinds of alternatives. For example, an

---

1. For a survey of the contributions of these three economists, as well as related work, see "Scientific background on the Sveriges Riksbank Prize in Economic Sciences in Memory of Alfred Nobel 2007, Mechanism Design Theory, Compiled by the Prize Committee of the Royal Swedish Academy of Sciences" at http://www.nobelprize.org/nobel_prizes/economics/laureates/2007/advanced-economicsciences2007.pdf.

alternative $x \in X$ could represent the attributes of a public good or service that is provided to the set of players, such as investment in education or in preserving the environment. Another example of an alternative $x \in X$ can be the allocation of a private good between the players, such as which of the n players gets the rights to a patent or a license to pollute. The flexibility of what X can be is a key feature of this framework, showing that it can be applied to many interesting and important settings. We call X the set of **public** alternatives because the chosen alternative affects all the players in N to the extent that even for a private good, if one player gets it then the consequence is that everyone else does not (e.g., an auction).

Each player i privately observes a signal (his type) $\theta_i \in \Theta_i$ which determines his preferences over outcomes just as we described in Chapter 12. The set of types Θ_i of player i can be finite or infinite. Given a realization of types θ_i for each i, let $\theta = (\theta_1, \theta_2, \ldots, \theta_n)$ be the **state of the world** as it describes the profile of all realized types for the n players. The state θ is drawn randomly from the **state space** $\Theta \equiv \Theta_1 \times \Theta_2 \times \cdots \times \Theta_n$, which is the set of all possible profiles of types $\theta \in \Theta$. The draw of θ is according to some prior distribution $\phi(\cdot)$ over Θ. ($\phi(\cdot)$ is given by probabilities if the state space Θ is finite and by a density, or cumulative distribution, if the state space Θ is continuous.) As before we assume that θ_i is player i's private information and that the prior $\phi(\cdot)$ is common knowledge.

Turning to payoffs, we will focus on a particularly convenient way to represent preferences over outcomes by assuming that every alternative has a "money-equivalent" value, and that preferences are additive in money. That is, if player i is given an amount of money equal to $m_i \in \mathbb{R}$ (if $m_i < 0$ then it means that money is taken away from i), and if the public alternative chosen is $x \in X$, then player i's payoff is given by

$$v_i(x, m_i, \theta_i) = u_i(x, \theta_i) + m_i, \qquad (14.1)$$

where we interpret $u_i(x, \theta_i)$ as the money-equivalent value of alternative $x \in X$ when i's type is θ_i. This class of preferences, in which money is added linearly to some function of other outcomes, is called **quasilinear preferences.**

As a result of this specification, we can think of **outcomes** as combinations of the choice of the public alternative together with monetary amounts that each player gets (or pays). Thus an outcome would be represented as $y = (x, m_1, \ldots, m_n)$, and each player i has preferences over outcomes given by (14.1). Notice that we restrict attention to the case of *private values* in which player i's payoff does not depend directly on other players' private information. In the more general *common values* case a player's payoff can depend on other players' information, which was the case in the adverse selection problem in Section 12.3 and in the example of common-values auctions in Section 13.2.

14.1.2 The Mechanism Designer

We now turn to the so-called central authority described earlier, whom we shall call the mechanism designer. The mechanism designer has the objective of achieving an outcome that depends on the types of the players. Before continuing we impose some restrictions on the set of outcomes by assuming that the mechanism designer does not have a source of funds to pay the players, which in turn implies that the monetary payments they make or receive have to be self-financed. That is, $\sum_{i=1}^{n} m_i \leq 0$. Notice that we allow for the sum of monetary transfers to be negative, which means that the mechanism designer can keep some of the money that he raises from the players. Thus the set of outcomes is restricted as follows:

$$Y = \{(x, m_1, \ldots, m_n) : x \in X, m_i \in \mathbb{R} \forall i \in N, \sum_{i=1}^{I} m_i \leq 0\}.$$

The mechanism designer's objective is given by a **choice rule** that takes the form

$$f(\theta) = (x(\theta), m_1(\theta), \ldots, m_n(\theta)),$$

where $x(\theta) \in X$ and $\sum_{i=1}^{I} m_i(\theta) \leq 0$. We will call $x(\theta)$ the **decision rule** and $(m_1(\theta), \ldots, m_n(\theta))$ the **transfer rule** because it determines the transfers made by or to each player. Two examples may be useful to illustrate our setup.

Example 14.1 (public good) Let $X = [0, \bar{x}]$ be the level of some public good, say, the size of a water treatment plant that will benefit some citizens and may displease others. The citizens are the group of players, N. Player i's willingness to pay (his monetary value or harm) from a level $x \in X$ of this public good is given by $u_i(x, \theta_i)$. The mechanism designer may be a utilitarian ruler who wants to choose the value of x that maximizes the sum of the players' valuations, so his decision rule $x(\theta)$ would maximize $\sum_{i=1}^{n} u_i(x, \theta_i)$.

Example 14.2 (allocation of a private good) A good, say a license to use a certain portion of the electromagnetic spectrum for cell phone coverage, can be allocated to one of a group of cellular carriers $i \in N$. Let $x_i \in \{0, 1\}$ indicate whether player i receives the good ($x_i = 1$) or not ($x_i = 0$). Then the possible set of alternatives is

$$X = \{(x_1, x_2, \ldots, x_n) \quad \text{such that } x_i \in \{0, 1\} \text{ and } \sum_{i=1}^{n} x_i = 1\}.$$

Player i's willingness to pay for owning the private good is given by $u_i(x, \theta_i) = \theta_i x_i$, where θ_i can be thought of as the value to each player of owning the portion of the spectrum (e.g., his future profits from providing service). The mechanism designer may again be a utilitarian ruler who wants to choose the value of x that maximizes the sum of the players' valuations, so his decision rule $x(\theta)$ would maximize $\sum_{i=1}^{n} u_i(x, \theta_i)$, which in this case would be achieved by allocating the license to the player who values it the most. Alternatively the mechanism designer may want to make the highest profit he can by collecting money from the player who gets the good, or who participates in some kind of auction to get it.

14.1.3 The Mechanism Game

In our setup the mechanism designer desires to implement a choice rule $f : \Theta \rightarrow Y$. However, he cannot do this directly because the choice rule depends on the unobserved state of the world θ. Thus in order to implement a state-contingent choice rule the mechanism designer has to ask players to reveal their types, and then he can prescribe an outcome on the basis of those types.

Perhaps, as you can imagine, players may not have incentives to reveal their types truthfully. In the example of a public good just described, some players who don't want the water treatment plant built might exaggerate their harm from such a plant, while others who want it built may exaggerate their benefit. The problem our mechanism designer faces can be described in two ways. First, what information would the players be willing to share if the mechanism designer would in turn implement his choice rule $f(\cdot)$? Second, would the mechanism designer be able to implement $f(\cdot)$ using some sophisticated set of rules that would end up revealing the players' private information?

This second question deserves some thought. Maybe our mechanism designer realizes that he cannot just extract the private information if he asks directly for the players to reveal their types. However, he may be able to design some clever game that the players will play, in which the rules of the game endow each player i with an action set A_i, and following the choice $a_i \in A_i$ by each player there is some outcome function $g(a_1, \ldots, a_n)$ that makes a choice of an outcome $y \in Y$. Player i's strategy is a contingent plan $s_i : \Theta_i \rightarrow A_i$. Because player i observes only his type θ_i and not other players' types, this is a game of incomplete information, and given the common prior over types, $\phi(\cdot)$, and the payoffs over outcomes, $v_i(g(s), \theta_i)$, this setup is just a Bayesian game. Formally we have

Definition 14.1 A **mechanism,** $\Gamma = \langle A_1, A_2, \ldots, A_n, g(\cdot) \rangle$ is a collection of n action sets $A_1, A_2, \ldots, A_n$ and an outcome function $g : A_1 \times A_2 \times \cdots \times A_n \rightarrow Y$. A pure strategy for player i in the mechanism Γ is a function that maps types into actions, $s_i : \Theta_i \rightarrow A_i$. The payoffs of the players are given by $v_i(g(s), \theta_i)$.

If the players are forced to play this game, then depending on their beliefs about the choices of other players, and on the realization of the players' types, each player i will choose a best response from his action set A_i. Because we are concerned with equilibrium behavior, we offer this definition:

Definition 14.2 The strategy profile $s^*(\cdot) = (s_1^*(\cdot), \ldots, s_n^*(\cdot))$ is a **Bayesian Nash equilibrium** of the mechanism $\Gamma = \langle A_1, \ldots, A_n, g(\cdot) \rangle$ if for every $i \in N$ and for every $\theta_i \in \Theta_i$

$$E_{\theta_{-i}}[v_i(g(s_i^*(\theta_i), s_{-i}^*(\theta_{-i})), \theta_i)|\theta_i]$$

$$\geq E_{\theta_{-i}}[v_i(g(a_i', s_{-i}^*(\theta_{-i})), \theta_i)|\theta_i] \quad \text{for all } a_i' \in A_i.$$

That is, if player i believes that other players are playing according to $s_{-i}^*(\theta_{-i})$ then he maximizes his expected payoff by following the behavior prescribed by $s_i^*(\theta_i)$ regardless of which type player i is.

Now imagine that our mechanism designer was clever enough to design a mechanism in which there is a profile of equilibrium strategies, $s_i^* : \Theta_i \rightarrow A_i$ such that given any realization of types θ, the outcome of the game is exactly what the mechanism designer wants to achieve. That is, for all $\theta \in \Theta$, $g(s_1^*(\theta_1), s_2^*(\theta_2), \ldots, s_n^*(\theta_n)) = f(\theta)$. In this case our mechanism designer would be very lucky indeed, because the mechanism he devised would result in the outcome he would have chosen had he known the types of the players. To define this scenario formally, we have

Definition 14.3 A mechanism Γ **implements** the choice rule $f(\cdot)$ if there exists a Bayesian Nash equilibrium of the mechanism Γ, $(s_1^*(\theta_1), s_2^*(\theta_2), \ldots, s_n^*(\theta_n))$, such that $g(s_1^*(\theta_1), s_2^*(\theta_2), \ldots, s_n^*(\theta_n)) = f(\theta)$ for all $\theta \in \Theta$.

The idea of a mechanism implementing a choice rule is illustrated in Figure 14.1. The mechanism designer would like to know θ and then choose the outcome $f(\theta)$. Using the mechanism with equilibrium strategies $s^*(\theta) = (s_1^*(\theta_1), s_2^*(\theta_2), \ldots, s_n^*(\theta_n))$, given any θ, the players choose actions $a = (a_1, a_2, \ldots, a_n)$ commensurate with

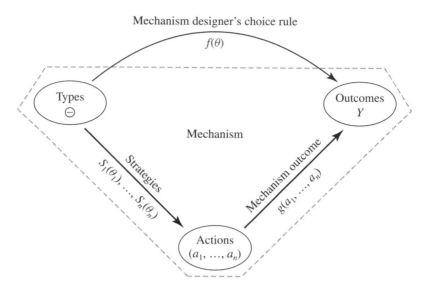

Mechanism designer's choice rule

$f(\theta)$

Types
Θ

Outcomes
Y

Mechanism

Strategies
$S_1(\theta_1), \ldots, S_n(\theta_n)$

Mechanism outcome
$g(a_1, \ldots, a_n)$

Actions
$(a_1, \ldots, a_n)$

FIGURE 14.1 A mechanism implementing $f(\cdot)$.

$s^*(\theta)$, and through the outcome function $g(\cdot)$ the mechanism does what the mechanism designer wanted to do, because for every θ a is chosen such that $g(a) = f(\theta)$.

Remark The formulated implementation concept introduced in definition 14.3 is sometimes called **partial implementation** because it requires that the desired outcome be *an* equilibrium, but allows for other, undesirable, equilibrium outcomes as well. Consider the following simple example of majority voting. Let $n = 3$, and assume there are two outcomes, $Y = \{a, b\}$, each being some alternative (e.g., a pair of candidates or a pair of policies). Suppose that we want to implement a choice rule $f(\cdot)$ such that $f(\theta) = a$ whenever at least two players strictly prefer a, and similarly for b. Suppose that we use the majority voting mechanism for this purpose: each of the three players casts a vote and the majority wins. The mechanism does have a desirable Bayesian Nash equilibrium in which each player votes his true preference. However, the majority voting mechanism also has an undesirable Nash equilibrium in which all players vote for b in all states. This equilibrium results in outcome b even in a state $\theta \in \Theta$ in which all players prefer a to b. The reason is that if *each player* believes that the other two will vote for b then he believes that b will win regardless of his vote, and hence he too is willing to vote for b. The study of implementation without so-called bad equilibria in which $f(\theta)$ is not implemented is called **full implementation.** We will not be concerned with the problem of undesirable equilibria and will content ourselves with the concept of partial implementation. For more on this topic see Mas-Colell et al. (1995, Chapter 23).

14.2   The Revelation Principle

We have just described what may be a useful tool in the hands of the mechanism designer. It may be that a clever designer can design a mechanism, which is a Bayesian game, and through it get the players to play equilibrium strategies that end up implementing exactly what the mechanism designer intended. This may be useful only if the mechanism designer cannot get the players to reveal their types when they realize that his intention is to implement $f(\cdot)$.

It seems that this scenario, in which the mechanism designer asks the players to reveal their types in order to implement $f(\cdot)$, can itself be regarded as a particular mechanism, and as such as a Bayesian game. The actions that players can choose are announcing any one of their types, and given a profile of announced types, the outcome is determined by $f(\cdot)$ operated on the announcements. Hence we can formally define this game as follows:

Definition 14.4 $\Gamma = \langle \Theta_1, \ldots, \Theta_n, f(\cdot) \rangle$ is a **direct revelation mechanism** for choice rule $f(\cdot)$ if $A_i = \Theta_i$ for all $i \in N$ and $g(\theta) = f(\theta)$ for all $\theta \in \Theta$.

It will also be useful to define a very important benchmark, in which this straight-forward direct revelation mechanism will actually have an equilibrium that implements the mechanism designer's intended outcome:

Definition 14.5 The choice rule $f(\cdot)$ is **truthfully implementable in Bayesian Nash equilibrium** if for all θ the direct revelation mechanism $\Gamma = \langle \Theta_1, \ldots, \Theta_n, f(\cdot) \rangle$ has a Bayesian Nash equilibrium $s_i^*(\theta_i) = \theta_i$ for all i. Equivalently, for all i,

$$E_{\theta_{-i}}[v_i(f(\theta_i, \theta_{-i}), \theta_i)|\theta_i] \geq E_{\theta_{-i}}[v_i(f(\hat{\theta}_i, \theta_{-i}), \theta_i)|\theta_i] \quad \text{for all } \hat{\theta}_i \in \Theta_i.$$

That is, $f(\cdot)$ is truthfully implementable in Bayesian Nash equilibrium if truth-telling is a Bayesian Nash equilibrium strategy in the direct revelation mechanism. If every player i believes that all other players are reporting their types truthfully, then player i is also willing to report truthfully.

Effectively if some choice function $f(\cdot)$ was truthfully implementable then it means that our mechanism designer does not have to resort to some sophisticated game to get the outcome he desires. The question then is, if the mechanism designer's choice function $f(\cdot)$ is *not* truthfully implementable, can he find some clever mechanism that implements it? The answer turns out to be negative:

Proposition 14.1 (The Revelation Principle for Bayesian Nash Implementation) *A choice rule $f(\cdot)$ is implementable in Bayesian Nash equilibrium if and only if it is truthfully implementable in Bayesian Nash equilibrium.*

Proof By definition, if $f(\cdot)$ is truthfully implementable in Bayesian Nash equilibrium then it is implementable in Bayesian Nash equilibrium using the direct revelation mechanism. To prove the converse, suppose that there exists some mechanism $\Gamma = (A_1, \ldots, A_n, g(\cdot))$ that implements $f(\cdot)$ using the equilibrium strategy profile $s^*(\cdot) = (s_1^*(\cdot), \ldots, s_n^*(\cdot))$ and $g(s^*(\cdot)) = f(\cdot)$, so that for every $i \in N$ and $\theta_i \in \Theta_i$,

$$E_{\theta_{-i}}[v_i(g(s_i^*(\theta_i), s_{-i}^*(\theta_{-i})), \theta_i)|\theta_i]$$

$$\geq E_{\theta_{-i}}[v_i(g(a_i', s_{-i}^*(\theta_{-i})), \theta_i)|\theta_i] \quad \text{for all } a_i' \in A_i, \tag{14.2}$$

which means that no player i wishes to deviate from $s_i^*(\cdot)$. One particular way to deviate for player i is by using $s_i^*(\cdot)$ but pretending that his type is $\hat{\theta}_i$ rather than θ_i, i.e., choosing the action $a'_i = s^*_i(\widehat{\theta}_i)$. Thus a particular case of (14.2) is

$$E_{\theta_{-i}}[v_i(g(s^*_i(\theta_i), s^*_{-i}(\theta_{-i})), \theta_i)|\theta_i]$$

$$\geq E_{\theta_{-i}}[v_i(g(s^*_i(\widehat{\theta}_i), s^*_{-i}(\theta_{-i})), \theta_i)|\theta_i] \quad \text{for every } \widehat{\theta}_i \in \Theta_i. \quad (14.3)$$

But because $g(s^*(\theta)) = f(\theta)$ for all $\theta \in \Theta$, (14.3) implies that, for all $i \in N$ and for all $\theta_i \in \Theta_i$,

$$E_{\theta_{-i}}[v_i(f(\theta_i, \theta_{-i}), \theta_i)|\theta_i] \geq E_{\theta_{-i}}[v_i(f(\widehat{\theta}_i, \theta_{-i}), \theta_i)|\theta_i] \quad \text{for all } \widehat{\theta}_i \in \Theta_i.$$

But this is just the condition for $f(\cdot)$ to be truthfully implementable in Bayesian Nash equilibrium. ∎

At first glance this proposition may seem quite remarkable. It basically says that if the mechanism designer cannot implement $f(\cdot)$ directly then there is no mechanism in the world that can. The result follows from the strong requirements of equilibrium analysis, which is at the core of game theory. In a nutshell, in equilibrium players are neither wrong nor surprised. Hence if players are playing a mechanism that results in the implementation of $f(\cdot)$, then by construction of equilibrium beliefs they must know that $f(\cdot)$ will be implemented, and hence the mechanism designer might as well implement it directly.

To be more precise, the reasoning works as follows: Imagine that there is some clever mechanism Γ that implements $f(\cdot)$ with the strategies $s^*_i(\cdot)$ for each player i. Now imagine that each player i writes a computer program to compute his equilibrium action according to $s^*_i(\cdot)$. That is, the player feeds his type $\theta_i \in \Theta_i$ into the program, and the program calculates the equilibrium action $s^*_i(\theta_i)$ and sends it to the mechanism designer. The player will not want to lie to his computer because, by construction, the program computes his best response strategy.

Because the mechanism designer *knows* that $s^*_i(\cdot)$ is the player's equilibrium strategy, he can save the players the trouble of programming their strategies by "importing" each player's program into his own computer, which calculates $g(s^*(\theta))$. This gives rise to the following new mechanism Γ^*: players announce their types, $A_i = \Theta_i$ for all i, to the mechanism designer; the new outcome function $g^*(\cdot)$ is computed by first feeding the players' type announcements θ into their equilibrium programs $s^*_i(\cdot)$; the mechanism designer computes the outcome $g(s^*_i(\theta))$ and implements it. That is, given announcements $\widehat{\theta}_1, \widehat{\theta}_2, \ldots, \widehat{\theta}_n$,

$$g^*(\widehat{\theta}_1, \widehat{\theta}_2, \ldots, \widehat{\theta}_n) = g(s^*_1(\widehat{\theta}_1), s^*_2(\widehat{\theta}_2), \ldots, s^*_n(\widehat{\theta}_n)) = f(\widehat{\theta}_1, \widehat{\theta}_2, \ldots, \widehat{\theta}_n),$$

because the mechanism Γ implemented $f(\cdot)$. But in the new mechanism every player will report his type θ_i truthfully to the designer by construction, and, playing the new mechanism, they all know that it implements $f(\cdot)$. This in turn implies that we have just created a direct revelation mechanism in which truth-telling is a Bayesian Nash equilibrium. Thus the idea behind the powerful revelation principle is a consequence of equilibrium analysis: in equilibrium the players know that the mechanism implements $f(\cdot)$, and they choose to stick to it, so they might as well announce their types truthfully and have the mechanism designer implement $f(\cdot)$ directly.

14.3 Dominant Strategies and Vickrey-Clarke-Groves Mechanisms

14.3.1 Dominant Strategy Implementation

One particular form of a Bayesian Nash equilibrium is an equilibrium in **dominant strategies:** that is, each player has a strategy that maximizes his expected payoff regardless of what strategies he expects others to play. Formally, we have

Definition 14.6 The strategy profile $s^*(\cdot) = (s_1^*(\cdot), \ldots, s_n^*(\cdot))$ is a **dominant strategy equilibrium** of the mechanism $\Gamma = \langle A_1, A_2, \ldots, A_n, g(\cdot) \rangle$ if for every $i \in N$ and for every $\theta_i \in \Theta_i$

$$v_i(g(s_i^*(\theta), a_{-i}), \theta) \geq v_i(g(a_i', a_{-i}), \theta) \quad \text{for all } a_i' \in A_i \text{ and for all } a_{-i} \in A_{-i}.$$

The appealing feature of this equilibrium concept is in the weak requirements it demands of the players: a player *need not forecast* what the others are doing, nor does he have to have correct expectations on the distribution of the other players' types. A related advantage is that "bad" equilibria are usually not a problem. (If a player has two dominant strategies they must be payoff-equivalent, and most games of interest do not have two outcomes with exactly the same payoffs.)

The question then is, can we find a mechanism Γ that implements $f(\cdot)$ in dominant strategies? By now you should be able to predict the answer. Because a dominant strategy equilibrium is just a special case of a Bayesian Nash equilibrium, the revelation principle applies. We conclude, therefore, that to check if $f(\cdot)$ is implementable in dominant strategies we need only check that $f(\cdot)$ is truthfully implementable in dominant strategies directly—that is, that for all $i \in N$ and all $\theta_i \in \Theta_i$,

$$v_i(f(\theta_i, \theta_{-i}), \theta_i) \geq v_i(f(\hat{\theta}_i, \theta_{-i}), \theta_i) \quad \text{for all } \hat{\theta}_i \in \Theta_i, \text{ and for all } \theta_{-i} \in \Theta_{-i}.$$

14.3.2 Vickrey-Clarke-Groves Mechanisms

Recall that we restricted our attention to preferences in which every alternative has a money-equivalent value, and that preferences are additive in money, so that $v_i(x, m_i, \theta_i) = u_i(x, \theta_i) + m_i$. These preferences are called quasilinear preferences because the payoff of each player is linear in money and has an added component that derives a type-dependent payoff from the choice of $x \in X$.

There is a nice feature of this quasilinear environment in which each player has quasilinear preferences. Imagine that there are two alternatives $x, x' \in X$ and a pair of players i and j with types θ_i and θ_j such that $u_i(x', \theta_i) > u_i(x, \theta_i)$ and $u_j(x, \theta_j) > u_j(x', \theta_j)$. That is, player i prefers the outcome x' over x, while player j prefers the outcome x over x'. Further imagine that

$$u_i(x', \theta_i) - u_i(x, \theta_i) > u_j(x, \theta_j) - u_j(x', \theta_j).$$

This implies that i prefers x' to x in monetary terms more than j prefers x to x'. The quasilinear environment implies that for any amount of money $k > 0$ that satisfies $u_i(x', \theta_i) - u_i(x, \theta_i) > k > u_j(x, \theta_j) - u_j(x', \theta_j)$, we can replace x with x' and transfer k from player i to player j so as to make both players better off. This is a consequence of the additivity in money in quasilinear preferences, and it gives rise to the following useful proposition:

Proposition 14.2 *In the quasilinear environment, given a state of the world $\theta \in \Theta$, an alternative $x^* \in X$ is* Pareto optimal *if and only if it is a solution to*

$$\max_{x \in X} \sum_{i=1}^{I} u_i(x, \theta_i).$$

That is, a solution is Pareto optimal if and only if it maximizes the sum of the $u_i(\cdot, \theta_i)$ functions. The proof is simple: if an alternative x did not maximize this sum then we can find an alternative x' that did, and then we can find money transfers among the players that would ensure that the gains of some players more than compensate for the losses of others. Notice that transfers of money between players are a "wash" as far as Pareto efficiency is considered because one person's gain is the other's loss. The only thing that matters for efficiency is getting the choice $x \in X$ right, and the transfers can be thought of as the way in which the surplus is divided among the players to settle things.

Recall that the choice rule $f(\theta)$ determines the choice of the alternative, $x(\theta)$, and the choice of monetary transfers, $m_1(\theta), \ldots, m_n(\theta)$. We call a decision rule $x^*(\cdot)$ the **first-best decision rule** if for all $\theta \in \Theta$, $x^*(\theta)$ is Pareto optimal. We now focus on the problem of a benevolent mechanism designer who wishes to implement the first-best decision rule in a quasilinear environment:

$$x^*(\theta) \in \arg \max_{x \in X} \sum_{i=1}^{I} u_i(x, \theta_i) \ \forall \theta \in \Theta.$$

The question of interest now becomes, can our mechanism designer find a transfer rule $(m_1(\cdot), \ldots, m_n(\cdot))$ such that the choice rule $(x^*(\cdot), m_1(\cdot), \ldots, m_n(\cdot))$ is implementable in dominant strategies? That is, if the players are faced with this Pareto optimal choice rule $(x^*(\cdot), m_1(\cdot), \ldots, m_n(\cdot))$, will truth-telling be a dominant strategy for each player in the direct revelation mechanism?

Let's see what incentives a player faces in a direct revelation game with a first-best choice rule $(x^*(\cdot), m_1(\cdot), \ldots, m_n(\cdot))$. Each player will choose to announce some $\hat{\theta}_i \in \Theta_i$, which generates an announcement profile, $\hat{\theta} = (\hat{\theta}_1, \ldots, \hat{\theta}_n)$. The implemented choice implies that each player's payoff is $v_i(y, \theta_i) = u_i(x^*(\hat{\theta}_i, \hat{\theta}_{-i}), \theta_i) + m_i(\hat{\theta}_i, \hat{\theta}_{-i})$, where $\hat{\theta}_i$ is player i's announcement, $\hat{\theta}_{-i}$ are the announcements of all the other players, and θ_i is player i's true type. Thus the incentives that influence player i's decision on what to announce come from two sources: how his announcement influences the choice $x^*(\hat{\theta}_i, \hat{\theta}_{-i})$, and how it affects the value of $m_i(\hat{\theta}_i, \hat{\theta}_{-i})$.

Consider for example a choice rule in which $m_i(\hat{\theta}_i, \hat{\theta}_{-i}) \equiv 0$, so that transfers are never made to player i and his incentives are driven by the way his announcement influences the choice of $x^*(\hat{\theta}_i, \hat{\theta}_{-i})$. It is easy to see that because $x^*(\hat{\theta}_i, \hat{\theta}_{-i})$ is the first-best decision rule, i may not want to announce his type truthfully. The reason is that $x^*(\hat{\theta}_i, \hat{\theta}_{-i})$ is chosen to maximize *total surplus,* but player i instead wants to maximize *his own payoff.* Thus given a belief that player i has about the announcements of the other players, player i will report a type $\hat{\theta}_i \in \Theta_i$ to achieve decision x that maximizes $u_i(x, \theta_i)$, and not one that maximizes the sum of all the players' payoffs.

Using some conventional jargon from economics, the problem of these misaligned incentives when no transfers are made is that the player does not account for the **externality** he imposes on others with his announcement. When player i's announcement induces some choice x' instead of the Pareto-optimal choice $x^*(\theta)$, then he will gain

$u_i(x', \theta_i) - u_i(x^*(\theta), \theta_i)$, at the expense of imposing a loss on the rest of society equal to $\sum_{j \neq i} \left(u_j(x^*(\theta), \theta_j) - u_j(x', \theta_j) \right)$. This problem could be solved by having a clever transfer rule $m_i(\hat{\theta}_i, \hat{\theta}_{-i})$ that lets player i **internalize the externality** and thus aligns his *private* incentives with *social* incentives. To achieve this we need to set the functions $m_i(\cdot)$ so that each player internalizes the externality that he imposes on the others. This is the idea behind the following important choice rule:

Definition 14.7 Given announcements $\hat{\theta}$, the choice rule $f(\hat{\theta}) = (x^*(\hat{\theta}), m_1(\hat{\theta}), \ldots, m_n(\hat{\theta}))$ is a **Vickrey-Clarke-Groves (VCG) mechanism** if $x^*(\cdot)$ is the first-best decision rule and if for all $i \in N$

$$m_i(\hat{\theta}) = \sum_{j \neq i} u_j(x^*(\hat{\theta}_i, \hat{\theta}_{-i}), \hat{\theta}_j) + h_i(\hat{\theta}_{-i}),$$

where $h_i(\hat{\theta}_{-i})$ is an arbitrary function of $\hat{\theta}_{-i}$.

This general family of choice rules was first described by Groves (1973). However, Vickrey (1961) and Clarke (1971) had studied particular instances of such mechanisms (which are described shortly). The appealing feature of VCG mechanisms is described in the following proposition:

Proposition 14.3 *Any VCG mechanism is truthfully implementable in dominant strategies.*

Proof In the VCG mechanism every player i solves

$$\max_{\hat{\theta}_i \in \Theta_i} u_i(x^*(\hat{\theta}_i, \hat{\theta}_{-i}), \theta_i) + \sum_{j \neq i} u_j(x^*(\hat{\theta}_i, \hat{\theta}_{-i}), \hat{\theta}_j) + h_i(\hat{\theta}_{-i}).$$

Note first that $h_i(\hat{\theta}_{-i})$ does not affect i's choice. Note also that the player's announcement $\hat{\theta}_i$ matters only through its effect on the decision $x^*(\hat{\theta}_i, \hat{\theta}_{-i})$. Thus we can ask which decision the player wants to implement: $\max_x u_i(x, \theta_i) + \sum_{j \neq i} u_j(x, \hat{\theta}_j)$. It follows that player i wants to implement a decision that maximizes total surplus according to his true type and the other players' announced types. That is, he would like to implement $x^*(\theta_i, \hat{\theta}_{-i})$. Player i can achieve this decision by announcing the truth $\hat{\theta}_i = \theta_i$ (by the definition of $x^*(\cdot)$). ∎

Intuitively a VCG mechanism operates as follows. By setting $m_i(\hat{\theta}) = \sum_{j \neq i} u_j(x^*(\hat{\theta}_i, \hat{\theta}_{-i}), \hat{\theta}_j) + h_i(\hat{\theta}_{-i})$, each player i's payoff includes three components: his own $u_i(x^*(\hat{\theta}), \theta_i)$, the component $\sum_{j \neq i} u_j(x^*(\hat{\theta}_i, \hat{\theta}_{-i}), \hat{\theta}_j)$, and the component $h_i(\hat{\theta}_{-i})$. Because $h_i(\theta_{-i})$ is not a function of i's announcement, it has no effect on i's incentives, and we can ignore it. Focusing on the second component, notice that it has a helpful interpretation: this part of the transfer calculates a value that is the sum of the other players' $u_j(\cdot, \hat{\theta}_j)$ functions from the chosen alternative, *if* their true types are the types that they announce. It is important to note, however, that it does not matter whether or not the other players are indeed telling the truth, but only that this part of the transfer to player i is calculated *as if* they are telling the truth. The consequence is that player i is faced with a choice that is *as if* the other players are telling the truth, and then his own payoff becomes the Pareto-optimal objective that the mechanism designer wishes to maximize, with the only decision variable being player i's announced type. Thus it is *as if* the player chooses the ultimate decision x, and through the sum component of m_i he internalizes the externality of his decision on the other players.

Clarke (1971) suggested a particular VCG mechanism known as the **pivotal mechanism.** It is obtained by setting

$$h_i(\hat{\theta}_{-i}) = - \sum_{j \neq i} u_j(x^*_{-i}(\hat{\theta}_{-i}), \hat{\theta}_j),$$

where

$$x^*_{-i}(\hat{\theta}_{-i}) \in \arg \max_{x \in X} \sum_{j \neq i} u_j(x, \hat{\theta}_j)$$

is the optimal choice of x for a society from which player i was absent. Thus

$$m_i(\hat{\theta}) = \sum_{j \neq i} u_j(x^*(\hat{\theta}_i, \hat{\theta}_{-i}), \hat{\theta}_j) - \sum_{j \neq i} u_j(x^*_{-i}(\hat{\theta}_{-i}), \hat{\theta}_j).$$

We may interpret the pivotal mechanism as follows. When player i makes his announcement, we can think of it as if he is joining the other players, and possibly affecting the outcome of what would have happened had he not been part of society. Thus there are two relevant cases:

Case 1: $x^*(\hat{\theta}_i, \hat{\theta}_{-i}) = x^*_{-i}(\hat{\theta}_{-i})$. In this case player i's announcement *does not change* what would have happened if he were not part of society. Then the mechanism specifies a transfer of zero to player i.

Case 2: $x^*(\hat{\theta}_i, \hat{\theta}_{-i}) \neq x^*_{-i}(\hat{\theta}_{-i})$. In this case player i is pivotal in the sense that his announcement *changes* what would have happened without him. His transfer ends up taxing him for the externality that his announcement imposes on the other players because the new choice is not as good for the other players as the choice that maximizes their payoffs without player i.

Example 14.3 (allocation of an indivisible private good) Consider the same setting as in example 14.2, where an object can be allocated to one of N players and the value of owning the private good for player i is given by $u_i(x, \theta_i) = \theta_i x_i$. The first-best allocation solves

$$\max_{(x_1, \ldots, x_n) \in \{0, 1\}^n} \sum_{i=1}^n \theta_i x_i \quad \text{subject to} \quad \sum_i x_i = 1,$$

which results in allocating the good to the player i^* with the highest valuation: $i^* \in \arg \max_i \theta_i$, and

$$x_i^*(\theta) = \begin{cases} 1 & \text{if } i = i^* \\ 0 & \text{otherwise.} \end{cases}$$

The pivotal mechanism then has transfers

$$m_i(\hat{\theta}) = \sum_{j \neq i} u_j(x^*(\hat{\theta}), \hat{\theta}_{-i}) - \sum_{j \neq i} u_j(x^*_{-i}(\hat{\theta}_{-i}), \hat{\theta}) = \begin{cases} -\{\max_{j \neq i^*} \hat{\theta}_j\} & \text{if } i = i^* \\ 0 & \text{otherwise.} \end{cases}$$

That is, every player $i \neq i^*$ is not pivotal: his presence does not affect the allocation, and therefore $m_i(\hat{\theta}) = 0$. On the other hand, player i^* is pivotal: without him, the object would go to the player with the second-highest valuation, and the total surplus would be $\max_{j \neq i^*} \theta_j$. This is exactly the externality player i^* imposes on the others by being present, and this is exactly how much he has to pay in the pivotal mechanism. Now notice that this mechanism is identical to the second-price sealed-bid auction. Recall that we have established that each player in this auction has a dominant strategy, which is bidding his true valuation. This feature of the mechanism was first studied by Vickrey (1961), and this mechanism is also known as the **Vickrey auction.**

14.4 Summary

- Many situations are characterized by a central designer who wishes to make a decision regarding the welfare of a group of players, where the optimal decision depends on the private information that the players have.

- A mechanism is a game that elicits information from the players to help the central (mechanism) designer make a decision according to a decision rule that the designer wishes to implement.

- The revelation principle states that if a decision rule can be implemented by the mechanism designer using some mechanism, then it can be implemented by the simple direct revelation game, in which each player announces his type and the mechanism designer uses his decision rule.

- A particularly useful mechanism is the VCG mechanism, which implements a Pareto-optimal decision rule in dominant strategies when players have quasi-linear preferences.

14.5 Exercises

This chapter offered a very brief introduction to mechanism design aimed at more technically trained readers. For a more complete treatment with numerous exercises and examples, see Fudenberg and Tirole (1991, Chapter 7), Mas-Colell et al. (1995, Chapter 23), and Jehle and Reny (2011, Chapter 9).

14.1 **Brothers:** Imagine two brothers who have preferences over the next possible family vacation. The possible options are traveling to Amsterdam (A), to Barcelona (B), or to Calcutta (C). Brother 1's preferences are fixed and his type is θ_1; he prefers A to B and B to C. Brother 2 can be one of two types, $\theta_2 \in \{\theta_2', \theta_2''\}$, where type θ_2' prefers C to B and B to A while type θ_2'' prefers B to A and A to C. Their parents wish to implement the following choice function: $f(\theta_1, \theta_2') = B$ and $f(\theta_1, \theta_2'') = A$.

    a.   If the parents use the direct revelation mechanism, is it a dominant strategy for brother 2 to reveal his type truthfully?

    b.   Does your answer to (a) change if the choice function is $f(\theta_1, \theta_2') = C$ and $f(\theta_1, \theta_2'') = A$?

    c.   Does your answer to (a) change if the choice function is $f(\theta_1, \theta_2') = B$ and $f(\theta_1, \theta_2'') = B$?

PART V

DYNAMIC GAMES OF INCOMPLETE INFORMATION

Sequential Rationality with Incomplete Information

As we argued in Chapter 7, static (normal-form) games do not capture important aspects of dynamic games in which some players respond to actions that other players have previously made. Furthermore, as we demonstrated with the introduction of backward induction and subgame-perfect equilibrium in Chapter 8, we need to pay attention to the familiar problem of credibility and *sequential rationality*. This chapter applies the idea of sequential rationality to dynamic games of incomplete information and introduces equilibrium concepts that capture these ideas. That is, we want to focus attention on equilibrium play in which players play best-response actions not only *on the equilibrium path* but also at points in the game that are not reached, which we referred to previously as *off the equilibrium path*.

As we saw in the examples in Section 12.2, in games of incomplete information some players will have information sets that correspond to the set of types that their opponents may have, because every player does not know which types Nature chose for the other players. Regardless of whether a player observes his opponents' past behavior (which in games of complete information would imply perfect information), there will always be uncertainty about *which types* the opponents are when incomplete information is present. This in turn implies that structurally there will be many information sets that are not singletons, and this will lead to many fewer proper subgames. As we will now see, this impedes the applicability of subgame perfection as a solution concept that guarantees sequential rationality. We will have to deal more rigorously with the idea that players hold beliefs, and that these beliefs need to be consistent with the environment (Nature) and the strategies of all other players.

15.1 The Problem with Subgame Perfection

To understand the role of proper subgames as introduced in Chapter 8, and how they were used to reach the concept of subgame-perfect equilibrium, it is illuminating to consider the following familiar entry game of complete information. Player 1 is a

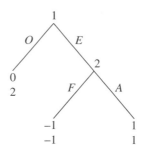

FIGURE 15.1 A simple entry game.

potential entrant to an industry that has a monopolistic incumbent, player 2. If player 1 stays out (O) then the incumbent earns a profit of 2, while the potential entrant gets 0. The entrant's other option is to enter (E), which gives the incumbent a chance to respond. If the incumbent chooses to accommodate entry (A), then both the entrant and the incumbent receive a payoff of 1. The incumbent's other option is to fight entry (F), in which case the payoff for each player is -1. The extensive form of this game is described in Figure 15.1. To find the Nash equilibria of this game it is useful to look at its matrix form as follows:

	F	A
O	0, 2	0, 2
E	$-1, -1$	1, 1

A quick observation reveals that the game has two pure-strategy Nash equilibria, which are (O, F) and (E, A). If, however, we consider the subgame-perfect equilibrium concept, backward induction clearly implies that following player 1's choice of entering, player 2 will strictly prefer to accommodate entry. Therefore player 1 should enter, anticipating a payoff of 1 rather than staying out and receiving 0. Thus subgame perfection implies sequential rationality and picks only one of the two Nash equilibria as the unique subgame-perfect equilibrium, (E, A), in which firm 1 enters and firm 2 accommodates entry.

Now consider a straightforward variant of this game that includes incomplete information. In particular imagine that the entrant may have a technology that is as good as that of the incumbent, in which case the game above describes the payoffs. However, the entrant may also have an inferior technology, in which case he would not gain by entering and the incumbent would lose less if fighting occurred. A particular case of this story can be captured by the following sequence of events:

1. Nature chooses the entrant's type, which can be weak (W) or competitive (C), so that $\theta_1 \in \{W, C\}$, and let $\Pr\{\theta_1 = C\} = p$. The entrant knows his type but the incumbent knows only the probability distribution over types.

2. The entrant chooses between E and O as before, and the incumbent observes the entrant's choice.

3. After observing the action of the entrant, and if the entrant enters, the incumbent can choose between A and F as before.

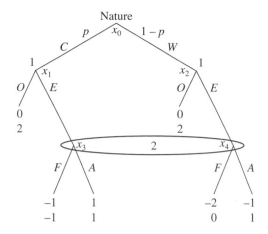

FIGURE 15.2 An entry game with incomplete information.

The payoffs are different for each realization of player 1's type, and they are given in the extensive form of the game in Figure 15.2.

When considering the normal form of this game, player 1 has four pure strategies that follow from the fact that a strategy for him is a type-dependent action, and there are two types and two actions. We define a strategy for player 1 as $s_1 = s_1^C s_1^W$, where $s_1^{\theta_1} \in \{O, E\}$ is what a type θ_1 of player 1 chooses. Thus the pure-strategy set of player 1 is

$$s_1 \in S_1 = \{OO, OE, EO, EE\}.$$

For example, $s_1 = OE$ means that player 1 chooses O when his type is C, and he chooses E when his type is W. Because player 2 has only one information set that follows entry, and two actions in that information set, he has two pure strategies, $s_2 \in \{A, F\}$.

To convert this extensive-form game to a normal-form matrix game, we need to compute the expected payoffs from each pair of pure strategies, where expectations are over the randomizations caused by Nature. Because player 1 has four pure strategies and player 2 has two, there will be eight entries in the normal-form matrix. For example, consider the pair of strategies $(s_1, s_2) = (OE, A)$. The payoffs of the game will be determined by one of the following two outcomes:

1. Nature chooses $\theta_1 = C$, in which case player 1 plays O and the payoffs are $(0, 2)$. This happens with probability $\Pr\{\theta_1 = C\} = p$.

2. Nature chooses $\theta_1 = W$, in which case player 1 plays E and player 2 plays A. The payoffs for this outcome are $(-1, 1)$, and this happens with probability $\Pr\{\theta_1 = W\} = 1 - p$.

From these two possible outcomes we can compute the expected payoffs for players 1 and 2 as follows:

$$Ev_1 = p \times 0 + (1 - p) \times (-1) = p - 1$$
$$Ev_2 = p \times 2 + (1 - p) \times (1) = 1 - p.$$

Similarly if the strategies are $(s_1, s_2) = (EE, F)$ then the expected payoffs for players 1 and 2 are

$$Ev_1 = p \times (-1) + (1-p) \times (-2) = p - 2$$
$$Ev_2 = p \times (-1) + (1-p) \times 0 = -p.$$

In this way we can complete the matrix and obtain the representation for this Bayesian game of incomplete information. To actually compute expected payoffs, set $p = 0.5$, in which case the matrix representation of the Bayesian game is as follows:

Player 2

		F	A
	OO	$\overline{0, 2}$	$\overline{0, 2}$
Player 1	OE	$-1, 1$	$\overline{-\frac{1}{2}, \frac{3}{2}}$
	EO	$-\frac{1}{2}, \frac{1}{2}$	$\overline{\frac{1}{2}, \frac{3}{2}}$
	EE	$-\frac{3}{2}, -\frac{1}{2}$	$\overline{0, 1}$

Following our analysis in Chapter 12, it is easy to find the pure-strategy Bayesian Nash equilibria: every Nash equilibrium of the matrix we have just calculated is a Bayesian Nash equilibrium of the Bayesian game. Therefore both (OO, F) and (EO, A) are pure-strategy Bayesian Nash equilibria of the Bayesian game.[1]

Interestingly these two Bayesian Nash equilibria are tightly related to the two Nash equilibria of the complete-information game in Figure 15.1. The equilibrium (OO, F) is one in which the incumbent "threatens" to fight, which causes the entrant to stay out regardless of his type, similar to the (O, F) equilibrium in the game of complete information in Figure 15.1. The equilibrium (EO, A) is one in which the incumbent accommodates entry, which causes the strong entrant to enter (getting 1 instead of 0) and the weak entrant to stay out (getting 0 instead of -1), similar to the (E, A) equilibrium in the game in Figure 15.1.

Not only are the equilibria similar, but there is a similar problem of credibility with the equilibrium (OO, F): player 2 threatens to fight, but if he finds himself in the information set that follows entry, he has a strict best response, which is to accommodate entry. Thus the Bayesian Nash equilibrium (OO, F) involves noncredible behavior of player 2 that is *not sequentially rational*.

Now comes the interesting question: which of these two equilibria survives as a subgame-perfect equilibrium in the extensive-form game? Recall that the definition of a subgame-perfect equilibrium is that in every *proper subgame*, the restriction of the strategies to that subgame must be a Nash equilibrium in the subgame. This, as we saw in Chapter 8, means that players are playing mutual best responses both on and off the equilibrium path. However, looking at the extensive-form game in Figure 15.2, it is easy to see that there is *only one proper subgame*, which is the complete game. Therefore, both (OO, F) and (EO, A) survive as subgame-perfect equilibria.

---

1. Notice that for player 1 *OE* is strictly dominated by *OO* and *EE* is strictly dominated by *EO*. Notice also that there are a continuum of mixed-strategy Bayesian Nash equilibria in which the incumbent plays *F* with probability $p \geq \frac{1}{2}$ and the entrant plays *OO*. There cannot be a mixed-strategy Bayesian Nash equilibrium in which the entrant mixes between *OO* and *EO* because then the incumbent's best response is *A*, in which case the entrant would strictly prefer to play *EO* over *OO*.

This example demonstrates that the very appealing concept of subgame-perfect equilibrium may have no bite for some games of incomplete information. At first this may seem somewhat puzzling. However, the problem is that the concept of subgame-perfect equilibrium restricts attention to best responses *within subgames,* but when there is incomplete information the induced information sets over types of other players often cause the only proper subgame to be the complete game.

The reason this happens is that in the modified entry game we analyzed, even though player 2 observes the actions of player 1, the fact that player 1 has several types implies that there is no proper subgame that starts with player 2 making a move. This is because whenever player 2 makes a move, he does not know the type of player 1, which implies that there are no proper subgames except for the whole game. Indeed this problem of action nodes being linked through information sets that represent the uncertainty players have over the types of their opponents will carry over to all games of incomplete information.

In order to extend the logic of subgame-perfect equilibrium to dynamic games of incomplete information, we need to impose a more rigorous structure on our solution concept in order for sequential rationality to be well defined. Our goal is to identify a structure of analysis that will cause the elimination of equilibria that involve noncredible threats such as those in the modified entry game, which we do in the next section.

15.2   Perfect Bayesian Equilibrium

To address the problem with subgame perfection demonstrated in the previous section we need to identify a way in which to express the fact that following entry, it is not sequentially rational for player 2 to play fight. In particular we need to make statements about the sequential rationality of player 2 *within each of his information sets* even though the information set is not itself the first node of a proper subgame. We need to be able to make statements like "in this information set player 2 is not playing a best response, and therefore his behavior is not sequentially rational."

This kind of reasoning is precisely what is missing from the definition of subgame-perfect equilibrium—we were not able to isolate player 2 within his information set. However, to describe a player's best response *within* his information set, we will have to ask *what* the player is playing a best response to. The answer, of course, must follow from the definition of what a best response is: we must include *beliefs* in the analysis. This will allow us to consider the beliefs of player 2 in his information sets and then analyze his best response to these beliefs.

More generally, to introduce sequential rationality at information sets that are not singletons, we need to insist that players form beliefs in *every* information set they have. These will include information sets that are reached with some positive probability given the proposed actions of the players as well as information sets that are not reached at all. Recall from Section 7.3 the notion of being on and off the equilibrium path, which is introduced again for Bayesian games as follows:

Definition 15.1   Let $\sigma^* = (\sigma_1^*, \ldots, \sigma_n^*)$ be a Bayesian Nash equilibrium profile of strategies in a game of incomplete information. We say that an information set is **on the equilibrium path** if given σ^* and given the distribution of types, it is reached with positive probability. We say that an information set is **off the equilibrium path** if given σ^* and the distribution of types, it is reached with zero probability.

Notice that an important part of being on or off the equilibrium path is having an equilibrium profile of strategies to start with. This is why the definition includes a proposed Bayesian Nash equilibrium profile, σ^*, from which we can discuss the information sets that are on the equilibrium path *given* σ^*, and those that are off the equilibrium path *given* σ^*. To illustrate this point consider the entry game previously described and illustrated in Figure 15.2.

Assume that player 1 chooses the strategy *EO,* implying that he enters when he is competitive and stays out when he is weak. In this case all information sets are reached with positive probability. First, given the structure of the game, both information sets of player 1 are always reached (with probabilities p and $1 - p$, respectively). Second, the information set of player 2 is reached because the competitive type of player 1 will enter. This implies that it is reached with probability p.

Now assume instead that player 1 chooses the strategy *OO,* implying that he never enters. In this case the information set of player 2 is not reached with positive probability. This illustrates a more general fact about games of incomplete information: the actions of the player with private information will have an effect on which information sets of the *uninformed* player are reached with positive probability.

We now formally introduce the notion that players must have beliefs in each of their information sets:

Definition 15.2 A **system of beliefs** μ of an extensive-form game assigns a probability distribution over decision nodes to every information set. That is, for every information set $h \in H$ and every decision node $x \in h$, $\mu(x) \in [0, 1]$ is the probability that player i who moves in information set h assigns to his being at x, where $\sum_{x \in h} \mu(x) = 1$ for every $h \in H$.

For example, in the entry game previously described the beliefs of player 1 are trivially defined because his information sets are singletons, while the belief of player 2 is a probability distribution over the two nodes in his unique information set. Denote by $\mu(x_3)$ player 2's belief that he is at the node corresponding to player 1 being competitive (C) and playing E, and let $1 - \mu(x_3)$ be his belief that he is at the node corresponding to player 1 being weak (W) and playing E. We are now ready to lay out our first requirement of sequential rationality in games of incomplete information:

Requirement 15.1 Every player will have a well-defined belief over where he is in each of his information sets. That is, the game will have a *system of beliefs.*

Now that we have established the need to have a system of beliefs, we need to ask the following question: how should the beliefs in a system of beliefs be determined? That is, can players have any beliefs they want, or should the beliefs be restricted by some elements that are not controlled by the player himself?

Just as we required the beliefs of players about the strategies of their opponents to be correct in order to define a Nash equilibrium, here too we will add a very similar requirement. In games of incomplete information, two constraints will influence whether a player's beliefs are correct. The first is the behavior of the other players, which we can consider as an *endogenous constraint on beliefs.* It is endogenous in the sense that it is determined by the strategies of the players, which are the "variables" that the players control. The second constraint on beliefs comes from the choices of Nature through the distribution of types. This is an *exogenous constraint on beliefs.* It is exogenous in the sense that it is determined by Nature, which is not something that the players control but rather part of the environment.

Let's illustrate this using the entry game. Following requirement 15.1, we defined player 2's belief in his information set as

$$\mu(x_3) = \Pr\{\text{player 1 is competitive} \mid E\}$$

$$1 - \mu(x_3) = \Pr\{\text{player 1 is weak} \mid E\}.$$

Imagine now that player 2 believes that player 1 is choosing the strategy *EO*, so that if he is competitive he enters and if he is weak he stays out. What should player 2 believe if his information set is reached? The only consistent belief would be that player 1 is competitive, and therefore it must be that $\mu(x_3) = 1$. This follows because with probability p Nature chooses a competitive type for player 1, and with the strategy *EO* a competitive player 1 always enters. The probability of reaching the node "weak followed by entry" inside the information set of player 2 is 0 because with probability $1 - p$ Nature chooses a weak type for player 1, and with the strategy *EO* a weak player 1 never enters. Any other belief would not be consistent with the belief that player 1 is playing *EO*.

Player 1's behavior can of course be more complex than choosing *EO*. More generally in the entry game, the probability of reaching the node "player 1 is type *C* and he chose *E*" inside the information set of player 2 is p times the probability that a type *C* player 1 chose *E*. By definition, $\mu(x_3)$ is the belief that player 2 assigns to "player 1 is type *C* *conditional on* entry occurring." Hence we need to consider the beliefs of player 2 as being conditional on the fact that his information set was indeed reached. Assume that player 1 is playing the following strategy: If $\theta_1 = C$ then he plays *E* with probability σ_C and *O* with probability $1 - \sigma_C$. Similarly if $\theta_1 = W$ then he plays *E* with probability σ_W and *O* with probability $1 - \sigma_W$. Now because Nature chooses $\Pr\{\theta_1 = C\} = p$, by Bayes' rule we must have[2]

$$\mu(x_3) = \Pr\{\text{player 1 is competitive} \mid \text{entry occurred}\}$$

$$= \frac{\Pr\{\text{competitive and entry occurred}\}}{\Pr\{\text{competitive and entry occurred}\} + \Pr\{\text{weak and entry occurred}\}}$$

$$= \frac{p\sigma_C}{p\sigma_C + (1 - p)\sigma_W}.$$

For example, the pure strategy *EO* is just a special case of this mixed (behavioral) strategy that has $\sigma_C = 1$ and $\sigma_W = 0$, and if this is the proposed strategy then of course $\mu(x_3) = 1$. Following this argument, we are ready to state our second requirement for sequential rationality: given a conjectured profile of strategies, and given Nature's choices, we require players' beliefs to be correct in information sets that are reached with positive probability. Formally, we have

Requirement 15.2 Let $\sigma^* = (\sigma_1^*, \ldots, \sigma_n^*)$ be a Bayesian Nash equilibrium profile of strategies. We require that in all information sets beliefs that are *on the equilibrium path* be consistent with *Bayes' rule*.

This is precisely how beliefs are formed; they include both the *exogenous constraints* that follow from Nature's probability distribution and the *endogenous*

---

2. See Section 19.4.3 for a detailed description of Bayes' rule and how it relates to dynamic games of incomplete information.

constraints that follow from beliefs about the other players' strategies from $\sigma^* = (\sigma_1^*, \ldots, \sigma_n^*)$.

Now consider the pure strategy OO (or $\sigma_C = \sigma_W = 0$). The probability of reaching player 2's information set is *not positive* because there is no way in which it is reached. That is, no type of player 1 plays E, and this implies that E is never chosen. If player 2 *believes* that player 1 chooses OO, and suddenly finds himself in his information set that follows entry, then Bayes' rule *does not apply* because given the suggested strategy both the numerator and the denominator in Bayes' rule are zero, and thus $\mu(x_3)$ is not well defined. What then should determine $\mu(x_3)$? In other words, if we cannot apply Bayes' rule because of the beliefs over strategies, what will we use to determine beliefs? For this we introduce the third requirement.

Requirement 15.3 At information sets that are *off the equilibrium path* any belief can be assigned to which Bayes' rule does not apply.

This means that when the moves of Nature combined with the belief over the strategies of the other players do not impose constraints on beliefs, then indeed beliefs could be whatever the player chooses them to be. Looking back at the case in which player 1 chooses the pure strategy OO in the entry game, then $\mu(x_3)$ can be any number in the interval $[0, 1]$ because it is not constrained by Bayes' rule.

All of the first three requirements were imposed to introduce well-defined beliefs for every player at each of his information sets. Now we come to the final requirement of sequential rationality:

Requirement 15.4 Given their beliefs, players' strategies must be *sequentially rational*. That is, in every information set players will play a best response to their beliefs.

To write down requirement 15.4 formally, consider player i with beliefs over information sets derived from the belief system μ, given player i's opponents playing σ_{-i}. Requirement 15.4 says that if h is an information set for player i then it must be true that he is playing a strategy σ_i that satisfies

$$E[v_i(\sigma_i, \sigma_{-i}, \theta_i)|h, \mu] \geq E[v_i(s_i', \sigma_{-i}, \theta_i)|h, \mu] \quad \text{for all } s_i' \in S_i,$$

where expectations are given over the beliefs of player i using μ.

We can now incorporate requirements 15.1–15.4 to show that the noncredible equilibrium in the entry game with incomplete information is fragile. To see this, notice that once we specify a belief $\mu(x_3)$ for player 2, then *for any* $\mu(x_3) \in [0, 1]$, it is a best response for player 2 to play A. This means that once we endow player 2 with a well-defined belief (requirement 15.1) then despite the fact that these beliefs are not restricted if player 1 chooses OO (by requirement 15.3), the Bayesian Nash equilibrium (OO, F) has player 2 not playing a best response to *any belief*, which violates requirement 15.4.

Now that we have defined beliefs together with sequential rationality, we need to combine all these components to define a coherent equilibrium concept:

Definition 15.3 A Bayesian Nash equilibrium profile $\sigma^* = (\sigma_1^*, \ldots, \sigma_n^*)$ together with a system of beliefs μ constitutes a **perfect Bayesian equilibrium** for an n-player game if they satisfy requirements 15.1–15.4.

This definition puts together our four requirements in a way that will guarantee sequential rationality. The next obvious question is, how do we find a perfect Bayesian equilibrium for a game? One way is to first find all the profiles of strategies in the Bayesian game that are Bayesian Nash equilibria; then we can systematically check for each Bayesian Nash equilibrium to see whether we can find a system of beliefs so that together they constitute a perfect Bayesian equilibrium.

A first step in this process is the following observation. If a game has a Bayesian Nash equilibrium that has *all the information sets* being reached on the equilibrium path (all information sets are reached with positive probability), then beliefs will be uniquely pinned down by Bayes' rule owing to requirement 15.2. This observation leads to the following result:

Proposition 15.1 *If a profile of (possibly mixed) strategies $\sigma^* = (\sigma_1^*, \ldots, \sigma_n^*)$ is a Bayesian Nash equilibrium of a Bayesian game Γ, and if σ^* induces all the information sets to be reached with positive probability, then σ^*, together with the belief system μ^* uniquely derived from σ^* and the distribution of types, constitutes a perfect Bayesian equilibrium for Γ.*

Proof Assume that the proposition was false, that is, σ^* is a Bayesian Nash equilibrium but (σ^*, μ^*), for which all the information sets are reached with positive probability, do not together form a perfect Bayesian equilibrium. This implies that for some player i playing according to σ_i^* is not a best response in (at least) one of his information sets, say h_i, to the beliefs derived from μ^*. Let σ_i' be a strategy for player i that chooses a best response to μ^* in the information set h_i and coincides with σ_i^* in every other information set of player i. But because μ^* is derived from σ^*, σ_i' will also be better than σ_i^* against σ_{-i}^*, contradicting the fact that σ^* is a Bayesian Nash equilibrium. ∎

As we soon turn to some applications of the perfect Bayesian equilibrium concept, this proposition will come in handy in analyzing an important class of games. It is easy to see how it applies to the entry game. As we demonstrated earlier, the game has two Bayesian Nash equilibria: (EO, A) and (OO, F). Because the Bayesian Nash equilibrium (EO, A) induces every information set to be reached, it follows that it can be supported as a perfect Bayesian equilibrium, and by Bayes' rule the belief that player 2 must have in his information is that he faces a competitive type of entrant with probability 1. As we also saw, (OO, F) cannot be supported as a perfect Bayesian equilibrium.

Remark As mentioned earlier, requirement 15.3 imposes no restrictions off the equilibrium path, which is one reason that some texts refer to this solution concept as **weak perfect Bayesian equilibrium.** It is worth pointing out that a more stringent requirement 15.3 is that beliefs off the equilibrium path are defined by Bayes' rule where possible. How can it be that Bayes' rule has bite when an information set is not reached? Consider the example in Figure 15.3, which does not have Nature move, but by definition can have beliefs assigned to each information set, and thus can accommodate requirement 15.3.

In this game, imagine that a profile of strategies has player 1 playing L, player 2 playing a mixed strategy $\sigma_A = \Pr\{A\}$, and player 3 playing a mixed strategy $\sigma_C = \Pr\{C\}$. If these are the beliefs of player 4 then in this case the information set following player 2 is reached with probability 1, and the belief must be $\mu_A = \sigma_A$. However, the information set following player 3 is not reached with positive probability, and requirement 15.3 as already stated allows for any beliefs in this information set. A more careful scrutiny would imply the following logic: If player 4 realizes that his

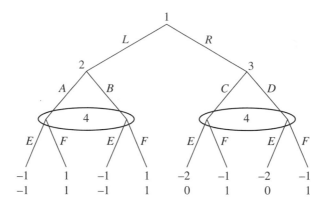

FIGURE 15.3   Example: requirement 15.3 has restrictions.

information set following player 3 was reached then he must assume that player 1 deviated from his strategy of playing L. But why should he believe that player 3 also deviated from his strategy σ_C? Indeed a natural belief would be that only player 1 deviated, and therefore we must have $\mu_C = \sigma_C$. This more stringent statement of requirement 15.3 implies that sometimes information sets that are reached with zero probability can still have constraints on beliefs. The next section introduces a refinement of perfect Bayesian equilibrium that explicitly deals with this issue.

15.3   Sequential Equilibrium

Perfect Bayesian equilibrium has become the most widely used solution concept for dynamic games with incomplete information. There are, however, examples of games in which the perfect Bayesian equilibrium solution concept allows for equilibria that seem unreasonable. The reason for this is that restriction 3 of the perfect Bayesian equilibrium concept places no restrictions on beliefs that are off the equilibrium path.

To see this, consider the game shown in Figure 15.4. If player 1 plays D with a positive probability then by requirement 15.2 the beliefs of player 2 are completely determined by Bayes' rule, which implies that $\mu_2(x_1) = \mu_2(x_2) = \frac{1}{2}$. With these beliefs player 2 must play L. If player 2 plays L then player 1's best response is

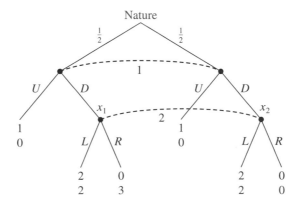

FIGURE 15.4   A game with an unreasonable perfect Bayesian equilibrium.

to play D, which implies that the pair of strategies (D, L) together with the implied beliefs $\mu_2(x_1) = \mu_2(x_2) = \frac{1}{2}$ form a perfect Bayesian equilibrium. Note, however, that the pair of strategies (U, R) can also be supported as a perfect Bayesian equilibrium. If player 1 plays U then by requirements 15.2 and 15.3 beliefs are not restricted in player 2's information set. In particular, player 2 can believe that $\mu_2(x_1) > \frac{2}{3}$, in which case he believes that by playing R his expected payoff will be $3\mu_2(x_1) > 2$, for which playing R is a best response. As a consequence, playing U for player 1 is also a best response!

This example may seem unusual, but it turns out to be quite common. It is for this reason that many applications of dynamic games of incomplete information use stronger concepts than perfect Bayesian equilibrium. These are often referred to as "equilibrium refinements," as their goal is to refine the set of outcomes that can be supported in equilibrium. The key idea behind these refinements is to put restrictions on the sorts of beliefs that players can hold in information sets that are off the equilibrium path.

One commonly used refinement is called *sequential equilibrium*, and it was introduced by Kreps and Wilson (1982). They define the notion of "consistent" beliefs as follows:

Definition 15.4 A profile of strategies $\sigma^* = (\sigma_1^*, \ldots, \sigma_n^*)$, together with a system of beliefs μ^*, is **consistent** if there exists a sequence of nondegenerate mixed strategies, $\{\sigma^k\}_{k=1}^{\infty}$, and a sequence of beliefs that are derived from each σ^k according to Bayes' rule, $\{\mu^k\}_{k=1}^{\infty}$, such that $\lim_{k \to \infty}(\sigma^k, \mu^k) = (\sigma^*, \mu^*)$.

To see the power behind this definition, observe that in the game described in Figure 15.4 the only consistent beliefs for player 2 are $\mu_2(x_1) = \mu_2(x_2) = \frac{1}{2}$. The reason is that any belief that is part of a consistent pair of strategies and beliefs must be derived from a sequence of strategies that cause every information set to be reached with positive probability on the equilibrium path. This follows from the requirement that $\{\sigma^k\}_{k=1}^{\infty}$ be a sequence of *nondegenerate* mixed strategies, which implies that each player is mixing among all his actions with positive probability. As we argued earlier, if player 1 plays D with positive probability then the beliefs of player 2 are completely determined by Bayes' rule to be $\mu_2(x_1) = \mu_2(x_2) = \frac{1}{2}$, so in any sequence of the form required by consistency the limit of these beliefs must be $\mu_2(x_1) = \mu_2(x_2) = \frac{1}{2}$.

Definition 15.5 A profile of strategies $\sigma^* = (\sigma_1^*, \ldots, \sigma_n^*)$, together with a system of beliefs μ^*, is a **sequential equilibrium** if (σ^*, μ^*) is a consistent perfect Bayesian equilibrium.

This definition implies of course that every sequential equilibrium is a perfect Bayesian equilibrium, but the reverse is not true. Indeed of the two perfect Bayesian equilibria we identified earlier, (U, R) is not part of a sequential equilibrium because the only beliefs that can be part of a consistent strategy profile–belief pair are $\mu_2(x_1) = \mu_2(x_2) = \frac{1}{2}$, which implies that (D, L) together with the beliefs $\mu_2(x_1) = \mu_2(x_2) = \frac{1}{2}$ is the unique sequential equilibrium.

Sequential equilibrium is a bit harder to apply than perfect Bayesian equilibrium in practice, so as with most applications in the social sciences we will stick to perfect Bayesian equilibrium as our solution concept. It is interesting to note that there are games in which even sequential equilibrium allows for unreasonable outcomes, which suggests that one might want to consider a stronger refinement for a solution concept. Indeed we will see one such refinement for signaling games in Chapter 16.

15.4   Summary

- Because games of incomplete information have information sets that are associated with Nature's choices of types, it will often be the case that the only proper subgame is the whole game. As a consequence, subgame-perfect equilibrium will rarely restrict the set of Bayesian Nash equilibria to those that are sequentially rational.

- By requiring that players form beliefs in every information set, and requiring these beliefs to be consistent with Bayes' rule, we can apply the concept of sequential rationality to Bayesian games.

- In a perfect Bayesian equilibrium, beliefs are constrained on the equilibrium path but not off the equilibrium path. It is important, however, that beliefs off the equilibrium path support equilibrium behavior.

- In some games the concept of perfect Bayesian equilibrium will not rule out play that seems sequentially irrational. Equilibrium refinements, such as sequential equilibrium, have been developed to address these situations.

15.5   Exercises

15.1   **Equilibrium Selection:** Consider the extensive-form game in Figure 15.5.

    a.   Find all the Bayesian Nash equilibria of this game.

    b.   Which of the Bayesian Nash equilibria are also perfect Bayesian equilibria? Why?

15.2   **Not All That Glitters Revisited:** Recall exercise 12.5 in Chapter 12 and your analysis of the unique Bayesian Nash equilibrium for that game. The secretary of commerce is contemplating the introduction of a certification program. This will allow a prospector to get a certificate that exactly states *and publicizes* the amount of gold in the mine and eliminates the problem of asymmetric information. If the program is implemented the game would be modified as follows: The prospector first decides whether or not to get a government certificate and then decides for what price to ask. Thus each type x chooses a pair $(y(x), p(y))$, where $y(x) \in \{N, x\}$ denotes his choice of certificate (N for no certificate and x for the true value on the certificate)

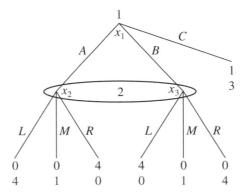

FIGURE 15.5   Exercise 15.1.

and $p(y)$ denotes his asking price given his certification choice. Assume that the cost of obtaining a certificate is $\frac{1}{2}$.

a. Imagine that a prospector of type x gets a certificate. What is the unique subgame-perfect equilibrium *in the subgame that starts after the certification?*

b. Show that *if there exists a perfect Bayesian equilibrium* in which some type $x' < 1$ chooses to certify, it *must be true* that all *higher* types $x'' \in (x', 1]$ also choose to certify in that perfect Bayesian equilibrium. Show that *in any perfect Bayesian equilibrium* type $x = 1$ must choose to be certified.

c. From (b) we know that *if there is a perfect Bayesian equilibrium with some certification* then it must take on the following form: an upper interval of types will choose to be certified, and the complementary lower interval will not. That is, there is some threshold type $x^* \in [0, 1]$ such that types in the interval $(x^*, 1]$ will certify, types in the interval $[0, x^*)$ will not, and type x^* will be indifferent (assume without loss that type x^* certifies). Show that *if such a perfect Bayesian equilibrium with certification exists,* then the lower interval $[0, x^*)$ will self-mine, and there is *no price* at which they can trade their mines to the company in equilibrium. Following your findings, find the (*almost*) unique pure-strategy perfect Bayesian equilibrium and explain why it is unique. (The meaning of "almost unique" is that the threshold type x^* can choose to certify with any probability, so excluding the behavior of this type, the equilibrium behavior is unique.)

d. If the price of the government certification program is sufficient to cover its cost, would you argue that the economy is better off in the Pareto sense (some are better off while no one is worse off) compared to the status before government intervention? If so, who is better off?

15.3 **Trust Thy Neighbor:** A student, player 1, has to hand in a problem set at the other end of (a large) campus but needs to rush into an exam. She has two options. She can deliver the problem set after the exam (call this option L) and incur a late penalty. Alternatively she can give the problem set to player 2, a random student who happens to be next to player 1 (call this option S). Player 2 can then either deliver the problem set on time (call this option D) or throw it away in the nearest trash can (call this option T). For player 1 the payoff is 1 if the problem set is delivered on time, -1 if it is thrown away, and 0 if it is delivered late. The payoffs for player 2 are x if he delivers and y if he throws it away.

a. Draw the game tree that represents this game. Using equilibrium analysis, what conditions on x and y would justify player 1 trusting player 2 to deliver the problem set?

b. Now assume that some proportion p of the population of students are "nice guys" (N), for which $x = 1$ and $y = 0$, while a proportion $1 - p$ of the population are "jerks" (J), for which $x = 0$ and $y = 1$. Thus we can think that before the game is played, Nature first chooses the type of player 2 that player 1 will meet, according to the probabilities previously described. Draw the game tree.

c. Now assume that for (b) $p = \frac{3}{4}$. What are the pure-strategy Bayesian Nash equilibria of this game? Are there Bayesian Nash equilibria that are not perfect Bayesian equilibria?

    d.   For which values of p are all the Bayesian Nash equilibria also perfect Bayesian equilibria?

15.4   **Campaigning:** Two politicians, an incumbent (player 1) and a potential rival (player 2), are running for the local mayoralty. The incumbent has either a broad base of support (B) or a small base of support (S), each occurring with probability $\frac{1}{2}$. The incumbent knows his level of support but the potential rival does not. The incumbent first chooses how much soft money to spend on campaign financing: a low quantity (L) or a high quantity (H), a decision that is observed by the potential rival. The rival can then decide to run (R) or not to run (N). If the incumbent chooses a level of campaign financing L then given the support base and the reaction of the potential rival, the payoffs are given by the following *payoff matrix* (these are payoffs that represent the expectations from winning, campaigning, and so on; this is not a matrix game):

		Player 2's response	
		R	N
Support base	B	6, 4	10, 0
	S	4, 4	6, 0

The **cost in payoffs** that an incumbent incurs for choosing H instead of L is 2 if he has a broad base of support and 4 if he has a small base of support (that is, these are costs that are deducted from the payoffs in the payoff matrix that are conditional on the type). A rival who runs against an incumbent with a broad base of support who chose H will obtain a payoff of -10, while a rival who runs against an incumbent with a small base of support who chose H will obtain a payoff of -4. If the rival chooses not to run then he obtains 0, as in the payoff matrix.

    a.   Draw the extensive form of this game and identify the proper subgames. Draw the matrix that represents the normal form of the extensive form.

    b.   If the rival could commit in advance to a certain pure strategy that he would follow regardless of the incumbent's choice of financing, anticipating that the incumbent would then choose his best response, what would that strategy be? What would be the incumbent's best response to this strategy? Is the pair of strategies you found a Bayesian Nash equilibrium?

    c.   Can the pair of strategies you found in (b) be part of a perfect Bayesian equilibrium?

    d.   Are there other pairs of strategies that can be part of a perfect Bayesian equilibrium?

15.5   **Entry (again):** The market for widgets has an incumbent firm. The total value of having 100% of the market is 10, which the incumbent receives if no one enters. A potential entrant arrives, and it can be one of two types: tough (T) or weak (W). A weak entrant can choose one of three options: small entry (S), big entry (B), or exit (X). A tough entrant can choose only between S or B (it is inconceivable that it would choose X). There is no cost for a tough entrant to enter at any level. However, it costs a weak entrant 6 to enter at any

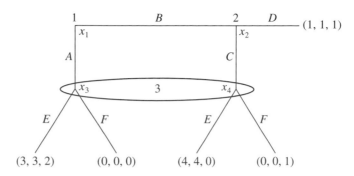

FIGURE 15.6 Selten's horse, exercise 15.6.

level. Exiting costs nothing to the entrant. The entrant knows his type, but the incumbent knows only the prior distribution: $\Pr\{T\} = \frac{1}{2}$.

In response to any level of entry, the incumbent can choose to accommodate (A) or fight (F). Accommodating an entrant imposes no costs. Independent of the entrant's type, accommodating small entry gives the incumbent 60% of the market and the entrant 40%, while accommodating big entry gives the incumbent 40% of the market and the entrant 60%. Fighting a tough entrant increases the incumbent's market share by 20% (relative to accommodating) but imposes a cost of 4 on the incumbent. Fighting a weak entrant that chose S increases the market share of the incumbent to 100% but imposes a cost of 2 on the incumbent. Fighting a weak entrant that chose B increases the market share of the incumbent to 100% but imposes a cost of 8 on the incumbent.

    a. Draw this game in extensive form.

    b. Using a matrix representation, find all the pure-strategy Bayesian Nash equilibria for this game.

    c. Which one of the Bayesian Nash equilibria is preferred by the incumbent? Can it be supported as a perfect Bayesian equilibrium?

    d. Find all the perfect Bayesian equilibria of this game.

15.6 **Selten's Horse:** Consider the three-person game described in Figure 15.6, known as Selten's horse (for the obvious reason).

    a. What are the pure-strategy Bayesian Nash equilibria of this game?

    b. Which of the Bayesian Nash equilibria that you found in (a) are perfect Bayesian equilibria?

    c. Which of the Bayesian Nash equilibria that you found in (a) are sequential equilibria?

16

Signaling Games

In games of incomplete information there is at least one player who is uninformed about the type of another player. In some instances it will be to the benefit of players to reveal their types to their opponents. For instance, if a potential rival to an incumbent firm or an incumbent politician knows that he is strong, he may want to reveal that information to the incumbent, to suggest "I am strong and hence you should not waste time and energy fighting me." Of course even a weak player would like to try to convince his opponent that he is strong, so merely stating "I am strong" will not do. There has to be some credible means, beyond such "cheap talk," through which the player can *signal* his type and make his opponent believe him.

Games in which such signaling is possible in equilibrium are called *signaling games;* they originated in the Nobel Prize–winning contribution of Michael Spence (1973), which he developed in his Ph.D. thesis. Spence investigated the role of education as an instrument that signals information to potential employers about a person's intrinsic abilities, but not necessarily what he has learned.

Signaling games share a structure that includes the following four components:

1. Nature chooses a type for player 1 that player 2 does not know, but cares about (common values).

2. Player 1 has a rich action set in the sense that there are at least as many actions as there are types, and each action imposes a different cost on each type.

3. Player 1 chooses an action first, and player 2 then responds after observing player 1's choice.

4. Given player 2's belief about player 1's strategy, player 2 updates his belief after observing player 1's choice. Player 2 then makes his choice as a best response to his updated beliefs.

These games are called signaling games because of the potential signal that player 1's actions can convey to player 2. If in equilibrium each type of player 1 is playing a different choice then *in equilibrium* the action of player 1 will fully reveal player 1's type to player 2. That is, even though player 2 does not *know* the type of player 1, in equilibrium player 2 *fully learns* the type of player 1 through his actions.

Of course, it need not be the case that player 1's type is revealed. If, for instance, in equilibrium all the types of player 1 choose the same action then player 2 cannot update his beliefs at all. Because of this variation in the signaling potential of player 1's strategies, these games have two important classes of perfect Bayesian equilibria:

1. *Pooling equilibria:* These are equilibria in which *all the types of player 1 choose the same action,* thus revealing nothing to player 2. Player 2's beliefs must be derived from Bayes' rule *only* in the information sets that are reached with *positive probability.* All other information sets are reached with probability zero, and in these information sets player 2 must have beliefs that support his own strategy. The sequentially rational strategy of player 2 given his beliefs is what keeps player 1 from deviating from his pooling strategy.

2. *Separating equilibria:* These are equilibria in which *each type of player 1 chooses a different action,* thus revealing his type in equilibrium to player 2. Player 2's beliefs are thus well defined by Bayes' rule in all the information sets that are reached with positive probability. If there are more actions than types for player 1, then player 2 must have beliefs in the information sets that are not reached (the actions that no type of player 1 chooses), which in turn must support the strategy of player 2. Player 2's strategy supports the strategy of player 1.

The choice of terms is not coincidental. In a **pooling equilibrium** all the types of player 1 *pool together* in the action set, and thus player 2 can learn nothing from the action of player 1. Player 2's posterior belief after player 1 moves must be equal to his prior belief over the distribution of Nature's choices of types for player 1. In a **separating equilibrium** each type of player 1 *separates* from the others by choosing a unique action that no other type chooses. Thus after observing what player 1 did, player 2 can infer *exactly* what type player 1 is.

Remark   There is a third class of equilibria called **hybrid** or **semi-separating equilibria,** in which different types choose different *mixed strategies.* As a consequence some information sets that belong to the uninformed player can be reached by different types with different probabilities. Thus Bayes' rule implies that in these information sets player 2 can learn something about player 1 but cannot always infer exactly which type he is. We will explore these kinds of equilibria in Chapter 17. See Fudenberg and Tirole (1991, Chapter 8) for a more advanced treatment.

The incomplete-information entry game that we analyzed in the previous chapter can be used to illustrate these two classes. In the Bayesian Nash equilibrium (OO, F), both types of player 1 chose "out," so player 2 learns nothing about player 1's type (in this case he has no active action following player 1's decision to stay out). Thus (OO, F) is a pooling equilibrium (though it is not a perfect Bayesian equilibrium, as demonstrated earlier). In the Bayesian Nash equilibrium (EO, A), which is also a perfect Bayesian equilibrium, player 1's action perfectly reveals his type: if player 2 sees entry, he believes with probability 1 that player 1 is strong, while if player 1 chooses to stay out then player 2 believes with probability 1 that player 1 is weak. Therefore (EO, A) is a separating equilibrium.

16.1   Education Signaling: The MBA Game

This section analyzes a very simple version of an education signaling game in the spirit of Spence's work that sheds some light on the signaling value of education. To focus attention on the signaling value of education, we will ignore any productive value that education may provide. That is, we assume that a person learns nothing productive from education but has to "suffer" the loss of time and the hard work of studying to get a diploma, in this case an MBA degree.[1] The game proceeds in the following steps:

1. Nature chooses player 1's skill (productivity at work), which can be high (H) or low (L), and only player 1 knows his skill. Thus his type set is $\Theta = \{H, L\}$. The probability that player 1's type is H is given by $\Pr\{\theta = H\} = p > 0$, and it is common knowledge that this is Nature's prior distribution.

2. After player 1 learns his type, he can choose whether to get an MBA degree (D) or be content with his undergraduate-level degree (U), so that his action set is $A_1 = \{D, U\}$. Getting an MBA requires some effort that is *type dependent*. Player 1 incurs a private cost c_θ if he gets an MBA, and a cost of 0 if he does not. We assume that high-skilled types find it easier to study, captured by the assumption that $c_H < c_L$. We assume in particular that $c_H = 2$ and $c_L = 5$.

3. Player 2 is an employer, who can assign player 1 to one of two jobs. Specifically player 2 can assign player 1 to be either a manager (M) or a blue-collar worker (B), so that his action set is $A_2 = \{M, B\}$. The employer will retain the profit from the project and must pay a wage to the worker depending on the job assignment. The market wage for a manager is w_M and that for a blue-collar worker is w_B, where $w_M > w_B$. We assume in particular that $w_M = 10$ and $w_B = 6$.

4. Player 2's payoff (the employer's profit) is determined by the combination of skill and job assignments. It is assumed that the MBA degree adds nothing to productivity. A high-skilled worker is relatively better at managing, while a low-skilled worker is relatively better at blue-collar work. The employer's net profits from the possible skill-assignment matches are given in the following table:

		Assignment	
		M	B
	H	10	5
Skill			
	L	0	3

Given the information about the game that is laid out in (1)–(4), the complete game tree is represented in Figure 16.1. Because player 2 does not know player 1's type and only observes his choice, there are two information sets. The two nodes that follow the choice U are in one information set, and the two nodes that follow the choice D are in the second information set. In the analysis that follows, we refer to the first information set as I_U and to the second as I_D.

---

1. It is clearly true that obtaining an MBA degree from a prestigious university acts as a strong signal of one's ability. I can attest, however, that—at least at the University of California, Berkeley, where I currently teach—there is also a serious productive component to the MBA program, from which students certainly benefit.

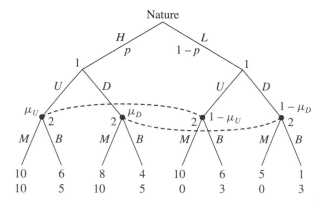

FIGURE 16.1 The MBA game.

First, to define beliefs, let μ_U denote the belief of player 2 that player 1's type is H conditional on player 1 choosing U, and similarly let μ_D denote the belief of player 2 that player 1's type is H conditional on player 1 choosing D. These beliefs will be determined by the distribution of Nature's choice, together with the beliefs that player 2 holds about the strategy that player 1 is playing. For equilibrium analysis, these beliefs will be determined according to requirements 15.2 and 15.3 described in Section 15.2.

In general if player 1 is using a mixed strategy in which type H chooses U with probability σ^H and type L chooses U with probability σ^L, and if both σ^H and σ^L are strictly between 0 and 1 (i.e., the two types are choosing nondegenerate mixed strategies) then requirement 15.2 implies that by Bayes' rule

$$\mu_U = \frac{p\sigma^H}{p\sigma^H + (1-p)\sigma^L} \tag{16.1}$$

and

$$\mu_D = \frac{p(1-\sigma^H)}{p(1-\sigma^H) + (1-p)(1-\sigma^L)}. \tag{16.2}$$

Notice that if both $\sigma^H = \sigma^L = 1$ (both types are choosing U) then beliefs are well defined only by (16.1) from Bayes' rule, so that $\mu_U = p$, while from (16.2) beliefs are not well defined by Bayes' rule, so we have the freedom to choose μ_D. Similarly if both $\sigma^H = \sigma^L = 0$ (both types are choosing D) then μ_U is not well defined while $\mu_D = p$.

We are now ready to proceed to find the perfect Bayesian equilibria in the MBA game. Because each player has two information sets with two actions in each of these sets, each player has four pure strategies. Let player 1's strategy be denoted $a_1^H a_1^L$, where $a_1^\theta \in \{U, D\}$ denotes what player 1 does if he is type $\theta \in \{H, L\}$. Similarly let $a_2^U a_2^D$ denote player 2's strategy, where $a_2^k \in \{M, B\}$ denotes what player 2 does if he observes that player 1 chose $k \in \{U, D\}$.

To make our analysis more straightforward, assume that Nature chooses player 1's type according to $p = \frac{1}{4}$, so that we can derive the matrix that is the normal-form representation of the MBA Bayesian game. As we have demonstrated earlier for the entry game, the payoffs in the matrix are calculated by taking each pair of pure strategies, observing which paths are played with the different probabilities that are due to Nature's choice, and then writing down the derived expected payoffs from this pair of strategies. For example, if (UD, MB) are the pair of strategies then with probability $\frac{1}{4}$ Nature chooses type H for player 1 who chooses U, and in response player 2 chooses M, yielding a payoff pair of $(10, 10)$. This follows because player 1 gets a wage of 10 and incurs no cost of obtaining an MBA, while player 2 assigns a high-skill worker to a managerial job, so he obtains a payoff of 10 as well. With probability $\frac{3}{4}$ Nature chooses type L for player 1 who chooses D, and in response player 2 chooses B, yielding a payoff pair of $(1, 3)$. This follows because player 1's net payoff is $6 - 5 = 1$ (wage equal to 6 and a cost of studying equal to 5) and player 2's net payoff is 3 (assigning a low-skill worker to a blue-collar job). The expected pair of payoffs for the players from the strategy (UD, MB) is therefore

$$(v_1, v_2) = \frac{1}{4}(10, 10) + \frac{3}{4}(1, 3) = (3.25, 4.75).$$

Similarly we can calculate the expected payoffs for all the other 15 entries in the Bayesian game matrix. Notice that when player 1 plays the same action for the different types (rows 1 and 4) then part of player 2's strategy is never used, so there are repeat entries which reduce the number of calculations needed. The matrix representation is

		Player 2			
		MM	MB	BM	BB
	UU	<u>10</u>, 2.5	<u>10</u>, 2.5	<u>6</u>, 3.5	<u>6</u>, $\overline{3.5}$
Player 1	UD	6.25, 2.5	3.25, $\overline{4.75}$	5.25, 1.25	2.25, 3.5
	DU	<u>9.5</u>, 2.5	<u>8.5</u>, 1.25	<u>6.5</u>, $\overline{4.75}$	<u>4.5</u>, 3.5
	DD	5.75, 2.5	1.75, $\overline{3.5}$	5.75, 2.5	1.75, $\overline{3.5}$

If we follow the method of underlining player 1's best responses for each column and overlining player 2's best responses for each row, we immediately observe that there are two pure-strategy Bayesian Nash equilibria: (UU, BB) and (DU, BM). To see whether these can be part of a perfect Bayesian equilibrium, we need to find a system of beliefs that support the proposed behavior, and that together with these strategies satisfy requirements 15.1–15.4.

From proposition 15.1 it follows that (DU, BM) can be part of a perfect Bayesian equilibrium because all of the information sets are reached with positive probability. In particular the derived beliefs from (DU, BM) are $\mu_U = 0$ and $\mu_D = 1$.[2] It follows from the Bayesian game matrix that player 2 is playing a best response to these beliefs in each of his information sets, and that player 1 is playing a best response in each of his. So (DU, BM) together with $\mu_U = 0$ and $\mu_D = 1$ constitute a perfect Bayesian equilibrium.

What about the pair of strategies (UU, BB)? From (16.1) and (16.2), unique beliefs are derived only for information set I_U because I_D is reached with zero probability. In particular $\mu_U = \frac{1}{4}$ and μ_D is not well defined. It is easy to check that

---

2. Using equations (16.1) and (16.2) that we derived from Bayes' rule, this is the case where $\sigma^H = 0$ and $\sigma^L = 1$, and the resulting μ_U and μ_D follow.

player 2 choosing B is a best response in information set I_U to the belief $\mu_U = \frac{1}{4}$. Therefore to see whether (UU, BB) can be part of a perfect Bayesian equilibrium we need to see if there are beliefs μ_D that support B as a best response for player 2 in information set I_D.

For B to be a best response in the information set I_D, it must be the case that given the belief μ_D the expected payoff from B is higher than the expected payoff from M. This can be written down as

$$5\mu_D + 3(1 - \mu_D) \geq 10\mu_D + 0(1 - \mu_D),$$

which is true if and only if $\mu_D \leq \frac{3}{8}$. This implies that we *can* support (UU, BB) as part of a perfect Bayesian equilibrium. In particular (UU, BB), together with belief $\mu_U = \frac{1}{4}$ and any belief satisfying $\mu_D \in \left[0, \frac{3}{8}\right]$, constitutes a perfect Bayesian equilibrium.

We conclude that in the first perfect Bayesian equilibrium with strategies (DU, BM), different types of player 1 choose different actions, thus using their actions to *reveal* to player 2 their true types. In other words, this is a separating perfect Bayesian equilibrium. In the second perfect Bayesian equilibrium, with strategies (UU, BB), both types of player 1 do the same thing, and thus player 2 learns nothing from player 1's action; this is a pooling perfect Bayesian equilibrium.

16.2 Limit Pricing and Entry Deterrence

We turn to another application of the analysis of dynamic games of incomplete information, this time to an important antitrust issue. Bain (1949) argued that an incumbent firm can engage in *limit pricing* to deter entry, and thus try to gain control as a long-run monopolist. This practice is defined as one in which a firm prices below marginal costs, which is obviously a losing strategy in the short run. But if by doing this the firm runs competitors out of business, deters other firms from entering the market, and thus becomes a monopolist, then this might be a winning strategy in the long run. Limit pricing is considered anticompetitive behavior and is illegal in the United States.

However, this argument should make one feel a bit uncomfortable: Is it reasonable to punish firms for charging low prices? How can we convince ourselves that Bain's intuition is correct and that a firm may indeed choose to incur short-run losses to generate long-run profits? These questions were answered by Milgrom and Roberts (1982). We present a simplified version of their analysis in the following game.

Consider a market for some good that will last for two periods with demand $P = 5 - q$ in each period. In period $t = 1$ only one firm, the incumbent (player 1), is in the market, and this incumbent will continue to produce in period $t = 2$. A potential entrant (player 2) observes what happened in period $t = 1$ and then makes a choice of whether to enter the market at a fixed cost of $\frac{1}{2}$ and compete against the incumbent in period $t = 2$, or to remain out of the market. If player 2 stays out of the market then player 1 remains a monopolist in period 2, while if player 2 enters the market then they compete in a Cournot duopoly game. Each firm has marginal costs c_i, where it is common knowledge that the entrant's marginal costs are $c_2 = 2$. There is asymmetric information with respect to the costs (the type) of the incumbent. In particular the entrant knows only that $c_1 \in \{1, 2\}$, so that the incumbent can be either more efficient than the entrant (have low [L] costs of $c_1 = 1$) or as efficient (have high [H] costs of $c_1 = 2$). It is common knowledge that $\Pr\{c_1 = 1\} = \frac{1}{2}$, but only the incumbent knows his true costs.

The timing of this game, which captures the story, is described as follows:

1. Nature chooses the type of player 1, $c_1 \in \{1, 2\}$, each type with equal probability.

2. Player 1 observes his costs and decides how much to produce in the first period, q_1^1, and the price in the first period is then $P = 5 - q_1^1$.

3. Player 2 observes q_1^1 and decides whether or not to enter at a fixed cost of $F = \frac{1}{2}$.

4. If player 2 stays out then in period $t = 2$ player 1 chooses q_1^2 and the price is $P = 5 - q_1^2$.

5. If player 2 enters then in period $t = 2$ each player i simultaneously chooses q_i^2 and the price is $P = 5 - q_1^2 - q_2^2$ (Cournot competition).

Given the continuous action spaces of this game (continuous quantity levels) it is not too useful to write down a game tree. As a consequence solving for the perfect Bayesian equilibria of this game is not as simple as for the MBA game. Because the strategy sets are continuous we cannot look for all the Bayesian Nash equilibria in a matrix and then check whether they can be supported as perfect Bayesian equilibria. Thus to solve for the perfect Bayesian equilibria of this game we must go to the heart of what sequential rationality means and solve the game using techniques similar to backward induction. We proceed in two steps: we look first for separating perfect Bayesian equilibria and then for pooling perfect Bayesian equilibria.

16.2.1   Separating Equilibria

We proceed to check for the existence and nature of separating perfect Bayesian equilibria using a sequence of steps:

Step 1:   What will q_1^2 be if firm 2 stays out?

The answer is very simple. If firm 2 stays out then firm 1 is a monopoly and maximizes profits,

$$\max_{q_1^2}(5 - q_1^2 - c_1)q_1^2.$$

This yields the solution using the first-order condition $5 - 2q_1^2 - c_1 = 0$ (which in this case is necessary and sufficient, as you can check), and this in turn gives us firm 2's quantity:

$$q_1^2 = \frac{5 - c_1}{2} = \begin{cases} 2 & \text{if } c_1 = 1 \\ 1.5 & \text{if } c_1 = 2. \end{cases}$$

Step 2:   What will q_1^2 and q_2^2 be if firm 2 enters and firm 1's costs are known to all?

In this case both firms play a Cournot game. Firm 1 maximizes its profits, given by

$$\max_{q_1^2} v_1^2 = (5 - q_1^2 - q_2^2 - c_1)q_1^2,$$

which yields the best-response function using the first-order condition $5 - 2q_1^2 - q_2^2 - c_1 = 0$ (which is again necessary and sufficient),

$$BR_1(q_2^2) = \frac{5 - q_2^2 - c_1}{2}. \tag{BR1}$$

Firm 2 solves a problem that is similar except that it also includes the fixed cost of entry in the objective function,

$$\max_{q_2^2} v_2^2 = (5 - q_1^2 - q_2^2 - c_2)q_2^2 - \frac{1}{2},$$

which yields the best-response function for firm 2 using the first-order condition $5 - 2q_2^2 - q_1^2 - c_2 = 0$ (which is again necessary and sufficient),

$$BR_2(q_1^2) = \frac{5 - q_1^2 - c_2}{2} = \frac{3 - q_1^2}{2}. \tag{BR2}$$

Because firm 1 can be one of two types, there are two cases to be considered if c_1 is known to firm 2 as well:

1. If firm 1 is a high-cost type then in this case $c_1 = 2$, and (BR1) and (BR2) imply that $q_1^2 = q_2^2 = 1$, $P = 3$, and profits for the two firms in the second period are $v_1^2 = (3 - 2)1 = 1$, while $v_2^2 = (3 - 2)1 - \frac{1}{2} = \frac{1}{2}$.
2. If firm 1 is a low-cost type then in this case $c_1 = 1$, and (BR1) and (BR2) imply that $q_1^2 = \frac{5}{3}$, $q_2^2 = \frac{2}{3}$, $P = \frac{8}{3}$, and profits for the two firms in the second period are $v_1^2 = \left(\frac{8}{3} - 1\right)\frac{5}{3} = \frac{25}{9}$, while $v_2^2 = \left(\frac{8}{3} - 2\right)\frac{2}{3} - \frac{1}{2} = -\frac{1}{18}$.

Step 3: What quantities for each type will support firm 1's separating (signaling) behavior?

This is where things become more complex. To answer this question we consider the logic of separating versus pooling equilibria. In *any* separating equilibrium different types of firm 1 choose different actions, implying that in any such equilibrium firm 2 perfectly learns the type of firm 1. To continue our analysis consider a separating perfect Bayesian equilibrium with the following notation: $q_1^{t\theta}$ denotes the choice of firm 1 in period $t \in \{1, 2\}$ when he is of type $\theta \in \{H, L\}$. That is, $q_1^{t\theta}$ is the strategy of type θ in period t. Let $q_2^2(q_1^1)$ denote the strategy of firm 2 in period $t = 2$, which is a function of q_1^1 being played by firm 1 in period 1. Finally let $\mu(q_1^1) = \Pr\{c_1 = 2|q_1^1\}$ be firm 2's posterior belief in period 2 after observing that q_1^1 was played by firm 1 in period 1. Using this notation we can write a perfect Bayesian equilibrium as[3]

$$\left\{ \underbrace{q_1^{1L}, q_1^{1H}}_{\text{Firm 1, } t=1}, \quad \underbrace{q_1^{2L}, q_1^{2H}}_{\text{Firm 1, } t=2}, \quad \underbrace{q_2^2(q_1^1)}_{\text{Firm 2, } t=2}, \quad \underbrace{\mu(q_1^1)}_{\text{Firm 2's beliefs}} \right\}.$$

---

3. To be precise, $q_1^{2\theta}$ must be a function of whether or not firm 2 decides to enter, and firm 2's strategy should include two components: first, whether or not to enter depending on q_1^1, and then upon entry what the choice $q_2^2(q_1^1)$ is. We are simplifying by collapsing both of these choices of firm 2 into the quantity choice, and we will let $q_2^2(q_1^1) = 0$ be the choice of no entry.

We can now establish a series of claims:

Claim 16.1 *In any separating perfect Bayesian equilibrium, when firm 2 believes that* $c_1 = 1$ *then it will stay out, and firm 1 will choose* $q_1^{2L} = 2$, *while if firm 2 believes that* $c_1 = 2$ *it will enter, and both firms produce* $q_1^{2H} = q_2^2 = 1$.

The proof follows from the foregoing analysis of the monopoly problem (step 1) and the Cournot problem (step 2), and from the fact that we require firm 2's behavior to be sequentially rational. This is one of the nice features of a separating perfect Bayesian equilibrium: even though firm 2 *does not know* the type of firm 1, in a separating equilibrium types are revealed, and firm 2 must act *as if he knew* the type of firm 1.

Claim 16.2 *In any separating perfect Bayesian equilibrium, in the first period a high-cost incumbent must produce* $q_1^{1H} = 1.5$.

Proof From claim 16.1, in any separating perfect Bayesian equilibrium it must be that following q_1^{1H} firm 2 will enter, and following q_1^{1L} firm 2 will stay out and let firm 1 be a monopolist (otherwise it would not be a separating perfect Bayesian equilibrium). To see that in a separating perfect Bayesian equilibrium it must be that $q_1^{1H} = 1.5$, assume in negation that this is not the case. Then in period 1 firm 1 is making less than monopoly profits when its marginal costs are $c_1 = 2$, and in period 2 firm 1 is making Cournot profits. Now consider a deviation of firm 1, when $c_1 = 2$, to the monopoly quantity of $q_1^{1H} = 1.5$. In period 1 profits will be higher, which means that for this deviation not to be profitable it must be that firm 1 gets *less than Cournot profits* in the second period. But this cannot happen because either: (1) firm 2's beliefs after the deviation remain $\Pr\{c_1 = 2\} = 1$, in which case they will play the same Cournot game, or (2) firm 2 changes its beliefs to $\Pr\{c_1 = 2\} < 1$, in which case firm 1 will make higher-than-Cournot profits (either firm 2 will stay out or it will play Cournot against an unknown rival and produce less than 1, depending on its beliefs). Thus we conclude that if $q_1^{1H} \neq 1.5$ then firm 1 has a profitable deviation to $q_1^{1H} = 1.5$. ∎

We have therefore established from our analysis that if a separating perfect Bayesian equilibrium exists then it must satisfy $q_1^{2L} = 2$, $q_1^{2H} = 1$, $q_2^2(q_1^{1L}) = 0$, $q_2^2(q_1^{1H}) = 1$, $\mu(q_1^{1L}) = 0$, and $\mu(q_1^{1H}) = 1$. From claim 16.2 we established that it must satisfy $q_1^{1H} = 1.5$. We are left to find two more elements: we must define beliefs for all other quantities $q_1^1 \notin \{q_1^{1H}, q_1^{1L}\}$, and we have to find q_1^{1L}. If we find these values in a way for which strategies and beliefs satisfy requirements 15.1–15.4 from Section 15.2 then we have found a separating perfect Bayesian equilibrium.

Step 4: Setting off-the-equilibrium-path beliefs.

To set off-the-equilibrium-path beliefs that will support behavior on the equilibrium path we will use a "trick" that is common for games with continuous strategy sets and is similar to what we did for the MBA game. Recall that we want the separating perfect Bayesian equilibrium to work in such a way that each type of firm 1 $\theta \in \{L, H\}$ will stick to his strategy $q_1^{1\theta}$, rather than deviating to some other quantity q_1^1. To do this, we can make the continuation game following any deviation from either q_1^{1L} or q_1^{1H} to be *as undesirable as possible for firm 1*. How can we achieve this? Precisely by causing firm 2 to enter following any such deviation, which is guaranteed to happen when firm 2 believes that firm 1 has high costs. Indeed when firm 2 acts in this way then firm 1 faces the most severe second-period competition, and this is the most undesirable outcome for firm 1. Hence the easiest way to prevent deviations and keep firm 1 *on the equilibrium path* is by setting beliefs that make *off-the-equilibrium-path* behavior very unattractive to firm 1. This will be satisfied if

$$\mu(q_1^1) = \begin{cases} 1 & \text{if } q_1^1 \neq q_1^{1L} \\ 0 & \text{if } q_1^1 = q_1^{1L}. \end{cases}$$

These beliefs cause a unique best response for firm 2, which is not to enter if $q_1^1 = q_1^{1L}$ and to enter and produce $q_2^2 = 1$ if $q_1^1 \neq q_1^{1L}$.

Step 5: What should q_1^{1L} be?

Once we have calculated all of the equilibrium components, for q_1^{1L} to satisfy the missing piece of the puzzle it must satisfy the following two important conditions:

1. When firm 1 is an L type, it prefers to choose q_1^{1L} over any other quantity, in particular over q_1^{1H}.

2. When firm 1 is an H type, it prefers to choose q_1^{1H} over q_1^{1L}.

The first condition is that type L is playing a best response. The second condition just says that an H type does not want to imitate an L type. What about other quantities? Using the belief system we defined earlier, claim 16.2 already implies that an H type prefers $q_1^{1H} = 1.5$ to *any other nonmonopoly profit* that induces entry. The reason we have to care about imitating an L type is because q_1^{1L} *prevents entry*, allowing firm 1 to remain a monopolist in the second period.

We call these two conditions—that each type prefers choosing his designated action rather than imitating some other type—**incentive compatibility constraints.** The meaning is precisely that in equilibrium each type has an incentive to choose his prescribed strategy and not choose the prescribed strategies for the other types. Similar incentive compatibility constraints held for the separating perfect Bayesian equilibrium that we found in the MBA game, and that is a general set of constraints for signaling games.

To solve for q_1^{1L} we can start with a natural candidate: the *monopoly quantity* for the L type. From the analysis of step 1, this is equal to $q_1^{1L} = 2$, and the first-period profits would be 4. If this choice is part of a separating perfect Bayesian equilibrium it will induce firm 2 to stay out, and thus firm 1 again gets a profit of 4 in period $t = 2$. We must check now for incentive compatibility of the two types of firm 1. Starting with the L type, would the L type of firm 1 want to deviate and choose q_1^{1H} or any other quantity? The answer is clearly no, because if it chooses q_1^{1H} or any quantity different than q_1^{1L} then not only will its first period profits be lower than monopoly profits, but this will induce entry and lower firm 1's profits in the second period. Hence the L type's incentive compatibility constraint is satisfied.

Turning to incentive compatibility of the H type, if that type follows its strategy then it produces 1.5 in the first period and gets monopoly profits of 2.25, which induces entry that results in Cournot profits of 1 in the second period. Total profits would therefore be 3.25. From the foregoing analysis, the H type would not deviate to any other entry-inducing quantity because that would lower its first-period profits without deterring entry. We are left to check whether the H type would want to deviate and choose $q_1^{1L} = 2$ instead of $q_1^{1H} = 1.5$.

If the H type indeed deviates and chooses q_1^{1L} then this will deter entry and it can obtain monopoly profits in the second period by producing at a level of q_1^{1H}. As a consequence total profits for the H type will be:

$$\pi = \overbrace{(5 - q_1^{1L} - c_H)q_1^{1L}}^{\text{First period}} + \overbrace{(5 - q_1^{1H} - c_H)q_1^{1H}}^{\text{Second period}}$$
$$= (5 - 2 - 2)2 + (5 - 1.5 - 2)1.5$$
$$= 4.25 > 3.25!$$

This shows that the incentive compatibility constraint of the H type is violated. This happens because by imitating an L type, the H type is sacrificing some of its monopoly profits in the first period to obtain full monopoly profits in the second period rather than the significantly lower Cournot equilibrium profits. It turns out that the sacrifice is well worth the extra gains, implying that $q_1^{1L} = 2$ causes a violation of the H type's incentive compatibility constraint and thus cannot be part of a separating perfect Bayesian equilibrium.[4]

The intuition from the previous unsuccessful attempt at finding q_1^{1L} is that for incentive compatibility to hold it must be the case that the H type will *suffer enough* from deviating. This can be done by making the deviation less attractive, which means we need to make q_1^{1L} larger (because the monopoly quantity for the L type is 1.5, the larger is q_1^{1L}, the lower will be the profits from this choice).

It turns out that we do not need to make any guesses about q_1^{1L} because our analysis has already pinned down *all* the other components of *any* separating perfect Bayesian equilibrium. Hence to find a perfect Bayesian equilibrium we need only to find a level of q_1^{1L} that satisfies the two incentive compatibility constraints. First, we need to make sure that an H type does not want to deviate to q_1^{1L}. This condition will be satisfied if the first-period profits from deviating, plus the second-period monopoly profits the firm gets (which equal 2.25 because firm 2 stays out after q_1^{1L}), are no more than what the firm gets from producing its monopoly quantity in the first period followed by Cournot competition in the second (which adds up to 3.25). This condition can be written as

$$(5 - q_1^{1L} - 2)q_1^{1L} + 2.25 \leq 3.25,$$

which is reduced to

$$q_1^{1L} \geq \frac{3}{2} + \sqrt{\frac{5}{4}} \simeq 2.62.$$

This is therefore the *lower bound* on q_1^{1L} that is needed to deter the H type from deviating from q_1^{1H} to q_1^{1L}. Similarly we have to make sure that the L type will not want to deviate from q_1^{1L} to some other quantity. This would put an upper bound on q_1^{1L}. To find this upper bound we need to figure out what would be the *best deviation* for an L type of firm 1 in the first period. Given the belief system we are imposing, any deviation from q_1^{1L} will induce entry, and firm 2 will produce $q_2^2 = 1$ from the previous analysis. So a type L that deviates should play a best response to $q_2^2 = 1$ in

---

4. Note that if we introduce discounting into this model this conclusion might change. You are asked to consider this in exercise 16.2.

the second period. This is calculated by

$$\max_{q_1^2}(5 - q_1^2 - 1 - 1)q_1^2,$$

which is maximized at $q_1^2 = 1.5$, yielding a second-period profit of 2.25. Because *any deviation* of an L type will be followed by this best response, an L type gains the most from deviating to his monopoly quantity of $q_1^1 = 2$ with a first-period profit equal to 4. This implies that the best deviation for firm 1 when it is an L type yields a profit of 6.25, and as a consequence incentive compatibility of the L type is guaranteed if

$$(5 - q_1^{1L} - 1)q_1^{1L} + 4 \geq 6.25,$$

which reduces to

$$q_1^L \lesssim 3.32.$$

Summing up, this analysis allows us to state the following conclusion on separating perfect Bayesian equilibria of this game:

Conclusion: In the limit pricing–entry deterrence game there is a continuum of separating perfect Bayesian equilibria. Any profile of strategies

$$\left\{ q_1^{1L}, q_1^{1H}, q_1^{2L}, q_1^{2H}, q_2^2(q_1^1) \right\}$$

together with beliefs $\mu(q_1^1)$ that satisfies

$$2.62 \leq q_1^L \leq 3.32$$

$$q_1^{1H} = 1.5$$

$$q_1^{2L} = 2$$

$$q_1^{2H} = 1$$

$$\mu(q_1^1) = \begin{cases} 1 & \text{if } q_1^1 \neq q_1^{1L} \\ 0 & \text{if } q_1^1 = q_1^{1L} \end{cases}$$

$$q_2^2(q_1^1) = \begin{cases} 1 & \text{if } q_1^1 \neq q_1^{1L} \\ 0 & \text{if } q_1^1 = q_1^{1L} \end{cases}$$

is a separating perfect Bayesian equilibrium.

It is worth noting something special about all these perfect Bayesian equilibria: for a low-cost firm to deter entry, it must choose a quantity that is higher than its short-run profit-maximizing (monopoly) profits. Why? Because to scare off opponents *credibly*, it must choose something that a high-cost firm would not find profitable to do. In other words, for a signal to be credible, it must be the case that no other type would want to use the signal.

Note that in the example the low-cost (L-type) firm is charging a price that is above marginal costs in the separating perfect Bayesian equilibrium. A modified example, however, would admit an equilibrium in which the first-period price is below marginal costs. For example, if the second-period market demand were much higher than the first-period demand (i.e., a growing market) then it may be the case that the first-period separating quantity for the low type needs to be so high that the price would be below marginal cost. Another alternative would be to extend the game to many more periods. If by deterring entry in the first period the firm would be a monopoly for several periods thereafter, then the lower bound we derived on q_1^{1L} would have to be higher, as would the upper bound. This would eventually drive the first-period price to be below marginal cost. This would be the extent of the sacrifice needed to make sure that the signal is credible. Thus Bain's intuition survives the test of a formal model that is built to capture the essence of anticompetitive limit pricing.

16.2.2 Pooling Equilibria

Pooling equilibria are easier to find in this game. A pooling equilibrium will be defined by $\{q_1^{1*}, q_1^{2L}, q_1^{2H}, q_2^2(q_1^1), \mu(q_1^1)\}$ because both types of firm 1 choose the same quantity q_1^{1*} in the first period. This immediately implies that beliefs must satisfy $\mu(q_1^{1*}) = \frac{1}{2}$ because it is common knowledge that Nature chooses $\Pr\{c_1 = 1\} = \frac{1}{2}$, and pooling behavior implies that firm 2 does not update its beliefs. This implies that firm 2, if it would enter following q_1^{1*}, will play a static game of incomplete information. We therefore need to find the Bayesian Nash equilibrium for this second-period game, which is an incomplete-information version of the Cournot model. Thus we find q_1^{2L}, q_1^{2H}, and $q_2^2(q_1^{1*})$ by simultaneously solving three maximization problems: one for the entrant and two for the incumbent's two types. The L-type incumbent solves

$$\max_{q_1^{2L}} (5 - q_1^{2L} - q_2^2(q_1^{1*}) - 1)q_1^{2L},$$

which yields the best-response function from the first-order condition

$$q_1^{2L} = \frac{4 - q_2^2(q_1^{1*})}{2}.$$

Similarly an H-type incumbent maximizes

$$\max_{q_1^{2H}} (5 - q_1^{2H} - q_2^2(q_1^{1*}) - 2)q_1^{2H},$$

for which the best response is

$$q_1^{2H} = \frac{3 - q_2^2(q_1^{1*})}{2}.$$

And finally the entrant, firm 2, maximizes

$$\max_{q_2^2(q_1^{1*})} \frac{1}{2}\left[(5 - q_1^{2L} - q_2^2(q_1^{1*}) - 2)q_2^2(q_1^{1*})\right]$$

$$+ \frac{1}{2}\left[(5 - q_1^{2H} - q_2^2(q_1^{1*}) - 2)q_2^2(q_1^{1*})\right] - \frac{1}{2},$$

which results in the best-response function

$$q_2^2(q_1^{1*}) = \frac{3 - \frac{1}{2}(q_1^{2L} + q_1^{2H})}{2}.$$

The Bayesian Nash equilibrium of this incomplete-information Cournot game can be found by solving the three best-response functions (which are just three first-order conditions) simultaneously, which yields

$$q_2^2(q_1^{1*}) = \tfrac{5}{6}, q_1^{2L} = \tfrac{19}{12}, q_1^{2H} = \tfrac{13}{12}.$$

This in turn results in second-period expected profits for firm 2 equal to $v_2^2 = \tfrac{7}{18} > 0$, for the H type of firm 1 equal to $v_1^{2H} = \left(\tfrac{13}{12}\right)^2$, and for the L type of firm 1 equal to $v_1^{2L} = \left(\tfrac{19}{12}\right)^2$. This implies that firm 2 will choose to enter in a pooling equilibrium in which it does not learn the type of firm 1.

To proceed we have to set beliefs for off-the-equilibrium-path choices of firm 1 $(q_1^1 \neq q_1^{1*})$, which will determine the best-response function of firm 2 $q_2^2(q_1^1)$ in information sets that are off the equilibrium path. We will then need to find the value of q_1^{1*} to complete the equilibrium. As we have seen in our analysis of the separating perfect Bayesian equilibrium, the best way to keep the different types of firm 1 on the equilibrium path is by making deviation off the equilibrium path very undesirable. This is done by setting the "worse" beliefs that firm 2 could have about firm 1, which are $\mu(q_1^1) = 1$ for $q_1^1 \neq q_1^{1*}$. Using these off-the-equilibrium-path beliefs together with on-the-equilibrium-path beliefs $\mu(q_1^{1*}) = \tfrac{1}{2}$ immediately implies the sequentially rational best-response function of firm 2,

$$q_2^2(q_1^1) = \begin{cases} 1 & \text{if } q_1^1 \neq q_1^{1*} \\ \tfrac{5}{6} & \text{if } q_1^1 = q_1^{1*}. \end{cases}$$

We are left to find a level of q_1^{1*} from which neither type of firm 1 would want to deviate. In the analysis of the separating equilibrium we calculated the "best" first-period deviation of each type, which is to choose its monopoly quantities followed by the best response to $q_2^2 = 1$ in the second period. These yielded a profit of 3.25 to the H type and 6.25 to the L type, as we calculated earlier. We therefore need to satisfy two inequalities for q_1^{1*} to be a best response for both types. For the H type,

$$(5 - q_1^{1*} - 2)q_1^{1*} + \left(\tfrac{13}{12}\right)^2 \geq 3.25,$$

which reduces to

$$1.083 \leq q_1^{1*} \leq 1.917. \tag{ICH}$$

Similarly for the L type,

$$(5 - q_1^{1*} - 1)q_1^{1*} + \left(\tfrac{19}{12}\right)^2 \geq 6.25,$$

which reduces to

$$1.493 \leq q_1^{1*} \leq 2.507. \tag{ICL}$$

Because q_1^{1*} must satisfy both incentive constraints (ICH) and (ICL), we can summarize our analysis as follows:[5]

Conclusion: In the limit pricing–entry deterrence game there is a continuum of pooling perfect Bayesian equilibria. Any profile of strategies $\{q_1^{1*}, q_1^{2L}, q_1^{2H}, q_2^2(q_1^1)\}$ together with beliefs $\mu(q_1^1)$ that satisfies

$$1.493 \leq q_1^{1*} \leq 1.917$$

$$q_1^{2L} = \tfrac{19}{12}$$

$$q_1^{2H} = \tfrac{13}{12}$$

$$q_2^2(q_1^1) = \begin{cases} 1 & \text{if } q_1^1 \neq q_1^{1*} \\ \tfrac{5}{6} & \text{if } q_1^1 = q_1^{1*} \end{cases}$$

$$\mu(q_1^1) = \begin{cases} 1 & \text{if } q_1^1 \neq q_1^{1*} \\ \tfrac{1}{2} & \text{if } q_1^1 = q_1^{1*} \end{cases}$$

is a pooling perfect Bayesian equilibrium.

Unlike the separating equilibrium in which the incumbent's action reveals its type to the entrant, this is not the case here. In fact it is not that convincing to state that different types of the incumbent will choose the same actions, which makes the pooling equilibrium look somewhat artificial and less appealing. As we will see in the next section, there are some convincing arguments that not only favor the separating equilibria as more reasonable but also often select one of the separating equilibria as the most reasonable.

Signaling games that result in some form of entry deterrence have also been applied to explain a phenomenon studied in political science: the huge efforts that strong political incumbents exert to raise campaign contributions. Epstein and Zemsky (1995) offer a formal game theoretic explanation that incumbents raise these huge sums of money as a signal of strength to deter potential challengers.

16.3 Refinements of Perfect Bayesian Equilibrium in Signaling Games

In both the MBA game and the limit pricing–entry deterrence game we had a plethora of perfect Bayesian equilibria. This outcome suggests that when we applied sequential rationality to the Bayesian Nash equilibrium concept we still did not manage to get rid of many equilibria. It implies that for the games we analyzed the predictive power of the perfect Bayesian equilibrium solution concept is not as sharp as we would hope.

Let's consider the MBA game first and focus our attention on the pooling equilibrium in which both types of worker choose U, and then regardless of the education

---

5. Note that the inequalities are intuitive. For each type the best deviation is to its monopoly quantity, which is 1.5 for the H type and 2 for the L type. Thus for each type the range is an interval that is symmetric around its monopoly quantity.

choice the employer assigns the worker to B. Now consider the following deviation and the speech that an H type could deliver:

> I am an H type. To convince you I am going to deviate and choose D. If you believe me, and put me in the M job instead of a B job, I will get 8 instead of 6. The reason you should believe me is that if I were an L type who chose D and you were to promote me then I would get 5 instead of 6. Therefore you should believe me when I tell you that I am an H type because no L type in his right mind would do what I am about to do.

What should the employer think? The argument makes sense because if the candidate were an L type then there is no way in which he could gain from this move, and in fact he will lose. In contrast an H type can gain if he is believed by the employer. This logic suggests that the employer should be convinced by this deviation *combined with the speech*. Now if we take this a step further, the employer can make these kinds of logical deductions himself:

> Let me see which type can gain from this deviation. If neither can or if both can, I will keep my off-the-equilibrium-path beliefs as before. But if only one type of worker can benefit and other types can only lose, then I should update my beliefs accordingly and act upon these new, more "sophisticated" beliefs.

This logical process leads to the **intuitive criterion** that was developed by Cho and Kreps (1987). For any given set of beliefs of player 2, player 1 (who has private information) can use his action to send a message to player 2 in the spirit of "only an x type would benefit from this move; therefore I am an x type." The intuitive criterion is a way of ruling out less reasonable equilibria, those for which such messages can be profitably sent, and hence it acts to *refine* the equilibrium predictions. That is, take a perfect Bayesian equilibrium and see if it survives the intuitive criterion. If it does not—that is, a player can make a deviation with such a convincing message—then it is ruled out by the intuitive criterion.

We now define the intuitive criterion more formally. Consider a signaling game in which player 1 has private information $\theta \in \Theta$ and chooses actions $a_1 \in A_1$ in the first period, after which player 2 observes his action, forms a posterior belief over player 1's type, and then chooses action $a_2 \in A_2$. Imagine that the set of player 1's types Θ is finite, and let $\widehat{\Theta} \subset \Theta$ be a subset of player 1's types. Let $BR_2(\widehat{\Theta}, a_1)$ be the set of best-response actions of player 2 if player 1 has chosen action $a_1 \in A_1$ and the belief μ of player 2 puts positive probability only on types in the set $\widehat{\Theta}$. Formally we have

$$BR_2(\widehat{\Theta}, a_1) = \cup_{\mu \in \Delta(\widehat{\Theta})} \arg \max_{a_2 \in A_2} \sum_{\theta \in \widehat{\Theta}} v_2(a_1, a_2; \theta) \mu(\theta).$$

Definition 16.1 A perfect Bayesian equilibrium σ^* fails the intuitive criterion if there exist $a_1 \in A_1$, $\theta \in \Theta$, and $\widehat{\Theta} \subset \Theta$ such that

1. $v_1(\sigma^*; \theta) > \max_{a_2 \in BR_2(\Theta, a_1)} v_1(a_1, a_2; \theta)$ for all $\theta \in \widehat{\Theta}$
2. $v_1(\sigma^*; \theta) < \min_{a_2 \in BR_2(\Theta/\widehat{\Theta}, a_1)} v_2(a_1, a_2; \theta)$

From the definition we see that a perfect Bayesian equilibrium *fails* the intuitive criterion if two conditions hold. Condition (1) states that any type in the subset of types $\widehat{\Theta}$ would *never* choose to play a_1 because regardless of which type player 2 believes him to be, he would do strictly worse than if he stuck to the equilibrium. Condition (2) states that type θ will do strictly better than the equilibrium by playing a_1 if he can convince player 2 that his type is *not* in $\widehat{\Theta}$.

We can now go back to the MBA game and see why the pooling equilibrium fails the intuitive criterion. In this equilibrium both types expect a payoff of 6. Using the definition for the pooling equilibrium, let $a_1 = D$, $\widehat{\Theta} = \{L\}$, and $\theta = H$. Condition (1) is satisfied because by choosing D the L type will be worse off: the most he can get is 5 (10 from a manager's wage less 5 for the cost of education). Condition (2) is satisfied because if player 2 is convinced that player 1 is indeed an H type then the H type will receive 8 (10 from a manager's wage less 2 for the cost of education) instead of 6. The separating equilibrium, however, does not fail the intuitive criterion because no type can be made better off by convincing player 2 that he is truly that type.

For the limit pricing–entry deterrence game the analysis is more complex. To see why the pooling equilibrium fails, we first need to define $BR_2(\Theta, a_1)$ for any a_1 (which for the limit pricing game was q_1^1). We saw from our analysis of the separating equilibrium that at one extreme belief, when the entrant believes that he is facing the high-cost incumbent, $\mu(q_1^1) = 1$, he will enter and in equilibrium produce $q_2^2 = 1$. At the other extreme, when he believes that the incumbent has low costs, $\mu(q_1^1) = 0$, then he will stay out (effectively producing $q_2^2 = 0$). We also saw that in the intermediate case of a pooling equilibrium, the entrant's belief is $\mu(q_1^1) = \frac{1}{2}$, and in equilibrium he chooses to produce $q_2^2 = \frac{5}{6}$. It should not be hard to realize that the lower the belief $\mu(q_1^1)$, the lower will be the equilibrium quantity chosen by player 2, until some lower bound on beliefs is reached for which the incomplete-information Cournot Bayesian Nash best response will be some level $q_2^2 = \underline{q} < \frac{5}{6}$ that yields player 2 an expected payoff of zero. Any lower quantity cannot be part of an equilibrium in which player 2 chooses to enter. Hence for the limit pricing–entry deterrence game we will have $BR_2(\Theta, a_1) = \{0\} \cup [\underline{q}, 1]$ regardless of a_1.

Now to see that pooling fails the intuitive criterion, let $a_1 = 3$, $\widehat{\Theta} = \{H\}$, and $\theta = L$. Condition (1) is satisfied because the pooling equilibrium yields the H type a payoff of at least 3.25, while by deviating to $q_1^1 = 3$ the H type will at most get monopoly profits in period 2 equal to 2.25. So this deviation is always worse for him because

$$(5 - 3 - 2)3 + 2.25 = 2.25 < 3.25.$$

Condition (2) is satisfied because if player 2 is convinced that player 1 is indeed an L type then player 2 will stay out and the L type will receive

$$(5 - 3 - 1)3 + 4 = 7 > 4 + \left(\tfrac{19}{12}\right)^2. \tag{16.3}$$

Notice that the right side of (16.3) is the highest pooling equilibrium payoff that the type can receive, because it includes his monopoly profits in the first period followed by the pooling equilibrium profits in the second. We conclude that the pooling equilibrium fails the intuitive criterion.

In the limit pricing–entry deterrence game we are still left with a continuum of separating equilibria, in which $2.62 \leq q_1^L \leq 3.32$ was undetermined. It turns out that of all these equilibria, the only one that does not fail the intuitive criterion is the **best separating equilibrium** for player 1, which is $q_1^{1L} = 2.62$. The intuition is that an L type would never benefit from choosing $q_1^1 = 2.62$, so the H type can use this deviation from any other separating equilibrium quantity. You are asked to verify this in exercise 16.7. For signaling games of the type we have seen, the intuitive criterion will always select the best separating equilibrium as the unique prediction.

16.4 Summary

- In games of incomplete information some types of players would benefit from conveying their private information to the other players.

- Announcements or cheap talk alone cannot support this in equilibrium, because then disadvantaged types would pretend to be advantaged and try to announce "I am this type" to gain the anticipated benefits. This strategy cannot be part of an equilibrium because by definition players cannot be fooled in equilibrium.

- For advantaged types to be able to separate themselves credibly from disadvantaged types there must be some signaling action that costs less for the advantaged types than it does for the disadvantaged types.

- Signaling games will often have many perfect Bayesian and sequential equilibria because of the flexibility of off-the-equilibrium-path beliefs. Refinements such as the intuitive criterion help pin down equilibria, often resulting in the least-cost separating equilibrium.

16.5 Exercises

16.1 **Separating Equilibrium:** Consider the entry game described in exercise 15.5. Is it true that in this game any separating perfect Bayesian equilibrium imposes unique beliefs for the incumbent in all information sets?

16.2 **Limit Pricing Revisited:** Consider the limit pricing–entry deterrence game described in Section 16.2 and imagine that second-period profits are discounted using a discount factor $\delta \leq 1$. Furthermore consider the strategies in which each type of player 1 chooses its monopoly quantity in the first period so that $q_1^{1L} = 2$ and $q_1^{1L} = 1.5$. For which values of δ will this be part of a separating perfect Bayesian equilibrium?

16.3 **More Limit Pricing:** An incumbent firm (player 1) is either a low-cost type $\theta_1 = \theta_L$ or a high-cost type $\theta_1 = \theta_H$, each with equal probability. In period $t = 1$ the incumbent is a monopolist and sets one of two prices p_L or p_H, and its profits in this period depend on its type and the price it chooses, given by the following table:

Type	Profit from p_L	Profit from p_H
θ_L	6	8
θ_H	1	5

After observing the period $t = 1$ price, a potential entrant (player 2), which does not know the incumbent's type but knows the distribution of types, can choose to enter the market (E) or stay out (O) in period $t = 2$. The payoffs of both players in period 2 depend on the entrant's choice and on the incumbent's type and are given by the following table:

Incumbent's type	Entrant's choice	Incumbent's payoff	Entrant's payoff
θ_L	E	0	-2
θ_L	O	8	0
θ_H	E	0	1
θ_H	O	5	0

At the beginning of the game the incumbent discounts profits for period $t = 2$ using a discount factor $\delta \leq 1$.

    a. Draw the extensive-form game tree of this game and write down the corresponding matrix.

    b. For $\delta = 1$ find a pooling perfect Bayesian equilibrium of the game in which both types of player 1 choose p_L in period $t = 1$.

    c. Find the range of discount factors for which a separating perfect Bayesian equilibrium of the game exists in which type θ_L chooses p_L and type θ_H chooses p_H in period $t = 1$.

16.4   **School Choice:** A prospective student, player 1, is deciding whether to go to Brown Bear University (B) or Silly Tree University (S). Both are top-rated universities that accept only the very best students. The difference is that Silly Tree offers a more holistic and comfortable lifestyle, while Brown Bear requires harder work and imposes more "studying" costs on its students. For this reason a player who goes to Brown Bear learns to become more self-sufficient and becomes more productive in the workforce. The cost of learning and level of final productivity depend on the type of player 1, who can be either excellent (E) or just really good (G) (because no other type would be accepted into these universities). Player 1 knows his type, but all others in society know only that a proportion p of young adults are type E. The cost of learning and the level of productivity from each choice are given as follows:

Type	University choice	Learning cost	Productivity
E	B	2	12
E	S	0	4
G	B	8	10
G	S	2	2

Once player 1 finishes school, he is hired by a firm (player 2), which can place him in one of two jobs: low-tech (L) or high-tech (H). The wage for the L job is $w_L = 2$ and that for the H job is $w_H = 6$. The payoffs to player 1 are the wages less the cost of education. The firm's profits depend on both the job assignment and the type of employee. If the employee is assigned to an H job, the net profits to the firm are equal to the productivity of the employed player 1 less the wage he is paid. If the assignment is to an L job, the net profits are half the productivity of the employed player 1 less the wage he is paid.

    a. Draw this game in extensive form.

    b. Assume that $p = \frac{1}{2}$. Represent the matrix form of the Bayesian game.

    c. Find all the pure-strategy Bayesian Nash equilibria.

    d. Find all the pure-strategy perfect Bayesian equilibria.

    e. In brief terms, what is the intuition that explains the comparison between your results in (d) and (c)?

16.5 **Drug Ads:** A pharmaceutical company (player 1) introduces a new cold medicine. The medicine may either be highly effective (H) or have little effect (L). The company knows the effectiveness of the drug, but a representative consumer (player 2) knows only that the prior probability that it is highly effective is $\frac{1}{2}$. The company can choose either to advertise the drug excessively (A), at a cost $c > 0$, or not to advertise (N), which costs nothing. The representative consumer decides whether or not to buy the product after observing whether the company advertised the drug. The net payoff to the representative consumer from buying the drug is 1 if it is highly effective and -1 if it has little effect, and his payoff from not buying the drug is 0. If the drug is highly effective then if consumers buy the drug once they will learn of its efficacy and buy it many more times, in which case the company earns a high payoff equal to $R > c$. If instead consumers learn that the drug has little effect then the company will sell the drug to them only once, and the company's returns are equal to $r > 0$. If the representative consumer does not buy the drug then the revenue of the company is 0. Assume that $R > c > r > 0$.

    a. Write down the extensive-form game tree.

    b. Find a separating perfect Bayesian Nash equilibrium in which the company chooses a different action depending on the drug's efficacy.

    c. Find a pooling perfect Bayesian Nash equilibrium in which the company chooses the same action regardless of the drug's efficacy.

    d. What changes when $R > r > c > 0$?

16.6 **Publish or Perish:**[6] Imagine that any newly minted Ph.D. who starts a tenure-track assistant professor job (player 1) is one of two types: high-ability (θ_H) or low-ability (θ_L), where $\theta_H > \theta_L > 0$. The assistant professor knows his type, but the department that hires him (player 2) knows only that he has high ability with probability $p < \frac{1}{2}$. The assistant professor first chooses how hard to work, which is effectively how many papers to publish in period 1 (the pre-tenure period). After observing how many papers the assistant professor published, the department decides whether to grant him tenure (T) or not to do so (N). If the department chooses to grant tenure then the assistant professor's payoff is $v_1(q, T \mid \theta) = V - \frac{q}{\theta}$, where V is the value of being tenured. The department's payoff is 1 if it tenures a high-ability type and -1 if it tenures a low-ability type. If the department denies tenure, it gets a payoff of 0 and the assistant professor's payoff is $v_1(q, N \mid \theta) = -\frac{q}{\theta}$. Denote by $\mu(q)$ the department's belief that the professor is a high-ability type given that he published q papers.

    a. Write down a "rough" extensive form of this game.

    b. If there is a pooling perfect Bayesian equilibrium, will the assistant professor be tenured? Does he write any papers? What then is the unique outcome of the pooling perfect Bayesian equilibrium?

---

6. This exercise is based on one written by Rod Garrett and John Wooders.

    c.   Characterize the perfect Bayesian separating equilibria using the parameters of the game.

    d.   If there is more than one separating perfect Bayesian equilibrium, what is the smallest number of papers that an assistant professor needs to publish in order to get tenure?

16.7   **Intuitive Limit Pricing:** Show that of all the separating equilibria of the limit pricing–entry deterrence game in Section 16.2 only the best separating equilibrium does not fail the intuitive criterion.

16.8   **Beer or Quiche?** Cho and Kreps (1987) introduced what is now a famous two-player signaling game. First, Nature selects player 1, who knows his type, to be either a wimp (W), with probability $p = 0.1$, or surly (S), with probability $1 - p = 0.9$. Player 1 then chooses what to have for breakfast: beer (B) or quiche (Q). A surly type prefers beer while a wimp prefers quiche. Player 1's preferred breakfast gives him a payoff of 1 while his less-preferred choice gives him 0. After breakfast player 2 observes what player 1 ate but does not know whether he is a wimp or surly. Player 2 then chooses whether to duel (D) with player 1 or not to duel (N). Player 1, regardless of his type, prefers no duel, yielding him an extra payoff of 2, to a duel, which gives him 0. (For example, if player 1 eats his preferred breakfast and avoids a duel then his final payoff is 3, while if he eats his preferred breakfast and is forced into a duel then his final payoff is 1.) Player 2, however, prefers to duel if and only if player 1 is a wimp. If player 1 is surly then player 2's payoff is 0 from D and 1 from N. If player 1 is a wimp then player 2's payoff is 2 from D and 1 from N.

    a.   Draw the extensive form of this game.

    b.   What are the Bayesian Nash equilibria of this game?

    c.   What are the perfect Bayesian equilibria?

    d.   Of the equilibria you found in (c), which fail the intuitive criterion?

17

Building a Reputation

It is common to hear descriptions of some ruthless businesspeople as having "a reputation for driving a tough bargain" or "a reputation for being greedy." Others are referred to as having "a reputation for being trustworthy" or "a reputation for being nice." What does it really mean to have a reputation for being a certain *type* of person? Would people put in the effort to build a reputation for being someone they really are not? Some of the most interesting applications of dynamic games of incomplete information are in modeling and understanding how reputational concerns affect people's behavior.[1]

This chapter provides some of the central insights of the game theoretic literature that deals with incentives of players to build or maintain reputations for being someone they are not, in the sense of being nice, tough, or any other adjective that comes to mind and can enhance their reputations in the eyes of others. The insights described below have spawned a large literature, and the curious (and more technically inclined) reader is encouraged to consult Mailath and Samuelson (2006).

17.1   Cooperation in a Finitely Repeated Prisoner's Dilemma

The main idea that drives most game theoretic models of reputation building was first developed by Kreps, Milgrom, Roberts, and Wilson (1982) in a seminal paper that became known as the "gang of four" paper. In it they considered a finitely repeated Prisoner's Dilemma game for which backward induction implies that there is a unique subgame-perfect equilibrium in which the players always defect (Chapters 9 and 10). However, it is somewhat counterintuitive that players will play this way—is there no chance that if the game is finite but sufficiently long then by cooperating today they can sustain some quid-pro-quo cooperative behavior?

As the gang of four demonstrated, if there is even just a little bit of incomplete information about the players' types then for a long enough horizon of T periods there can actually be quite a lot of cooperation. To understand the idea behind their arguments, consider the following version of the Prisoner's Dilemma:

---

1. In full disclosure I must admit that I am biased toward this area of game theory. My Ph.D. thesis and the paper that got me my first academic job explored reputation as a tradable asset. The curious reader is referred to Tadelis (1999).

Player 2

		c	d
Player 1	C	1, 1	−1, 2
	D	2, −1	0, 0

Imagine that player 2 has standard "strategic" Prisoner's Dilemma payoffs given by the matrix, so that for him the dominant strategy is to defect (D). Player 1, however, is one of two types: he is either a strategic type or a grim-trigger type. The grim-trigger type uses the following simple strategy: In the first period he always cooperates (C); in every period $t > 1$ thereafter he continues to cooperate as long as player 2 has cooperated (c). If in any period t player 2 defects then from period $t + 1$ onward the grim-trigger type will defect (D). Let $p > 0$ be the probability that Nature chooses player 1 to be the grim-trigger type.

If this game is played only once then the outcome is obvious: both player 2 and the strategic player 1 will choose to defect as it is their dominant strategy. If the game is played for two periods, however, things become a bit more subtle. We have the following:

Claim 17.1 *If the game is played for two periods then there is a unique perfect Bayesian equilibrium. In period $t = 2$ player 2 and the strategic type of player 1 choose to defect. In period $t = 1$ the strategic type of player 1 chooses to defect while player 2 chooses to cooperate if and only if $p > \frac{1}{2}$.*

Proof It's clear that in period $t = 2$ both player 2 and the strategic player 1 will choose to defect regardless of what happened in period $t = 1$ and of player 2's beliefs about player 1's type. In period $t = 1$ the strategic type of player 1 will choose d because regardless of what he plays in period $t = 1$ he expects player 2 to play D in period $t = 2$. We are left to show that player 2 chooses to cooperate in period $t = 1$ if and only if $p > \frac{1}{2}$. The grim-trigger type of player 1 will cooperate in period $t = 1$, and in period $t = 2$ he will cooperate if and only if player 2 cooperated in period $t = 1$. As a consequence player 2's payoff from cooperating is

$$v_1(c) = \overbrace{p(1) + (1 - p)(-1)}^{t=1} + \overbrace{p(2)}^{t=2} = 4p - 1,$$

while his payoff from defecting is

$$v_1(d) = \overbrace{p(2) + (1 - p)(0)}^{t=1} + \overbrace{p(0)}^{t=2} = 2p.$$

It follows that player 2's best response is to play c in period $t = 1$ if and only if $p > \frac{1}{2}$. ∎

Claim 17.1 shows that when there is a high enough chance that player 1 is a grim-trigger type then player 2 will choose to cooperate in the first period. The reason is simply that if he defects, he guarantees himself a payoff of 0 in the second period, while if he cooperates, he may get a payoff of 2 from defecting. He is, however, taking a risk that he plays against a strategic type. Hence if that risk is not too high and if the expected reward is high enough then he will cooperate. Notice, however, that this has nothing to do with reputation—the strategic type of player 1 always plays defect and hence reveals himself immediately after the first period. With longer horizons, however, things change quite dramatically, as the next claim shows.

Claim 17.2 *If the game is played for three periods and if $p > \frac{1}{2}$ then in the unique perfect Bayesian equilibrium both player 2 and the strategic type of player 1 choose to cooperate in period $t = 1$.*

Proof We begin by considering how the game will proceed after period $t = 1$. Observe that if player 2 defects in the first period then both players will defect thereafter. Now imagine that player 2 cooperates in period $t = 1$ and let $\mu(GT)$ denote player 2's belief in period $t = 2$ that player 1 is a grim-trigger type. From claim 17.1 we know how the perfect Bayesian equilibrium will proceed in the last two periods, so that player 2 will choose c in period $t = 2$ if $\mu(GT) > \frac{1}{2}$ and d otherwise. We now will see if the perfect Bayesian equilibrium can support both player 2 and the strategic type of player 1 cooperating in period $t = 1$.

Now imagine that the strategic type of player 1 believes that player 2 will choose c in period $t = 1$. If the strategic type chooses C then he pools with the grim-trigger type, implying by Bayes' rule that $\mu(GT) = p > \frac{1}{2}$, and his payoff from choosing C is $v_2(C) = 1 + 2 + 0 = 3$. (He will gain 1 in period $t = 1$ from cooperating, and because $\mu(GT) > \frac{1}{2}$ then from claim 17.1 player 2 will choose c and the strategic type of player 1 will choose D in period $t = 2$, followed by (d, D) in period $t = 3$.) If the strategic type chooses d then he separates from the grim-trigger type, implying by Bayes' rule that $\mu(GT) = 0$, and his payoff from choosing D is $v_2(D) = 2 + 0 + 0$. (Because he reveals himself to be a strategic type then $\mu(GT) = 0$ and the play continues with both players choosing to defect in the following two periods.) This implies that a strategic type of player 1 should pool with the grim-trigger type and cooperate in the first period so as not to reveal his true type.

Turning to player 2, if he believes that both types of player 1 cooperate in $t = 1$ then his payoff from choosing c is $v_1(c) = 1 + p(1 + 2) + (1 - p)(-1) = 4p$ while his payoff from choosing d is $v_1(d) = 2 + 0 + 0$. Because $p > \frac{1}{2}$ this implies that player 2 should also cooperate in period $t = 1$. The reason this is the unique perfect Bayesian equilibrium is that even if player 2 believes that the strategic type of player 1 is playing D then he should still choose c because in this case $v_1(c) = p(1 + 1 + 2) + (1 - p)(-1) = 5p - 1$ while $v_1(d) = p(2) + (1 - p)(0) = 2p$, which is less than $v_1(c)$ if $p > \frac{1}{3}$. ∎

Claim 17.2 demonstrates that with three periods a strategic type of player 1 has incentives to behave *as if* he is a grim-trigger type, thus "building a reputation" for cooperation. Of course, for this to be the unique perfect Bayesian equilibrium we had to assume that $p > \frac{1}{2}$. It turns out, however, that the gang of four proved a much more striking result: if the game is repeated for long enough then the players will cooperate in almost all periods. More formally, we have the following:

Proposition 17.1 *Consider a T-period repeated Prisoner's Dilemma game in which T is large. The number of periods in which either player 2 or the strategic player 1 defects is bounded above by a constant $M < T$ that depends on p and not on T.*

A formal proof is not provided, and the interested reader can refer to the original paper by Kreps et al. (1982) or to Mailath and Samuelson (2006, Chapter 17). The intuition follows from the fact that if the strategic type of player 1 chooses D in some period $t < T$ then it becomes common knowledge that that player 1 is strategic, and backward induction implies that both players will defect until the end of the game. For this reason the strategic player 1 has an incentive to mimic the grim-trigger type and maintain a reputation for not being strategic. These reputational incentives only diminish as the game gets close to T. Anticipating these incentives, it is in the best interest of player 2 to cooperate as well. What makes the gang of four result remarkable is that it does not require specific assumptions on p. Even for very small values of p, for a large enough T cooperation will be sustained for all but a finite number of periods.

Remark The foregoing analysis relied on player 1 possibly being a grim-trigger type. One can think of many other types for player 1 that follow all kinds of conditional or unconditional strategies, in which case the equilibria will be different. Fudenberg and Maskin (1986) showed that if all conceivable types are possible for both players then a folk theorem result is obtained: any conceivable path of play is possible in equilibrium.

17.2 Driving a Tough Bargain

Most bargaining games take on a simple form in which the more one player gets, the less the other player gets, and the players need to reach some form of agreement in order to share the surplus. In Chapter 11 we analyzed models of alternating-offer bargaining in which one party makes offers to another, and the other player must accept the offer for gains to be split. We start with a simple example of a one-stage ultimatum game.

Consider a firm owner (player 1) who is in financial trouble and must renegotiate a debt agreement with his lender (player 2) in order to continue operating. Player 1 can offer player 2 one of two options: a modest default on debt (M) in which player 1 pays player 2 a large portion of the debt he owes, or a high default on debt (H) in which player 1 pays player 2 a small portion of the debt he owes. Player 2 can choose to accept (A) or reject (R) player 1's offer. If player 2 rejects, the firm will go bankrupt and the legal fees will leave both players with a payoff of 0. If player 2 accepts then the debt is renegotiated successfully, and the payoffs to the two players are shown in the extensive-form game depicted in Figure 17.1.

A pure strategy for player 1 is an offer $s_1 \in \{H, M\}$, and that for player 2 is a planned response to each of player 1's offers, $s_2 \in \{AA, AR, RH, RR\}$, where $s_2 = xy$ means that player 2 will choose x following an offer of H and y following an offer of M. It is easy to see that this game has a unique subgame-perfect equilibrium: (H, AA). You should easily be able to convince yourself that this game has Nash equilibria that are not subgame-perfect equilibria, for example (H, AR), which yields

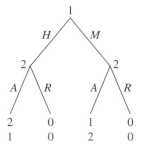

FIGURE 17.1 A perfect-information ultimatum game.

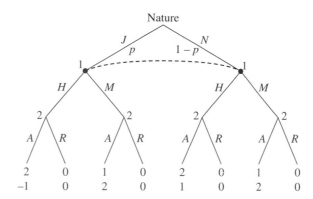

FIGURE 17.2 An incomplete-information ultimatum game.

the same outcome as (H, AA) but in which player 2 is not playing a best response if M was offered. The more interesting Nash equilibrium is (M, RA), in which player 2 gets his preferred outcomes, but this is supported by the sequentially irrational threat of rejecting a high default offer and receiving a payoff of 0 instead of 1. One can interpret this equilibrium as the one in which player 2 "drives a hard bargain," but because we believe that sequential rationality is an important feature of rational behavior, this outcome seems less likely.

Now consider a variation of this game to include some incomplete information about the type of player 2. In particular imagine that player 2 can be normal ($\theta_2 = N$) with payoffs as already described, or he can be a jerk ($\theta_2 = J$), who prefers both players to get nothing over getting the inferior outcome. Imagine further that Nature first chooses the type of player 2, who is a jerk with probability p, and assume that the payoff to the jerk of accepting an offer of H is -1 instead of 1, whereas all other payoffs are the same as in Figure 17.1. Player 1 does not know the type of player 2, but he does know that player 2 is a jerk with probability p. This game of incomplete information is described in Figure 17.2.

We can proceed to find the perfect Bayesian equilibrium of this game in one of two ways. First, we can write down the matrix representation of this game, find the Bayesian Nash equilibria, and then see which of these can be supported as a perfect Bayesian equilibrium. For this we would need to write down a 2×16 matrix, which is not too difficult; but in this case it is much simpler to use backward induction.

Consider player 2 at each of the four nodes (information sets) that follow an offer from player 1. In any perfect Bayesian equilibrium, player 2 must be playing a best response to his beliefs in every information set. Thus for each node player 2 must play a best response *at that node*, which immediately implies that a normal player 2 will accept any offer, while a jerk will accept M and reject H. Given this behavior of player 2, and the fact that the uninformed player 1 plays after Nature makes its choices, the beliefs of player 1 are uniquely determined by Nature's choice, so he must believe that $\Pr\{\theta_2 = J\} = p$. This implies that player 1 will strictly prefer to offer H if and only if the following inequality holds:

$$\underbrace{p \times 0 + (1 - p) \times 2}_{\text{Offer } H} > \underbrace{p \times 1 + (1 - p) \times 1}_{\text{Offer } M}$$

or

$$p < \tfrac{1}{2},$$

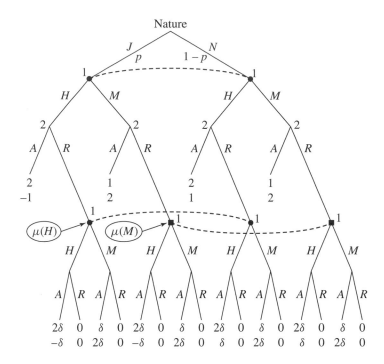

FIGURE 17.3   Two-period bargaining with incomplete information.

and he will prefer to offer M otherwise. The intuition is simple—if there is a good chance that player 2 is normal ($p < \frac{1}{2}$) then player 1 is better offering H and risking rejection than offering M and obtaining acceptance for sure.

Now we take our incomplete-information ultimatum game a step further and allow for the possibility of reputation building by introducing the following sequence of events. Player 1 makes an offer and player 2 responds as already described. If player 1's initial offer is accepted then the game ends as before. If the initial offer is rejected by player 2 then the decision is passed on to a bankruptcy judge, and player 1 can appeal for another chance to renegotiate the debt. The judge will always grant such a request, but this process causes delay. When player 1 is granted permission to make a second offer then the game proceeds as in Figure 17.2 with the same payoffs. The delay will result in discounted payoffs, with a discount factor of $\delta < 1$. The extensive form of this game is depicted in Figure 17.3.

Let $\mu(s_1)$, $s_1 \in \{H, M\}$, denote player 1's belief in each of his information sets that follow rejection. That is, $\mu(H) \in [0, 1]$ is player 1's belief that player 2 is a jerk *conditional on* player 2's rejection of an offer H, and $\mu(M) \in [0, 1]$ is his belief that player 2 is a jerk *conditional on* player 2's rejection of an offer M.

If we attempt to turn this into a normal-form matrix game this will result in an 8×4096 matrix! (Player 1 has 3 information sets and player 2 has 12 information sets with two actions each, so $2^3 = 8$ and $2^{12} = 4096$ pure strategies.) However, by using backward induction over all the information sets that are singletons we can drastically simplify the relevant pure strategies that can be part of any perfect Bayesian equilibrium. In the second stage a normal player 2 will accept any offer, and a jerk will accept M and reject H. Taking into account this best response of player 2 at the final stage after a second offer the game reduces to the one depicted in Figure 17.4, which we call game 2.

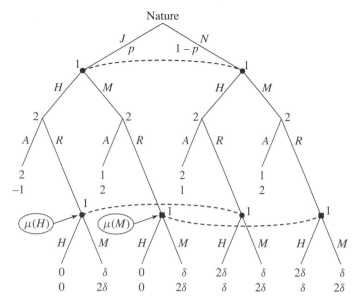

FIGURE 17.4 Game 2.

The reduced game 2 will still be a bit large to write down as a matrix because it implies 8 rows and 16 columns, but it seems that we cannot proceed with backward induction owing to the information sets of player 1 after a rejection in the first round of bargaining. However, a more careful observation of game 2 suggests that we can perform another step of backward induction. To do this consider the nodes in the first round at which player 2 has to move after he is offered M. If he accepts then he receives a payoff of 2, while if he rejects then the game moves into the next stage of bargaining. Note, however, that in the next stage the *most* player 2 can receive is $2\delta < 2$. Sequential rationality therefore implies that player 2 must accept an offer of M in the first stage, which allows us to reduce the game further because in any perfect Bayesian equilibrium player 2 must accept M in the first round *regardless of his type*. The further-reduced game appears in Figure 17.5, which we call game 3.

Game 3 is very manageable because it can be represented in its normal form by a 4×4 matrix. To do this we define strategies for both players in this reduced form of the initial game as follows: Let $s_1^1 s_1^2 \in \{HH, HM, MH, MM\}$ be a pure strategy for player 1, where s_1^t is what player 1 offers in bargaining stage $t \in \{1, 2\}$. Similarly let $s_2^J s_2^N \in \{AA, AR, RA, RR\}$ be a pure strategy for player 2, where s_2^θ is what player 2 chooses in the first stage when offered H when his type is $\theta \in \{J, N\}$.

To complete the matrix with real numbers, however, we need to specify values for δ and p. Consider the case in which the future matters a lot, with $\delta = 0.9$, and in which the likelihood of being a jerk is small, $p = 0.1$. In this case, for example, the pair of expected payoffs from player 1 choosing HM and player 2 choosing AR will be

$$(v_1, v_2) = p(2, -1) + (1 - p)(\delta, 2\delta) = (1.01, 1.52).$$

Similarly we can compute the other combinations of strategies to obtain the following matrix:

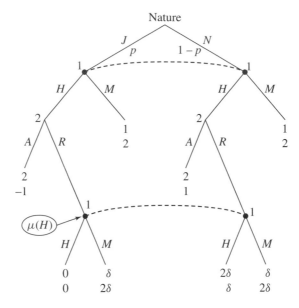

FIGURE 17.5 Game 3.

		Player 2		
	AA	*AR*	*RA*	*RR*
HH	2, 0.8	1.82, 0.71	1.8, 0.9	1.62, 0.81
HM	2, 0.8	1.01, 1.52	1.89, 1.08	0.9, 1.8
MH	1, 2	1, 2	1, 2	1, 2
MM	1, 2	1, 2	1, 2	1, 2

Player 1 labels the rows; "Player 2" heads the columns.

Any Nash equilibrium of this matrix will correspond to a Bayesian Nash equilibrium of the reduced Bayesian game, game 3. Furthermore any perfect Bayesian equilibrium of game 3 together with the sequentially rational best-response strategies of player 2 that we have already found will be part of a perfect Bayesian equilibrium in the original game. It is easy to see that for player 1 MH and MM are both strictly dominated by HH. The intuition is simple: For player 1, the strategy HH will replicate the one-stage game in which he commits to offer H and then player 2 has to respond with no chance of getting M. The *worst that can happen* for player 1 with this strategy is that both types reject the initial offer, but at the second stage the normal player 2 will accept. Because the likelihood of being a jerk is small, and there is little discounting, this yields a better payoff than offering M for sure (1.62 versus 1). If some type of player 2 accepts the initial offer, then things are even better for player 1.[2]

Once we eliminate MH and MM, it is easy to see that for player 2 the strategies AA and AR are strictly dominated by RR. To give intuition, it is actually easier to consider the following set of dominance relations: AA is dominated by RA because

---

2. Notice that if discounting were more severe, or if the probability of being a jerk were significantly higher, then HH would not necessarily dominate.

given that H is offered in the first stage, the jerk should reject rather than accept. Similarly AR is dominated by RR. Thus we are left with the following simple 2×2 matrix:

	Player 2	
	RA	RR
HH	1.8, 0.9	1.62, 0.81
HM	1.89, 1.08	0.9, 1.8

Player 1 is to the left of the HH/HM rows.

for which it is easy to see that there are no pure-strategy Nash equilibria.

As Nash's existence theorem shows (Section 6.4), any finite game must have a Nash equilibrium. To find it we need to find a mixed strategy $\sigma_1(HH) \in (0, 1)$ of choosing HH by player 1 that will make player 2 indifferent between RA and RR, and a mixed strategy $\sigma_2(RA) \in (0, 1)$ of choosing RA by player 2 that will make player 1 indifferent between HH and HM. To find $\sigma_1(HH)$ we solve

$$\overbrace{\sigma_1(HH)0.9 + (1 - \sigma_1(HH))1.08}^{\text{Player 2's payoff from choosing } RA} = \overbrace{\sigma_1(HH)0.81 + (1 - \sigma_1(HH))1.8}^{\text{Player 2's payoff from choosing } RR},$$

which yields $\sigma_1(HH) = \frac{8}{9}$. Similarly to find $\sigma_2(RA)$, we solve

$$\overbrace{\sigma_2(RA)1.8 + (1 - \sigma_2(RA))1.62}^{\text{Player 1's payoff from choosing } HH} = \overbrace{\sigma_2(RA)1.89 + (1 - \sigma_2(RA))0.9}^{\text{Player 1's payoff from choosing } HM}$$

which yields $\sigma_2(RA) = \frac{8}{9}$ as well.

Because this is the unique Bayesian Nash equilibrium of the reduced game 3, and because all information sets are reached with positive probability, we know that these mixed strategies are also part of a perfect Bayesian equilibrium with the induced beliefs using Bayes' rule. Hence the *unique* perfect Bayesian equilibrium in game 3 is as follows:

Player 1: Play HH with probability $\frac{8}{9}$ and HM with probability $\frac{1}{9}$.

Player 2: Play RA with probability $\frac{8}{9}$ and RR with probability $\frac{1}{9}$.

Beliefs: $\mu(H) = \frac{0.1}{0.1 + 0.9\frac{1}{9}} = 0.5$.[3]

We now look back to the original game and incorporate all that we have analyzed into a perfect Bayesian equilibrium. The following pair of strategies together with the induced beliefs forms a perfect Bayesian equilibrium of the original game:

Player 1:

- In the first stage play H.
- After rejection following an H offer, play H with probability $\frac{8}{9}$ and M with probability $\frac{1}{9}$.
- After rejection following an M offer, play H with probability 1.

---

3. This follows because the information set of rejection is reached for sure if player 2 is a jerk, hence the 0.1, and is reached with probability $\frac{1}{9}$ if player 2 is normal, hence the $0.9\frac{1}{9}$. The rest follows from Bayes' rule.

Player 2:

- **Type J:** following H play R and following M play A in any stage.
- **Type N:** In the first stage following M play A, and following H play R with probability $\frac{1}{9}$ and A with probability $\frac{8}{9}$. In the second stage play A following any offer.

Beliefs: $\mu(H) = 0.5$ and $\mu(M) = \mu^*$, where $\mu^* \in \left[0, \frac{1}{2}\right]$ is the off-the-equilibrium-path belief.

It is important to notice that the combination of backward induction and the analysis of the reduced game leaves undetermined what player 1 will do in the information set that occurs after the rejection of an M offer in the first stage. However, as we noticed earlier, nothing that player 1 does in this information set will affect the behavior of player 2 when M is offered, so all we have to do is assign some belief to player 1 in that information set and have him play a best response to that belief. We suggested that he play H in that information set, which implies that he must believe that $\Pr\{\theta_2 = J | M \text{ rejected}\} < \frac{1}{2}$, which imposes the restriction on μ^*. Alternatively we could suggest that he play M in that information set, which implies that he must believe that $\Pr\{\theta_2 = J | M \text{ rejected}\} > \frac{1}{2}$, and therefore we would have had to impose the opposite restriction.

The interpretation of this type of equilibrium, which is called a semi-separating equilibrium, is also interesting. One type, in this case the jerk, is playing a pure strategy by accepting M and rejecting H, while the normal type always accepts M and *sometimes but not always* accepts H. Player 1 will therefore sometimes learn player 2's type, which happens when player 2 accepts H in the first period, and sometimes only partially learn his type after a rejection of M. By the fact that the normal type is rejecting H only with some probability, after a rejection of H player 1's posterior that player 2 is a jerk goes up. Hence the term "semi-separating"—the normal type sometimes separates and sometimes does not.

To understand further the intuition that causes the semi-separating equilibrium to be the unique perfect Bayesian equilibrium it is useful to understand why other forms of behavior *cannot* be an equilibrium. We consider three natural cases and highlight what problem each poses as an equilibrium candidate:

Case 1: Pooling on accept. In this case following an offer H by player 1, both types of player 2 pool and choose to accept. This cannot be an equilibrium because accepting H is a dominated strategy for the jerk. He gets 0 instead of -1 and can guarantee himself at least 0 in the second period, giving an expected payoff of at least 0 by rejecting.

Case 2: Pooling on reject. In this case following an offer H by player 1, both types of player 2 pool and choose to reject. If this were part of an equilibrium then in the second stage player 1 could not update his prior on the type of player 2 and would continue to believe that $\Pr\{\theta_2 = J\} = 0.1$, in which case he would offer H again. However, if this is the continuation that a normal player 2 faces, he is better off accepting the H offer in the first period than in the second because of discounting. Thus this case cannot be an equilibrium.

Case 3: Separating. In this case following an offer H by player 1, a normal player 2 accepts the offer while a jerk does not. If this were part of an equilibrium, then in the second stage player 1 would update his prior on the type of player 2 and believe that $\Pr\{\theta_2 = J\} = 1$, in which case he would offer M. However, if this is the continuation in which a normal player 2 believes, he is better off rejecting the H offer in the first period and getting M in the second because the discounting is not too severe. Thus this case cannot be an equilibrium.

Now that we understand the problems of pooling or complete separation, it is easier to understand the intuition behind the mixed-strategy equilibrium that we found. In this equilibrium the normal type of player 2 *sometimes* acts like a jerk, which causes player 1 to update his beliefs about the type of player 2 in a way that makes him believe that player 2 is a jerk with a higher probability. For this to work, player 1's updated beliefs must be set high enough to just make him indifferent in the second stage between offering H and M. Being indifferent, player 1 will then mix between offering H and M in such a way that makes the normal type of player 2 indifferent between rejecting and accepting H in the first stage. We can interpret this as player 2 sometimes building a reputation of being a jerk, and in this way gaining a chance of getting M in the second period.

17.3   A Reputation for Being "Nice"

Consider a variant of the Centipede Game that we analyzed in Section 8.3.1, as shown in Figure 17.6. Each player, starting with player 1, has 49 decision nodes of whether to continue or stop. We denote each of the 98 nodes by x_n, where n is the number of potential moves left in the game, so that x_1 is the last decision node, x_2 is the second-to-last, and so on until x_{98}, which is the root of the game. If the players continue until the end of the game then each gets a payoff of 50. However, backward induction implies that player 2 will stop (s) at the last node x_1. Anticipating this player 1 will stop (S) at node x_2, and this logic causes the game to unravel and end immediately with player 1 choosing S at node x_{98}.

Of course this prediction may seem preposterous. Can't the players see that they can make so much more by continuing the game? This critique is similar to the critique of players not ever cooperating in a long but finitely repeated Prisoner's Dilemma. It turns out, maybe not surprisingly by now, that the same logic illustrated by the argument of Kreps et al. (1982) will also help us explain how players can continue playing the Centipede Game.

Consider a variant of the game in Figure 17.6 in which player 2 exhibits the same payoffs but player 1 can be one of two types. With probability 0.99 player 1 is a selfish type with payoffs as in Figure 17.6, while with probability 0.01 he is a nice type who prefers not to cause the game to stop. To make things simple assume that if this type of player 1 continues then he gets an extra payoff of 2 from not stopping the game. This Centipede Game of incomplete information is depicted in Figure 17.7.

As in the perfect-information game in Figure 17.6 we index the nodes of each player along the path in which player 1 is a selfish type by x_{98}, x_{97}, and so on until x_1. Similarly we index the nodes of each player along the path in which player 1 is a nice

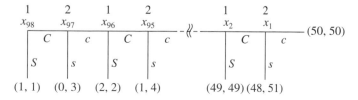

FIGURE 17.6   A Centipede Game.

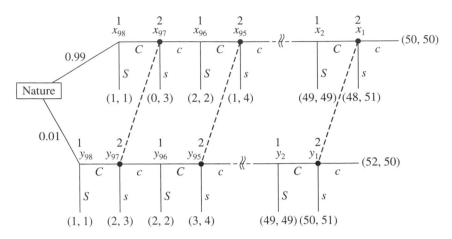

FIGURE 17.7 An incomplete-information Centipede Game.

type by y_{98}, y_{97}, and so on until y_1. In this game of incomplete information player 2 has 49 information sets, $I_1 = \{x_1, y_1\}$, $I_3 = \{x_3, y_3\}$, and so on until $I_{97} = \{x_{97}, y_{97}\}$. Let $\mu(y_n)$ denote the belief of player 2 that he is at node y_n conditional on being in information set I_n.

We know that a nice player 1 will always choose to continue. We will now establish a series of claims that will characterize the properties of the unique perfect Bayesian equilibrium of the incomplete-information Centipede Game depicted in Figure 17.7.

Claim 17.3 *In any perfect Bayesian equilibrium, if player 2 will choose to continue with probability 1 at information set I_n then player 1 will choose to continue with probability 1 at node x_{n+1}. Similarly if player 2 will choose to stop with probability 1 in information set I_n then a selfish player 1 will choose to stop with probability 1 at node x_{n+1}.*

Proof If player 2 chooses c with probability 1 at some information set I_n then a selfish player 1's best response at node x_n is to choose C with probability 1 because instead of getting the stop payoff at node $n + 1$ he can continue and let player 2 continue further to node $n - 1$, at which point the selfish player 1 can stop and get a strictly higher payoff. Clearly if player 2 chooses s with probability 1 at some information set I_n then a selfish player 1's best response at node x_{n+1} is to choose S with probability 1. ∎

Claim 17.3 simply says that if in equilibrium player 2 has a pure strategy in any of his information sets then the selfish player 1 will have one at the node that precedes this information set. In particular if player 2 wishes to continue for sure at some information set then player 1 will continue for sure at the nodes that precede that information set.

Claim 17.4 *In any perfect Bayesian equilibrium, if player 2 will choose to continue with probability 1 at information set I_n then both players will choose to continue with probability 1 from the beginning of the game at least until information set I_n. As a consequence, $\mu(y_n) = 0.01$.*

Proof This follows directly from claim 17.3. If player 2 will continue for sure at information set I_n then both types of player 1 will continue at nodes x_{n+1} and y_{n+1}. But then player 2 at information set I_{n+2} will guarantee himself to reach information set I_n by continuing at I_{n+2}, and if continuing from I_n is an equilibrium best response then his payoff must be higher than that from stopping at information set I_n. But from claim 17.1 this implies that player 1 will choose to continue with probability 1 at node x_{n+3}, and this inductive argument continues until node x_{98}. That $\mu(y_n) = 0.01$ follows from the fact that both the selfish and nice types of player 1 will continue until information I_n, which implies that player 2 cannot update his original prior (both types of player 1 pool). ∎

Claim 17.4 establishes an interesting finding which states that if there is some stage at which player 2 chooses to continue with probability 1 then the game must reach that stage for sure. The next result establishes that player 2 will never choose to stop the game prematurely.

Claim 17.5 *In any perfect Bayesian equilibrium player 2 will choose to continue with positive probability in all his information sets but his last one I_n, in which he will stop.*

Proof Suppose instead that player 2 chooses s with probability 1 at some information set $I_n, n > 1$. A selfish player 1's best response is to choose S with probability 1 at node x_{n+1}. If this is part of the equilibrium then Bayes' rule implies that $\mu(y_n) = 1$ because only a nice player 1 would continue in information set I_{n+1}. But then player 2's best response in information set I_n is to choose c with probability 1, a contradiction. ∎

Claim 17.5 implies that in any perfect Bayesian equilibrium each of player 2's 49 information sets is reached with positive probability because the nice player 1 plays C at each of his decision nodes and from claim 17.5 player 2 must play c with positive probability in all but his last information set. This conclusion is convenient because if all the information sets are reached with positive probability in any perfect Bayesian equilibrium then the beliefs are determined by Bayes' rule everywhere (recall requirement 15.2 in Section 15.2).

Claim 17.6 *In any perfect Bayesian equilibrium the selfish player 1 continues with positive probability at all nodes but his last node x_2.*

Proof Suppose instead that the selfish player 1 chooses S with probability 1 at some node x_n, $n > 2$. If this is part of the equilibrium then Bayes' rule implies that $\mu(y_{n-1}) = 1$ because only a nice player 1 would continue in information set I_n. This implies that player 2's best response in information set I_{n-1} is to choose c with probability 1. But from claim 17.3 the best response of a selfish player 1 is to choose C with probability 1 at x_n, a contradiction. ∎

Claim 17.6 highlights the crux of incomplete-information reputation models. If the private-information player reveals his selfish preferences at any point, then behavior that is not consistent with this immediately implies that player 1 is nice. This will encourage player 2 to continue down the game until the very end, which in turn gives player 1 incentives to continue himself and cheat player 2 at a later stage. However, this is something that equilibrium behavior precludes because *by definition* no player can be fooled in equilibrium.

We can now proceed to actually find the unique perfect Bayesian equilibrium of the game. Let $\sigma_1(x_n)$ denote the probability that a selfish player 1 chooses to stop at node

x_n, and let $\sigma_2(I_n)$ denote the probability that player 2 chooses to stop at information set I_n.

We know that player 2 chooses s in I_1 and hence, from claim 17.1, the selfish player 1 chooses S at node x_2. From claim 17.5 we know that in equilibrium player 2 mixes between s and c at I_3, implying that he must be indifferent between these choices, or equivalently,

$$50 = \mu(y_3)(51) + (1 - \mu(y_3))49,$$

which implies that in equilibrium

$$\mu(y_3) = \tfrac{1}{2}. \tag{17.1}$$

Let $v_2(s|I_n)$ denote player 2's expected payoff from stopping in information set I_n, $n > 3$. From claim 17.5 player 2 must be indifferent between stopping and receiving $v_2(s|I_n)$ and continuing and receiving a random outcome as follows. With probability $\mu(y_n)$ player 1 is nice and the game will proceed to I_{n-2}, where player 2 receives a payoff of $v_2(s|I_n) + 1$. With probability $1 - \mu(y_n)$ player 1 is selfish and chooses to stop with probability $\sigma_1(x_{n-1})$, giving player 2 a payoff of $v_2(s|I_n) - 1$, while player 1 continues with probability $1 - \sigma_1(x_{n-1})$, giving player 2 a payoff of $v_2(s|I_n) + 1$. This implies that

$$v_2(s|I_n) = \mu(y_n)(v_2(s|I_n) + 1)$$
$$+ (1 - \mu(y_n))[\sigma_1(x_{n-1})(v_2(s|I_n) - 1) + (1 - \sigma_1(x_{n-1}))(v_2(s|I_n) + 1)],$$

which simplifies to

$$\sigma_1(x_{n-1})(1 - \mu(y_n)) = \tfrac{1}{2}. \tag{17.2}$$

Now observe that if player 2 chooses to continue then he moves from information set I_n to information set I_{n-2} in one of two ways: either player 1 is nice and he continues for sure, or he is selfish and he continues with probability $(1 - \sigma_1(x_{n-1}))$. Hence Bayes' rule implies that

$$\mu(y_{n-2}) = \frac{\mu(y_n)}{\mu(y_n) + (1 - \mu(y_n))(1 - \sigma_1(x_{n-1}))} \tag{17.3}$$

$$= \frac{\mu(y_n)}{1 - \sigma_1(x_{n-1})(1 - \mu(y_n))}.$$

Notice that the left side of (17.2) appears in the denominator of (17.3), so substituting (17.2) into (17.3), we can rewrite (17.3) as

$$\mu(y_n) = \frac{\mu(y_{n-2})}{2}. \tag{17.4}$$

We can now use (17.1) together with an iterative application of (17.4) to obtain the following sequence of equilibrium beliefs for player 2 in each of his information sets I_n (n being odd):

$$\mu(y_3) = \tfrac{1}{2}, \ \mu(y_5) = \tfrac{1}{4}, \ \mu(y_7) = \tfrac{1}{8}, \ \ldots, \ \mu(y_n) = \left(\tfrac{1}{2}\right)^{\frac{n-1}{2}}, \ \ldots \tag{17.5}$$

However, recall that $\mu(y_n)$ is bounded below by 0.01, which is the prior probability that player 1 is nice. This implies that the equilibrium sequence can go only up to information set I_{13}, in which $\mu(y_{13}) = (0.5)^6 = 0.015625$ because $\mu(y_{15}) = (0.5)^7 < 0.01$, which is impossible given the prior probability. This observation together with claims 17.2–17.4 implies the following result:

Corollary 17.1 *The incomplete-information Centipede Game depicted in Figure 17.7 has a unique perfect Bayesian equilibrium in which player 2 continues in all his information sets I_n, $n > 13$, and the selfish player 1 continues at all nodes x_n, $n > 14$. At all but their last information sets both players mix between stop and continue, and at their last information set both players stop.*

To complete the full characterization of the perfect Bayesian equilibrium we must compute the values of $\sigma_1(x_n)$ and $\sigma_2(I_n)$, the probabilities that the selfish player 1 and player 2 choose to stop in their later information sets. Rearranging (17.2) implies that for any even n we have

$$\sigma_1(x_n) = \frac{1}{2(1 - \mu(y_{n+1}))},$$

which together with the sequence defined in (17.5) allows us to obtain

$$\sigma_1(x_n) = \frac{1}{2\left(1 - \left(\frac{1}{2}\right)^{\frac{n}{2}}\right)} \qquad \text{for all } n \leq 12 \text{ even.} \tag{17.6}$$

To compute $\sigma_2(I_{n-1})$ we use the fact that player 2's mixed strategy must cause player 1 to be indifferent between continuing and stopping. Letting $v_1(S|x_n)$ denote the expected payoff of a selfish player 1 from stopping at x_n, we have

$$v_1(S|x_n) = \sigma_2(I_{n-1})(v_1(S|x_n) - 1) + (1 - \sigma_2(x_{n-1}))(v_1(S|x_n) + 1),$$

which reduces to

$$\sigma_2(I_{n-1}) = \tfrac{1}{2} \qquad \text{for all } n \leq 14 \text{ even.} \tag{17.7}$$

This analysis highlights another interesting aspect of reputation games that we saw toward the end of Section 17.2. As corollary 17.1 indicates, backward induction shows that as we move back at least 13 decision nodes from the end of the game, the two players stop playing a mixed strategy and instead continue for sure. An implication of this is that if this game was much shorter, like the game in Section 8.3.1 that had 4 decision nodes, 2 for each player, then the unique perfect Bayesian equilibrium would be a *semi-separating equilibrium* as follows: players 1 and 2 would stop in their last information sets x_2 and I_2; from (17.6) we know that a selfish player 1 would start the game by stopping with probability $\sigma_1(x_4) = \frac{2}{3}$; and from (17.7) player 2 will choose $\sigma_2(I_3) = \frac{1}{2}$. In many reputation games that are not "long enough," the perfect Bayesian equilibrium will involve semi-separating behavior.

Remark This last observation has another side to it. Recall proposition 17.1, which states that the "number of periods in which either player 2 or the strategic player 1 defects is bounded above by a constant $M < T$ that depends on p and not on T." We can see that this is exactly what happens in the incomplete-information Centipede Game as well. In the example just considered, the probability of player 1 being a nice type is $p = 0.01$, which implies that there will be no more than 13 decision stages in which the players choose to stop with positive probability, and only one decision stage for each player in which they choose to stop for sure. Hence if the Centipede Game is of length $T > 13$ stages then for all stages $T - 13$ the two players will continue with probability 1.

17.4 Summary

- Games of incomplete information can shed light on incentives for rational strategic players to behave in ways that help them build a reputation for having certain behavioral characteristics.

- In equilibrium models players are, by definition, never fooled. However, if there is incomplete information then players will have rational uncertainty about whether players they face are set in their ways.

- This rational uncertainty is what gives strategic players an incentive to imitate behavioral "types" and act in ways that are not short-run best-response actions, but that in turn give rise to long-run benefits, thus providing reputational incentives.

- The incomplete information and the resulting reputational incentives cause finitely repeated games and other finite dynamic games not to unravel to the often grim backward-induction outcome, but instead to result in high-payoff behavior that can persist on very long time horizons.

- These game theoretic models help us understand concepts such as apparently "crazy" behavior that in turn results in long-run benefits for the player acting in this way.

17.5 Exercises

17.1 **Tit-for-Tat Reputation:** Consider the following Prisoner's Dilemma:

		Player 2	
		c	d
Player 1	C	1, 1	$-1, 2$
	D	2, -1	0, 0

Player 2 has payoffs given by the matrix while player 1 has one of two types: he is either a strategic type or a tit-for-tat (TFT) type. The TFT type uses the following strategy: in the first period he always cooperates; in every period $t > 1$ thereafter he plays what his opponent played in period $t - 1$. Let $p = \frac{1}{4}$ be the probability that Nature chooses player 1 to be the TFT type. Assume that the players do not discount future payoffs.

  a. What is the perfect Bayesian equilibrium if the game is played twice?
  b. If the game is played three times, is there an equilibrium in which the strategic type of player 1 defects in the first period?

c. What is the perfect Bayesian equilibrium if the game is played three times?

17.2 **Building Trust:** A graduating senior (player 1) has an idea for two start-up ventures, a small one and a large one. He can trust (T) his ideas with his roommate (player 2), who is the only one who can implement these ideas, or he can not trust (N) player 2, forget his dreams, and get a boring job, resulting in the status quo payoffs of $(0, 0)$ for players 1 and 2, respectively. There are two types of player 2: an honest type that *always* honors trust (H) and a normal type that can choose to honor trust (H) or abuse trust (A). It is common knowledge that a quarter of the people in the world are honest, but only player 2 knows whether he is honest or not.

The sequence of play is as follows: First, player 1 can choose T or N with respect to sharing his small idea. A choice of N (no trust) *ends the game* and the payoffs are $(0, 0)$. Following T player 2 can honor the trust, resulting in payoffs of $(1, 1)$, or abuse the trust and expropriate the idea, resulting in payoffs of $(-1, 2)$. (Recall that an honest player 2 *cannot* abuse trust.) If trust was offered in the first stage then first-stage payoffs are determined (i.e., the outcome is known to both players), and player 1 can then choose whether or not to trust player 2 with his big idea. (If trust was not offered in the first stage then the game ended.) No trust in the second stage will result in no payoffs in addition to those obtained in the first stage. Following trust in the second stage, if player 2 honors trust then the additional payoffs will be $(7, 7)$, while if he abuses trust and expropriates the idea the additional payoffs will be $(-5, 11)$. There is no discounting between these stages.

a. Draw the whole game tree and write down the pure strategies of each player for the whole game. How many pure strategies does each player have?

b. Using sequential rationality argue that following an abuse of trust in the first stage, there will be no trust in the second stage, effectively ending the game. Write down the reduced game tree that is implied by this argument, and the resulting pure strategies for each of the two players. From now on focus on the reduced game you derived, in which the game ends after an abuse in the first stage.

c. Can there be a pooling perfect Bayesian equilibrium in which the normal type in the first period honors trust for sure?

d. Can there be a separating perfect Bayesian equilibrium in which the normal type in the first period abuses trust for sure?

e. Find the perfect Bayesian equilibrium of this game. Is there a unique one?

17.3 **Reputational Incentives:** A seller (player 1) and buyer (player 2) can engage in trade for two periods. The seller is good with probability 0.5. A good seller produces a high-quality product with probability π and a low-quality product with probability $1 - \pi$. The seller is strategic with probability 0.5. A strategic seller chooses effort, $e \in [0, 1]$, at a personal cost of $c(e) = e^2$ and produces a high-quality product with probability $e\pi$ and a low-quality product with probability $1 - e\pi$. The buyer's payoff from a high-quality product is 1 and that from a low-quality product is 0. The game proceeds as follows: in each period t the seller makes an ultimatum price offer p_t to the buyer for his product. The buyer either accepts or rejects the offer. If he rejects the offer then they move to the second period. If he accepts the offer then a strategic seller chooses $e_t \geq 0$ and delivers the product. Neither the seller's type nor his choice of e_t (if he is strategic) is known to the buyer. After the product is delivered the buyer learns its quality.

    a.   If this game had only one period, what would be the effort level e and the price p in the unique perfect Bayesian equilibrium?

    b.   Imagine the game is repeated for two periods. Is there a perfect Bayesian equilibrium in which strategic sellers exert the level of effort e that you found in (a) in each period?

    c.   Find the unique perfect Bayesian equilibrium of the two-period game. Why is there a difference in the effort levels in each period?

18

Information Transmission and Cheap Talk

Chapter 16 showed that in some situations it will be to the benefit of players to reveal their types to their opponents. In the classic signaling example of Spence (1973), a high-productivity worker wishes to convey the information "I am high productivity and hence you should hire me for a high-paying job." As we argued, if the only means of communication available is cheap talk, then even a low-productivity worker will try to convince his opponent that he is a high-productivity type, so merely stating "I am high productivity" will not do. We concluded that if there is a credible and costly signal—in the case of the worker the signal was education—then it can act as a credible way for the player to signal his type and cause his opponent to believe him.

However, a credible and costly signal may not be available in every situation in which some types may benefit from revealing their information. This chapter describes the way in which game theoretic reasoning has been applied to situations in which one player can use costless communication to try to convey hidden information to an interested party. As in signaling games, in these **cheap-talk** or **information-transmission games** player 1 has private information and the payoffs exhibit common values, so that both players' payoffs depend on player 1's private information. Unlike in signaling games, however, player 1's action is a message that has no direct effect on payoffs.

Given the nature of cheap-talk games, player 1 is often referred to as the sender and player 2 as the receiver, and these games will typically proceed in the following four steps:

1. Nature selects a type of player 1 $\theta \in \Theta$ from some common-knowledge distribution p.
2. Player 1 learns θ and chooses some message (action) $a_1 \in A_1$.
3. Player 2 observes message a_1 and chooses action $a_2 \in A_2$.
4. Payoffs $v_1(a_2, \theta)$ and $v_2(a_2, \theta)$ are realized.

As we will now see, the inherent conflict of interest between the players will put limits on how much information the informed player can credibly communicate to the uninformed player in equilibrium.

18.1 Information Transmission: A Finite Example

This section illustrates a simple example of a cheap-talk game in which player 1 has a finite type space. A good friend of mine recently decided to accept a job in San Francisco and was contemplating where to live. He wants to trade off living close to me (we're really good friends) with having an easy commute. If traffic is generally bad he would like to live very close to work, that is, somewhere in San Francisco. If traffic is generally good then he would not mind commuting, and would like to live in Palo Alto, right next to me. If traffic is neither bad nor good, he would prefer living in between Palo Alto and San Francisco; the worse the traffic, the closer to San Francisco he would like to live. Naturally I care more about him living close to me than he does, and so I care less about how bad a commute he has. My friend relies on my intimate knowledge of commuting conditions so that he can make an informed choice.

To formalize the game we can think of me as player 1, who is the *sender* of information, and my friend as player 2, who is the *receiver* of information, and we can imagine that expected traffic conditions are given by $\theta \in \{1, 3, 5\}$, where 1 is bad, 3 is mediocre, and 5 is good. Player 1 knows the true value of θ, but player 2 knows only the prior distribution of θ. Player 1 transmits a message (his action) to player 2 about the traffic conditions. Player 2 then chooses an action (where to live) $a_2 \in A_2 = \{1, 2, 3, 4, 5\}$, where 1 is San Francisco, 5 is Palo Alto, and 2, 3, and 4 are towns in between the two cities in that order.

The preferences of player 2 are described by the following payoff function:

$$v_2(a_2, \theta) = 5 - (\theta - a_2)^2.$$

Notice that player 2 has a clear best response: given any level of traffic, he wants to choose his residence location equal to the traffic level. That is, his optimal choice is $a^*(\theta) = \theta$. To capture the fact that player 1 is biased toward having player 2 live closer to location 5, the preferences of player 1 are given by the following payoff function:

$$v_1(a_2, \theta) = 5 - (\theta + b - a_2)^2,$$

where $b > 0$ is the *bias* of player 1.

This is a dynamic game of incomplete information: player 1's type, or the state of the world θ, is known only to player 1, while player 2 knows only the distribution of θ. Player 1's type affects both his payoff and the payoff of player 2, making this a common-values game. Player 1's action set includes messages that he can transmit to player 2, and player 2's action set includes choosing where to live. To further fix ideas, imagine that player 1 is restricted to sending only one of three messages corresponding to one of these states of nature: $a_1 \in A_1 = \{1, 3, 5\}$.[1] For concreteness imagine that $b = 1.1$ and the prior is such that each of

---

1. It suffices if player 1 has at least as many messages as there are states of the world. If there are less, then the message space will not be rich enough to describe all the possible states, while if it is large enough then without loss we can consider only a number of messages that equals the number of states.

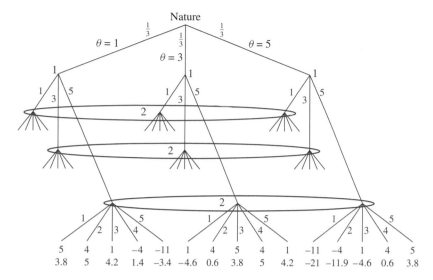

FIGURE 18.1 The commuting conditions information-transmission game.

the three states of the world occurs with probability $\frac{1}{3}$. The game is described in Figure 18.1.[2]

In a perfect Bayesian equilibrium player 1 will send (possibly) type-dependent messages to player 2, who in turn will form a posterior belief over the states of the world, given his belief about player 1's strategy, and then choose where to live. A mixed strategy for player 1 is a mapping $\sigma_1 : \Theta \rightarrow A_1$, so that for any state $\theta \in \Theta$ he chooses a probability distribution over his potential messages. A mixed strategy for player 2 is a mapping $\sigma_2 : A_1 \rightarrow \Delta A_2$, so that for each announcement of player 1 player 2 chooses a probability distribution over A_2.

It is easy to see that for any value of traffic conditions, player 1 prefers player 2 to choose a higher action a than player 2 himself would want to choose, that is, a location closer to 5. For example, if $\theta = 3$, then player 2's best choice is $a = 3$, whereas player 1's best choice is $a = 4$, and he even prefers $a = 5$ over $a = 3$.[3] This should alert you, and player 2, to a problem. If player 2 wants player 1 to share his expert opinion about traffic in the Bay area, can he rely on player 1 to reveal the truth, or will player 1 offer biased advice? In particular how much information can be credibly communicated in a perfect Bayesian equilibrium? It is useful to start with an easy observation:

Claim 18.1 *There is no perfect Bayesian equilibrium in which player 1 reports the true state of the world.*

Proof Assume in negation that player 1 truthfully reporting $a_1 = \theta$ is part of a perfect Bayesian equilibrium. It therefore must follow that when player 1 sends the message

---

2. The payoffs shown are rounded to the first decimal place. Furthermore not all the payoffs are shown, but because player 1's message does not influence the payoffs, the payoffs written down are just replicated for the other nodes in which θ and a_2 are the same.

3. We have $v_1(3, 3) = 5 - (3 + 1.1 - 3)^2 = 3.79$, whereas $v_1(4, 3) = 5 - (3 + 1.1 - 4)^2 = 4.99$ and $v_1(5, 3) = 5 - (3 + 1.1 - 5)^2 = 4.19$.

$a_1 \in \{1, 3, 5\}$ player 2 chooses $a_2 = a_1$. We saw that when $\theta = 3$ player 1 prefers $a_2 = 5$ over $a_2 = 3$. But if player 1 believes that player 2 will follow his advice then when $\theta = 3$ player 1's best response is $a_1 = 5$, a contradiction. ∎

The intuition behind this result is simple, and easily generalizes to all such information-transmission games in which there is a bias between the sender and receiver with respect to the receiver's optimal choice. If it is indeed the case that the sender is reporting information truthfully, then it is in the receiver's best interest to take the sender's information at face value. But then if the receiver is acting in this way the sender has an incentive to lie.[4] The next observation is also quite straightforward:

Claim 18.2 *There exists a **babbling equilibrium** in which player 1's message reveals no information and player 2 chooses an action to maximize his expected utility given his prior belief.*

Proof To construct the babbling (perfect Bayesian) equilibrium let player 1's strategy be to send a message $a_1 \in \{1, 3, 5\}$ with equal probability of $\frac{1}{3}$ each regardless of θ. This means that the message is completely uninformative: player 2 knows that regardless of the message, $\Pr\{\theta\} = \frac{1}{3}$ for all $\theta \in \{1, 3, 5\}$. This implies that player 2 maximizes his expected payoff,

$$\max_{a_2 \in \{1,2,3,4,5\}} Ev_2(a_2, \theta) = 5 - \tfrac{1}{3}(-(1 - a_2)^2) + \tfrac{1}{3}(-(3 - a_2)^2) + \tfrac{1}{3}(-(5 - a_2)^2),$$

which is maximized when $a_2 = 3$.[5] Because each of player 2's information sets is reached with positive probability, player 2's beliefs are well defined by Bayes' rule everywhere, and player 1 cannot change these beliefs by changing his strategy.[6] Hence player 1 is indifferent between each of the three messages and is therefore playing a best response. ∎

Claims 18.1 and 18.2 paint a rather disappointing picture for our simple game. Not only will truthful messages never be part of an equilibrium (claim 18.1), but there is an equilibrium in which player 1's valuable private information will have no effect on player 2's choice. The remaining question is whether there are other equilibria in which there is *some* valuable information transmitted from the sender to the receiver.

Notice that the reason truth-telling is not an equilibrium is because were it to be an equilibrium, player 1 could always achieve his most preferred outcome by misleading player 2 into choosing a "higher" action. The reason for this is precisely because player 1 is biased toward higher actions compared to player 2. This suggests that if we can

---

4. In the case of a finite number of states, for this argument to be true it must be the case that the bias is large enough. If we set $b = 0.1$ then it is easy to check that truth-telling will be part of a perfect Bayesian equilibrium. In the next section we will analyze the continuous type case in which any bias, no matter how small, will prevent truth-telling in equilibrium.

5. Because this is a discrete-choice problem we can't use a first-order condition and simply have to solve the expected payoff function for each of the five values of a_2.

6. There are many other ways to support an uninformative babbling equilibrium. For instance, player 1's strategy can be to choose the message $a_1 = 5$ for any θ. Player 2's on-the-equilibrium-path belief will be that $\Pr\{\theta|a_1 = 5\} = \frac{1}{3}$ for all $\theta \in \{1, 3, 5\}$. But now we need to add off-the-equilibrium-path beliefs for messages $a_1 \in \{1, 3\}$. It is easy to satisfy player 1's equilibrium incentive constraint by having player 2 believe that $\Pr\{\theta = 1|a_1 \neq 5\} = 1$, so that if player 1 deviates from $a_1 = 5$, player 2 will choose $a_1 = 1$, which is always worse than $a_2 = 3$.

construct an equilibrium in which some valuable information is transmitted, then we need to remove some of the temptation that player 1 has to announce higher numbers. This can be done if we somehow put an upper bound on how high a number player 2 will choose, as the following claim shows:

Claim 18.3 *There is a perfect Bayesian equilibrium in which player 1 partially reports the true state of the world. In particular player 1 truthfully reveals state $\theta = 1$ but pools information in states $\theta = 3$ and $\theta = 5$.*

Proof To construct the equilibrium let player 1's strategy be to send a message $a_1 = 1$ when $\theta = 1$ and $a_1 \in \{3, 5\}$ with equal probability of $\frac{1}{2}$ each if $\theta \in \{3, 5\}$. This implies that player 2's posterior belief is $\Pr\{\theta = 1 | a_1 = 1\} = 1$ while $\Pr\{\theta = 3 | a_1\} = \Pr\{\theta = 5 | a_1\} = 0.5$ for $a_1 \in \{3, 5\}$. As a consequence, if $a_1 = 1$ then player 2 will choose $a_2 = 1$, while if $a_1 \in \{3, 5\}$ then player 2 maximizes his expected payoff,

$$\max_{a_2 \in \{1,2,3,4,5\}} Ev_2(a_2, \theta) = 5 - \tfrac{1}{2}(-(3 - a_2)^2) + \tfrac{1}{2}(-(5 - a_2)^2),$$

which is maximized when $a_2 = 4$. To see that player 1's strategy is a best response it is easy to confirm that if $\theta \in \{3, 5\}$ then player 1 prefers $a_2 = 4$ over $a_2 = 1$, implying that he is indifferent between selecting $a_1 = 3$ and $a_1 = 5$ and strictly prefers either over $a_1 = 1$. If $\theta = 1$ then player 1 prefers $a_2 = 1$ over $a_2 = 4$ so he strictly prefers $a_1 = 1$ over his other possible actions. ∎

The equilibrium constructed in the proof of claim 18.3 supports partially truthful information transmission because of the way in which information is revealed by player 1 to player 2. We know from player 1's bias that he would always prefer that player 2 choose a somewhat higher action than player 2 would like, except for when player 2 wishes to choose the highest possible action. At the same time, for a low enough state, player 1 may prefer player 2's optimal action over much higher actions. For this reason, if we want player 1 to truthfully reveal the state $\theta = 1$, in which player 2 will choose $a_2 = 1$, it must be the case that if player 1 lies and announces a higher state then player 2's action will be significantly higher. That is, there must be a large enough "gap" between the actions of player 2 in equilibrium. This satisfies the equilibrium incentive constraint, requiring that when $\theta = 1$, player 1 strictly prefers to announce $a_2 = 1$. The way in which this is done is by pooling player 1's higher information types so that when any other message except $a_1 = 1$ is sent, player 2 will believe that the state of the world is either $\theta = 3$ or $\theta = 5$, in which case he prefers to choose $a_2 = 4$.

This simple finite-information transmission game is useful in illustrating the three main insights of cheap-talk equilibria: a fully truthful equilibrium never exists, a babbling equilibrium always exists, and if player 1's bias is not too large then some information can be truthfully revealed in equilibrium. The reason that player 1's bias must not be too large is that if it is it will not be possible to create a large enough gap in player 2's choices. You are left to show that this is the case in exercise 18.1. We now turn to a slightly more general and commonly used version of this game.

18.2 Information Transmission: The Continuous Case

This section presents the more elegant and extremely well-known model of cheap-talk games introduced by Crawford and Sobel (1982), whose insights are highlighted

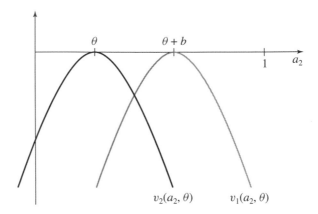

FIGURE 18.2   Payoffs in the cheap-talk game.

here. The model they proposed is more general, but the special-case version presented in this section has been widely applied in economics and political science.

The game is basically the same as the one described in the previous section, with the exception of $\theta \in \Theta = [0, 1]$ and the assumption that the state of the world θ is uniformly distributed on $[0, 1]$. Let player 2's action set include all real numbers, $a_2 \in \mathbb{R}$. Player 1's action set can be any arbitrary set of messages A_1, but it will be convenient to let $A_1 = [0, 1]$ so that the message space conforms with the state space Θ. Player 2's payoff is $v_2(a_2, \theta) = -(\theta - a_2)^2$, while player 1's payoff is $v_1(a_2, \theta) = -(\theta + b - a_2)^2$, which implies that for any given value of $\theta \in [0, 1]$, player 2's optimal choice is $a_2 = \theta$, while player 1's is $a_2 = \theta + b$. The payoff functions of the two players are depicted in Figure 18.2.

As in the finite example, both players would prefer a higher action to be taken when the state θ is higher, but player 1 has a constant bias, making him prefer even higher actions than player 2. This immediately implies that claim 18.1 generalizes to the continuous setting because of the same reasoning: If player 2 believes that player 1 is reporting θ truthfully, then player 2's best response is to choose $a_2 = \theta$. But if this is player 2's strategy then player 1 will report $a_1 = \theta + b$ for any $b \neq 0$. Hence there can never be a fully truthful equilibrium. Not surprisingly a babbling equilibrium still exists:

Claim 18.4   *There exists a babbling perfect Bayesian equilibrium in which player 1's message reveals no information and player 2 chooses an action to maximize his expected utility given his prior belief.*

Proof   We construct the equilibrium in a similar way to the finite case. Let player 1's strategy be to send a message $a_1 = a_1^B \in [0, 1]$ regardless of θ. This means that the message is completely uninformative and player 2 believes that θ is distributed uniformly on $[0, 1]$. This implies that, conditional on receiving the message a_1^B, player 2 maximizes his expected payoff,

$$\max_{a_2 \in \mathbb{R}} E v_2(a_2, \theta) = \int_0^1 -(\theta - a_2)^2 d\theta = -\tfrac{1}{3} + a_2 - a_2^2,$$

which is maximized when $a_2 = \tfrac{1}{2}$. Let player 2's off-equilibrium-path beliefs be $\Pr\{\theta = \tfrac{1}{2} | a_1 \neq a_1^B\} = 1$ so that his off-the-equilibrium-path best response to any other message is $a_2 = \frac{1}{2}$ as well. It is easy to see that player 1 is indifferent between any of his messages and hence choosing $a_1 = a_1^B$ is a best response. ∎

We see that the continuous-space cheap-talk model has the same two extreme results demonstrated for the discrete-space game: there is no truthful equilibrium and there is always a babbling equilibrium. The question then is, how much information can the sender, player 1, credibly transmit to the receiver, player 2? We begin by constructing a perfect Bayesian equilibrium in which player 1 uses one of two messages, a_1' and a_1'', and player 2 chooses a different action following each message, $a_2(a_1') < a_2(a_1'').$[7]

Claim 18.5 *In a two-message equilibrium player 1 must use a threshold strategy as follows: if $0 \leq \theta \leq \theta^*$ he chooses a_1', whereas if $\theta^* \leq \theta \leq 1$ he chooses a_1''.*

Proof For any θ player 1's payoffs from a_1' and a_1'' are as follows:

$$v_1(a_2(a_1'), \theta) = -(\theta + b - a_2(a_1'))^2$$

$$v_1(a_2(a_1''), \theta) = -(\theta + b - a_2(a_1''))^2,$$

which implies that the extra gain from choosing a_1'' over a_1' is equal to

$$\Delta v_1(\theta) = -(\theta + b - a_2(a_1''))^2 + (\theta + b - a_2(a_1'))^2.$$

The derivative of $\Delta v_1(\theta)$ is equal to $2(a_2(a_1'') - a_2(a_1')) > 0$ because $a_2(a_1') < a_2(a_1'')$. This implies that if type θ prefers to send message a_1'' over a_1' then every type $\theta' > \theta$ will also prefer a_1''. Similarly if type θ prefers to send message a_1' over a_1'' then so will every type $\theta' < \theta$. This in turn implies that if two messages are sent in equilibrium then there must be some threshold type θ^* as defined in claim 18.5. It follows that when $\theta = \theta^*$ player 1 must be indifferent between sending the two messages. ∎

Now that we know what restrictions apply to player 1's strategy in a two-message equilibrium, we can continue to characterize the strategy of player 2 in a two-message perfect Bayesian equilibrium as follows:

Claim 18.6 *In any two-message perfect Bayesian equilibrium in which player 1 is using a threshold θ^* strategy as described in claim 18.5, player 2's equilibrium best response is $a_2(a_1') = \frac{\theta^*}{2}$ and $a_2(a_1'') = \frac{1+\theta^*}{2}$.*

Proof This follows from player 2's posterior belief and from him playing a best response. In equilibrium player 2's posterior following a message a_1' is that θ is uniformly distributed on the interval $[0, \theta^*]$, and his posterior following a message a_1'' is that θ is uniformly distributed on the interval $[\theta^*, 1]$. Player 2 plays a best response if and only if he sets $a_2(a_1) = E[\theta|a_1]$, which proves the result. ∎

---

7. Two things are worth noting. First, it does not matter which message is larger because the messages have no inherent meaning. Second, there may be many more messages than just a_1' and a_1'', especially if the message space is continuous. There are two ways to proceed. First, we can take the message space A_1 and partition it into $a_1' \subset A_1$ and $a_1'' = A_1 \backslash a_1'$ so that there are no off-equilibrium-path beliefs. Second, we can define off-equilibrium-path beliefs that will support a_1' and a_1'' as the two equilibrium messages. Later in this section we will take the latter approach.

Using claims 18.5 and 18.6 we can now characterize the two-message perfect Bayesian equilibrium as follows:

Claim 18.7 *A two-message perfect Bayesian equilibrium exists if and only if $b < \frac{1}{4}$.*

Proof   From claim 18.5 we know that when $\theta = \theta^*$ player 1 must be indifferent between his two messages so that

$$v_1(a_2(a_1'), \theta^*) = v_1(a_2(a_1''), \theta^*),$$

which from claim 18.6 and from the fact that $\frac{\theta^*}{2} < \theta^* < \frac{1+\theta^*}{2}$ is equivalent to

$$\theta^* + b - \frac{\theta^*}{2} = -\left(\theta^* + b - \frac{1+\theta^*}{2}\right). \tag{18.1}$$

The solution to (18.1) is $\theta^* = \frac{1}{4} - b$, which can result in a positive value of θ^* only if $b < \frac{1}{4}$. To complete the specification of off-the-equilibrium-path beliefs, let player 2's beliefs be $\Pr\{\theta = \frac{\theta^*}{2} | a_1 \notin \{a_1', a_1''\}\} = 1$, so that he chooses $a_2 = \frac{\theta^*}{2}$, which causes player 1 to be indifferent between sending the message a_1' and any other message $a_1 \notin \{a_1', a_1''\}$, implying that his threshold strategy is a best response. ∎

It is interesting to observe the similarities between the equilibrium constructed for the finite model in claim 18.3 of Section 18.1 and the two-message equilibrium constructed in claim 18.7. First, note that the bias cannot be too big. If it is, then we cannot find a threshold type θ^* for which player 1 would not wish to deviate from message a_1' to message a_1''. Second, note that player 2 has more information when message a_1' is sent than when message a_1'' is sent, which is similar to the fact that in the discrete game player 2 knew in equilibrium when $\theta = 1$ but could not distinguish between $\theta = 3$ and $\theta = 5$. To see that player 2 has more information when a_1' is sent in the continuous game, from claim 18.7 we know that $\theta^* = \frac{1}{4} - b < \frac{1}{4}$, so that the interval $[0, \theta^*]$ is significantly shorter than the interval $[\theta^*, 1]$. This implies that in equilibrium player 2 has much less uncertainty about the value of θ when the message a_1' is sent.

The steps we used to find the condition $b < \frac{1}{4}$ under which a two-message equilibrium exists can be used to find more informative equilibria. For example, to construct a three-message equilibrium we first divide the interval $[0, 1]$ into three segments with a message for each: a_1' for $[0, \theta']$, a_1'' for $[\theta', \theta'']$, and a_1''' for $[\theta'', 1]$. Player 2's best response must be $a_2(a_1') = \frac{\theta'}{2}$, $a_2(a_1'') = \frac{\theta''+\theta'}{2}$, and $a_2(a_1''') = \frac{1+\theta''}{2}$. Finally player 1 must be indifferent between a_1' and a_1'' when $\theta = \theta'$, and indifferent between a_1'' and a_1''' when $\theta = \theta''$. This last condition yields two equations with two unknowns that will determine the equilibrium thresholds θ' and θ''.

You are left to solve this equilibrium in exercise 18.3 and show that it exists if and only if $b < \frac{1}{12}$. This should not be too surprising because the bias is what hampers player 1's ability to transmit more credible information in equilibrium. As the bias drops, we can construct finer and finer partitions of the interval $[0, 1]$ so that there is more meaningful communication between the sender and the receiver. In particular define $b_2 = \frac{1}{4}$ and $b_3 = \frac{1}{12}$. We have shown that if $b > b_2$ we have only a babbling equilibrium, if $b < b_2$ then we can construct a two-message equilibrium, and if $b < b_3$ then we can construct a three-message equilibrium. It turns out that the smaller the bias b, the more information can be transmitted in a perfect Bayesian equilibrium. Indeed there exist a series $b_2 > b_3 > b_4 > \cdots > b_M$ such that if $b < b_M$ then an equilibrium with M partitions exists.

Two further conclusions are worth noting. First, no matter how small the bias is, as long as $b > 0$ there is no fully truthful equilibrium, which is what we demonstrated in claim 18.1. In any equilibrium there must be some loss of information that depends on the magnitude of the bias b. Second, if $b < b_M$ then there are M different perfect Bayesian equilibria, starting with the babbling equilibrium up until the "most informative" equilibrium with M partitions.[8]

18.3 Application: Information and Legislative Organization

An interesting application of cheap-talk games was proposed by Gilligan and Krehbiel (1987), who develop a model that sheds light on the role of parliamentary committees and on the rules that govern the relative power of committees. The central idea of their theory is that policy makers are not informed about how different policies may affect outcomes. Committees are therefore formed to act as experts to transmit potentially useful information to the general body of the parliament, which then must make policy decisions.

The two players in this setting are a committee, player 1, which is required to advise the policy makers, player 2, who in turn set policy. Player 1 has private information about the state of the world, which can take on two values, $\theta \in \{-w, w\}$, where $w > 0$, while player 2's prior knowledge is only that each state is equally likely. Player 2 must choose a policy from the real numbers, $a_2 \in \mathbb{R}$. Player 2's preferences are given by $v_2(a_2, \theta) = -(\theta - a_2)^2$, which implies that player 2's optimal policy is $a_2 = \theta$. Player 1's preferences are given by $v_1(a_2, \theta) = -(\theta + b - a_2)^2$, with $b > 0$, implying that its optimal policy is $a_2 = \theta + b$. The sequence of events is as follows: first, player 1 sends a message to player 2, after which a policy decision is made.

As a benchmark, consider what player 2 would do if it had no further information beyond its prior. In this case it would maximize its expected payoff, equal to

$$\max_{a_2 \in \mathbb{R}} Ev_2(a_2, \theta) = -\tfrac{1}{2}(w - a_2)^2 - \tfrac{1}{2}(-w - a_2)^2 = -a_2^2 - w^2,$$

which is maximized when $a_2 = 0$. We refer to the policy choice of $a_2 = 0$ as the **status quo policy** because it would be selected in lieu of any information transmitted from the committee to the floor. The expected payoff of player 2 from the status quo is $v_2^{SQ} = -w^2$, and the expected payoff of player 1 (before learning the state of the world) is $v_1^{SQ} = -b^2 - w^2$.

The question posed by Gilligan and Krehbiel (1987) is whether different parliamentary rules can support more or less information sharing in equilibrium and make both players better off. Imagine that the parliament is considering one of two institutional rules: **open rule,** in which the floor may choose any policy it wants after the committee sends its message, or **closed rule,** in which the committee makes a policy recommendation and the floor must choose between it and the status quo policy $a_2 = 0$.

---

8. Note that the kind of refinements we discussed in Chapter 16 will not help refine the set of equilibria because there is no scope for "forward induction" types of arguments. This is a consequence of the sender's messages being cheap talk.

We begin the analysis with the open-rule institution because it is basically identical in structure to the cheap-talk games we analyzed in Sections 18.1 and 18.2. The question is, what are the conditions under which player 1 will truthfully reveal the state of the world, or put differently, when will there be a fully separating perfect Bayesian equilibrium? From the analysis of Sections 18.1 and 18.2 we know that a babbling equilibrium always exists in which player 1 reveals no information. This can be implemented (similar to claim 18.2) with the strategy in which player 1 chooses each of its possible messages with equal probability regardless of the value of θ, causing player 2 to choose the status quo regardless of player 1's message.

Now consider the possibility of a truthful separating equilibrium in which player 1 sends one message, a_1' when $\theta = -w$ and a_1'' when $\theta = w$. In such an equilibrium it must be the case that player 2 will choose $a_2 = \theta$ because it fully learns the state of the world by the nature of a separating equilibrium. This results in the following characterization:

Claim 18.8 *In an open-rule institution a fully truthful (separating) equilibrium in which $a_2 = \theta$ exists if and only if $b \leq w$.*

Proof Imagine that player 1 does fully reveal θ so that player 2 chooses $a_2 = \theta$. Player 1's expected utility in a truthful equilibrium is therefore $v_1^T = -b^2$. Because player 1 is biased toward higher policies, the deviation we are concerned with is whether, when $\theta = -w$, player 1 will send the message a_1'' instead of a_1'. In this case, instead of $v_1^T = -b^2$ player 1 will receive a deviation payoff of $v_1^D = -(-w + b - w)^2$. Hence player 1 will prefer not to deviate if and only if

$$-(-2w + b)^2 \leq -b^2,$$

which reduces to $b \leq w$. If this condition is satisfied then the following is a perfect Bayesian equilibrium: player 1 sends messages a_1' if $\theta = -w$ and a_1'' if $\theta = w$; player 2's beliefs are $\Pr\{\theta = w | a_1 = a_1''\} = 1$ and $\Pr\{\theta = w | a_1 \neq a_1''\} = 0$; player 2's policy choices are $a_2(a_1') = -w$ and $a_2(a_1'') = w$. ∎

Claim 18.8 reveals that the limits of the open-rule institution are that the committee's bias cannot be too large, because if $b > w$ then the only perfect Bayesian equilibrium is the babbling equilibrium, which implements the status quo.

We now turn to the analysis of the closed-rule institution. With a closed rule in place the committee will offer a policy, after which the floor will choose between the committee's recommendation and the status quo. Now consider a separating equilibrium in which player 1's proposed policy will be voted up or down and in which player 2 fully learns the state of the world θ. We know that player 1's bias will motivate it to offer a policy greater than that preferred by player 2. In particular player 1 would like to implement a policy of $a_2 = \theta + b$. We also know that player 2 can guarantee itself a payoff of $v_2^{SQ} = -w^2$ by choosing the status quo policy $a_2 = 0$. This observation results in the following:

Claim 18.9 *In a closed-rule institution a fully truthful (separating) equilibrium in which $a_2 = \theta + b$ exists if and only if $b \leq w$.*

Proof Imagine that player 1 proposes the policy $a_2 = \theta + b$. Player 2's payoff from accepting the proposal is $v_2(\theta + b, \theta) = -(\theta - (\theta + b))^2 = -b^2 \geq -w^2$ because $b \leq w$. This implies that the status quo is no better for player 2, and it will accept player 1's proposed policy. Player 1's payoff is the highest possible payoff it can receive, $v_2 = 0$. ∎

Claim 18.8 shows that when there is a separating truthful equilibrium under the open-rule institution then there is also one under the the closed-rule institution. It turns out that the closed-rule institution has more to offer.

Claim 18.10 *In a closed-rule institution a fully truthful (separating) equilibrium in which $a_2 = \theta + w$ exists when $b < 2w$.*

Proof Imagine that player 1 proposes the policy $a_2 = \theta + w$. Player 2's payoff from accepting player 1's policy is $v_2(\theta + w, \theta) = -(\theta - (\theta + w))^2 = -w^2$, implying that it is as well off as from the status quo and it will indeed accept the policy. We now need to check if player 1 will indeed propose these policies. Because player 1 is biased toward higher policies, the deviation with which we are concerned is whether when $\theta = -w$, player 1 will recommend $2w$ instead of 0. Player 1's payoff from recommending 0 when $\theta = -w$ is $v_1(0, w) = -(-w + b)^2$. If instead player 1 will deviate to proposal $2w$ then it receives a deviation payoff of $v_1^D = -(-w + b - 2w)^2$. Hence player 1 will prefer not to deviate if and only if

$$-(-w + b)^2 \geq -(-w + b - 2w)^2,$$

which reduces to $b \leq 2w$. ∎

Claim 18.10 shows that the closed-rule institution can support a separating perfect Bayesian equilibrium when the open rule cannot. The closed rule imposes fewer limits on the committee's bias for information to be revealed in equilibrium. Gilligan and Krehbiel (1987) therefore make the argument that empowering committees by tying the hands of the floor can result in more information transmission.

18.4   Summary

- Many situations are characterized by a decision maker who would like to know information to which a potential adviser with incongruent preferences is privy.

- Information-transmission or cheap-talk games offer a framework to explore these situations and consider how much information can be transferred from the adviser (sender) to the decision maker (receiver).

- Because the preferences of the two parties are not fully aligned, it will not be in the interest of the sender to reveal fully the private information that he has.

- If the sender's information space is very large then even a small amount of bias between his preferences and the receiver's preferences will result in some information not being revealed.

- Cheap-talk games have been successfully applied to shed light on institutional and organizational design.

18.5   Exercises

18.1   **Only Babbling:** Consider the pair of strategies used to construct the equilibrium in claim 18.3 in Section 18.1.

   a.   Find a threshold bias b^* such that if $b > b^*$ then the pair of strategies used in claim 18.3 is no longer an equilibrium.

     b. Show that if $b > b^*$ then the only perfect Bayesian equilibrium is the babbling equilibrium.

18.2 **Stockbrokers:** A stockbroker can give his client one of three recommendations regarding a certain stock: buy (B), hold (H), or sell (S). The stock can be one of three kinds: a winner (W), mediocre (M), or a loser (L). The stockbroker knows the type of stock but the client knows only that each type is equally likely. The game proceeds as follows: first, the stockbroker makes a recommendation $a_1 \in \{B, H, S\}$ to the client, after which the client chooses an action $a_2 \in \{B, H, S\}$ and payoffs are determined. The payoffs to the stockbroker (player 1) and client (player 2) depend on the type of stock and the action taken by the client (the pairs are (v_1, v_2), where v_i is player i's payoff) as follows:

		Player 2's action a_2		
		B	H	S
	W	$(2, 2)$	$(-1, -1)$	$(-2, -2)$
Type of stock θ	M	$(0, 1)$	$(1, 0)$	$(0, -1)$
	L	$(-2, 0)$	$(-1, 1)$	$(2, 0)$

     a. Find a babbling perfect Bayesian equilibrium of this game.
     b. Is there a fully truthful perfect Bayesian equilibrium in which the stockbroker makes the recommendation that, if followed, maximizes the client's payoff?
     c. What is the most informative perfect Bayesian equilibrium of this game?

18.3 **Three Messages:** Consider the three-message equilibrium described in Section 18.2. Find the threshold values θ' and θ'' and show that for this to be an equilibrium it must be that $b < \frac{1}{12}$.

18.4 **Committees:** Consider the committee cheap-talk model of Section 18.3 and consider the case of a closed rule under which the committee can propose only w or $-w$ to be considered against the status quo choice.

     a. Describe a babbling equilibrium for this game.
     b. For what values of b is a separating perfect Bayesian equilibrium possible?

19

Mathematical Appendix

This appendix is intended to refresh your memory in case you have forgotten some of the mathematical concepts that are used in this text. An attempt has been made to keep it as self-contained as possible even for those who may not have studied some of the mathematical tools used, but in those cases you may need to consult other texts as well.

19.1   Sets and Sequences

19.1.1   Basic Definitions

A set is a well-defined collection of objects. A set can include any collection of elements, such as numbers, cars, people, fruit, or even other sets. Two sets A and B are **equal** if and only if they include the exact same elements. One way to describe a set is by using a defining rule. For example, we can define C as the set of all countries that belong to the United Nations. Or we can define a set P to include all the positive integers. A second way to define a set is by providing an explicit list of its elements (or members). To do this we use curly brackets to define the set. For example, the set A of integers from 1 through 5 can be written as

$$A = \{1, 2, 3, 4, 5\},$$

or the set C of colors in the U.S. flag can be written as

$$C = \{\text{red, white, blue}\}.$$

If a set includes an infinite number of members, such as the set of natural numbers (or all positive integers), commonly denoted by $\mathbb{N}$, then we cannot list all of its members. Two common ways of defining the set of natural numbers are

$$\mathbb{N} = \{n : n > 0 \text{ is an integer}\}$$

or

$$\mathbb{N} = \{1, 2, 3, \ldots\},$$

where the notation ":" means "such that," and the notation "..." means "and so on and so forth."

Every member of a set must be unique, and the order in which they are listed is irrelevant.[1] The notation $x \in A$ is used to denote the fact "x is a member of the set A," while "y is not a member of A" is denoted by $y \notin A$. We denote by $\varnothing$ the **empty set,** which is a set that contains no elements.

A **sequence** is an ordered list of objects (or elements). Unlike with a set, the order matters, and the same object can appear multiple times in different positions in the sequence. The **length** of the sequence equals the number of ordered elements in the sequence, which can be infinite.

The notation $|A|$ is used to denote the **cardinality** of A, which loosely is the number of elements in A. For example, if $A = \{2, 4, 6, 8\}$ then $|A| = 4$. As already noted, given that $\varnothing$ denotes the empty set, the unique set that includes no elements, $|\varnothing| = 0$. The set $\mathbb{N}$ of natural numbers has an infinite number of members, but some sets have a higher cardinality, or a "higher" infinity. For example, the set of **real numbers** $\mathbb{R}$ (any number that is positive, negative, or zero) has a greater cardinality than the set $\mathbb{N}$. In fact, the set $A = [0, 1]$ of all real numbers between 0 and 1, including 0 and 1,[2] has a **continuum** of elements (numbers) and, unlike $\mathbb{N}$, its elements cannot be counted. For this reason we say that the set $\mathbb{N}$ is **countable** while $\mathbb{R}$, or for that matter any interval, is **uncountable.**

If every member of set B is also a member of set A, then we say that B is a **subset** of A, denoted by $B \subseteq A$. If B is a subset of A but A is not a subset of B, denoted $A \nsubseteq B$, then we say that B is a **proper subset** of A. For example, because every natural number is a real number, but the reverse is not true, then $\mathbb{N}$ is a proper subset of $\mathbb{R}$, which we denote by $\mathbb{N} \subset \mathbb{R}$.

19.1.2 Basic Set Operations

The **union** of two sets is the set that includes all the members that belong either to A or to B. We denote the union of the sets A and B by $A \cup B$. For example, if $A = \{1, 2\}$ and $B = \{x, y, z\}$ then $A \cup B = \{1, 2, 3, x, y, z\}$. The **intersection** of two sets is the set that includes all the members that belong both to A and to B. We denote the intersection of the sets A and B by $A \cap B$. For example, if $A = \{1, 2\}$ and $B = \{x, y, z\}$ then $A \cap B = \varnothing$ while if $C = \{a, b, c, d\}$ and $D = \{b, d, f, h\}$ then $C \cap D = \{b, d\}$. The **subtraction** of one set from another is possible. The notation $A \backslash B$ (in some texts, $A - B$) is the subtraction of B from A and is the set of all elements that are members of A but are not members of B. For example, if $A = \{a, b, 6, 9\}$ and $B = \{g, a, 12, 6\}$ then $A \backslash B = \{b, 9\}$. Similarly if $C = [0, 2]$ and $D = [1, 3]$ then $C \backslash D = [0, 1)$.

The **Cartesian product** of two sets A and B, denoted by $A \times B$, defines a new set that includes all the **ordered pairs** (a, b) such that $a \in A$ and $b \in B$. For example, if $A = \{1, 2\}$ and $B = \{5, 6\}$ then $A \times B = \{(1, 5), (1, 6), (2, 5), (2, 6)\}$. A common use of this concept in this book is the following: if S_i is the set of all possible strategies of player i and there are n players then the Cartesian product of the strategy sets is the set

$$S = S_1 \times S_2 \times \cdots \times S_n = \{(s_1, s_2, \ldots s_n) : s_i \in S_i \quad \text{for all } i = 1, 2, \ldots, n\}.$$

---

1. This is in contrast to a *sequence,* for which the order of members is important.
2. The set of real numbers between 0 and 1 that does not include 0 or 1 is denoted using parentheses and not square brackets, $(0, 1)$. In a similar way $[0, 1)$ is the set of real numbers between 0 and 1 including 0 but excluding 1.

A **partition** of a set A is a set of non-empty subsets of A, $P = \{A_1, A_2, \ldots A_n\}$, such that every element $a \in A$ is in exactly one of these subsets, $A_i \in P$. Equivalently a partition P of A satisfies three properties: (1) it does not contain the empty set; (2) the union of the elements of P is equal to A, that is, $\cup_{A_i \in P} A_i = A$; and (3) the intersection of any two distinct elements of P is empty, or $A_i \cap A_j = \varnothing$ for any $A_i, A_j \in P$ such that $A_i \neq A_j$. For example, one partition of the set $A = \{a, b, c, d, e\}$ is $P_1 = \{\{a, b\}, \{c, d\}, \{e\}\}$ while another is $P_2 = \{\{a\}, \{b, c, d\}, \{e\}\}$.

Sets are also useful to define the elements that are used to add up numbers. For example, if N is the set of players and each player $i \in N$ has x_i dollars, then to denote the amount of all the money in the hands of all the players, say x, we can write

$$x = \sum_{i \in N} x_i.$$

If the players are ordered as $i \in \{1, 2, \ldots, n\}$ then this is the same as writing $x = x_1 + x_2 + \cdots + x_n$.

19.2 Functions

19.2.1 Basic Definitions

A **function** defines a relationship between a set of inputs and a set of potential outputs with the property that each input is related to *exactly one output* (though one output can be related to more than one input). There are several ways to represent a function. Functions are often described using a formula that explains how to compute the output for any given input. For example, the function x^2 takes any number x and computes its square. Note that for any input $x \in \mathbb{R}$ there is a unique output, but for the output 4 (or any positive output for that matter) then both $x = 2$ and $x = -2$ can be the input that generated the output 4. Another common way to describe a function is using ordered pairs. The square function generates the ordered pairs $<-2, 4>$, $<2, 4>$, and infinitely many more such pairs. The combination of all ordered pairs that describe a function forms the **graph** of the function, and each ordered pair is a point on the graph.

Because it is generally impossible to list all the ordered pairs, we formally denote by $f : X \to Y$ a function f that relates inputs from the set X, called the **domain** of the function, to potential outputs from the set Y, called the **codomain** of the function. The set of all outputs that can occur is called the **range** or **image** of the function, which is often a subset of a set of codomains. We denote by $f(x) \in Y$ the output of the function when it is operated on the input $x \in X$. For example, the rule $f(x) = x^2$ transforms any input $x \in \mathbb{R}$ to its square, $x^2 \in \mathbb{R}$. The graph of $f(x) = x^2$ is the set of ordered pairs $\{<x, x^2> : x \in \mathbb{R}\}$, and its image is the set of all nonnegative numbers, $\mathbb{R}^+ = \{y \in \mathbb{R} : y \geq 0\}$, where $\mathbb{R}^+ \subset \mathbb{R}$. If the domain and codomain of f are both the set of real numbers, $\mathbb{R}$ (which will often be the case in this textbook), then we call f a **real-valued function.**

Inputs or outputs need not be numbers. A common function used in this textbook is a player's payoff, which relates profiles or collections of players' strategies to a payoff value. The inputs of payoff functions are profiles of strategies from the Cartesian product of players' strategy sets, $S = S_1 \times S_2 \times \cdots \times S_n$, to real numbers, and we therefore denote by $v_i : S \to \mathbb{R}$ the payoff function of player i which relates profiles $(s_1, s_2, \ldots, s_n)$ to real numbers.

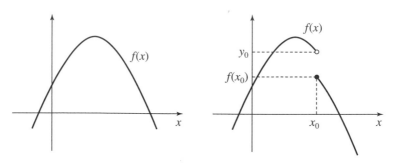

FIGURE 19.1 Continuous (left) and discontinuous (right) functions.

Given a function $f : X \to Y$, the **inverse function** of f is denoted by $f^{-1} : Y \to X$, and it maps outputs back into the inputs that generated them through f. For example, if we take the square root function $f(x) = \sqrt{x}$, so that $f : \mathbb{R}^+ \to \mathbb{R}^+$, then $f^{-1} : \mathbb{R}^+ \to \mathbb{R}^+$ is $f^{-1}(x) = x^2$. Note that many functions do not have an inverse. For example, $f(x) = x^2$ does not have an inverse function because more than one input generates the same output, and if we tried to invert the function then the inverse would try to relate an input to more than one output, which by definition is not a function. (Recall that a function relates each input to *exactly one output*.) Functions that do have an inverse are called **invertible,** and a function is invertible if and only if it has two properties: (1) It is **one-to-one** (also called an injection), which has the property that if $f(x) = f(z)$ then $x = z$ (this means that every output is associated with only one input). (2) It is an **onto function** (also called a surjection), which has the property that for every $y \in Y$ there exists an $x \in X$ such that $f(x) = y$, so that every element in the codomain has an associated element in the domain.

19.2.2   Continuity

A real-valued function is **continuous** if it has no "jumps." For instance, the function $f(x)$, which is depicted in the left panel of Figure 19.1, is continuous, while the function $f(x)$, which is depicted in the right panel, is discontinuous at the point x_0 because it has a jump at that point. Specifically as the value of the input x gets closer and closer to x_0 from below, then the value of $f(x)$ gets closer and closer to y_0, but at x_0, $f(x_0) \neq y_0$.

More formally a function f is continuous at some point $x_0 \in X$ if the limit of $f(x)$ as x approaches x_0 is equal to $f(x_0)$. We write this statement as $\lim_{x \to x_0} f(x) = f(x_0)$.

One of the most commonly known examples of a continuous function is a **linear function** that takes the form $f(x) = a + bx$, where a and b are real numbers, a is known as the **intercept,** and b is known as the **slope.** The intercept is the value of the function when $x = 0$, which is where the function intercepts the vertical (y) axis, and the slope of a function is the rate of change of the output as a function of changes in the input. For example, if $f(x) = 5 + 2x$ then the intercept is 5 and the slope is 2, because when $x = 0$ then $f(x) = 5$, and for every one unit of increase (decrease) in the value of x, the value of y increases (decreases) by 2 units. Two examples of linear functions are depicted in Figure 19.2. You should be able to convince yourself that any linear function is invertible.

Another common example of a continuous function is the **quadratic function** that takes the form $f(x) = ax^2 + bx + c$, where a, b, and c are real numbers. For

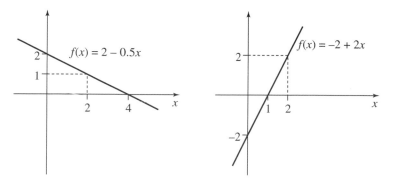

FIGURE 19.2 Linear functions.

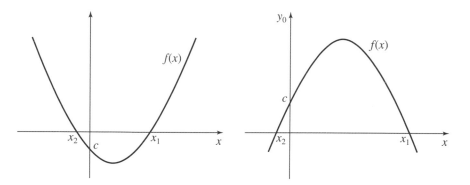

FIGURE 19.3 Quadratic functions.

this function the intercept is equal to c, but the slope, or rate of change in the output as a function of changes in the input, is not constant as in the linear function (which we will revisit in the next section). When $a > 0$ the function is "U-shaped," where as x becomes very negative or very positive the value of $f(x)$ becomes ever more positive. Such a function is shown in the left panel of Figure 19.3. When $a < 0$ the function is "inverse U-shaped," where as x becomes very negative or very positive the value of $f(x)$ becomes ever more negative. Such a function is shown in the right panel of Figure 19.3.

In Figure 19.3 notice that the functions cross the x axis twice, at x_1 and at x_2. These points are often referred to as **the roots** of the quadratic equation $f(x) = ax^2 + bx + c = 0$. The well-known formula for solving for the roots of a quadratic equation, assuming that $b^2 - 4ac \geq 0$, is given by

$$x_1 = \frac{-b + \sqrt{b^2 - 4ac}}{2a} \quad \text{and} \quad x_2 = \frac{-b - \sqrt{b^2 - 4ac}}{2a}.$$

19.3   Calculus and Optimization

19.3.1   Basic Definitions

In much of this textbook we will be maximizing functions that represent the payoffs of players under the assumption that these players wish to make themselves as well

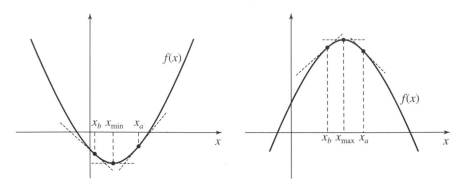

FIGURE 19.4 The maximum and minimum of a function.

off as possible. A real function f obtains a **maximum** at a point x if moving away from x in either direction causes the function's value to decrease. An example of such a point is x_{max} in the right panel of Figure 19.4. Similarly a real function f obtains a **minimum** at a point x if moving away from x in either direction causes the function's value to increase. An example of such a point is x_{min} in the left panel of Figure 19.4. An **optimum** is either a maximum or a minimum.

There is an important relationship between the slope of a function and whether or not the function is at an optimum. This can be seen by considering the points x_b, which is slightly below x_{max}, and x_a, which is slightly above x_{max} in the right panel of Figure 19.4. Because x_{max} is a maximum, as we move up from x_b toward x_{max}, the value of the function rises, which means that the slope of the function is positive at x_b. Similarly as we move up from x_{max} toward x_a, the value of the function falls, which means that the slope of the function is negative at x_b (if we were to reduce x at x_b then $f(x)$ would rise). At the maximum value of x_{max}, we cannot increase the value of the function if we move in any direction, and for this reason if the slope at x_{max} is well defined (a concept to be explained soon) then it must be zero. A similar argument can be applied to the figure's left panel.

19.3.2   Differentiation and Optimization

The **derivative** of a real-valued function $f : \mathbb{R} \to \mathbb{R}$ at a point x is equal to the slope of the tangent line to the graph of the function at x. Referring to the left panel of Figure 19.4, the derivative at the point x_b is negative because at x_b the tangent line to the graph has a negative slope (it is downward sloping). The derivative at the point x_a is positive because at x_a the tangent line to the graph has a positive slope. At x_{min} the derivative is equal to 0 (as it is at x_{max}).

The derivative of a function at a point x, or the slope at that point, can be expressed as follows:

$$\text{Slope at } x = \frac{\text{Change in } f(x)}{\text{Change in } x} = \frac{\Delta f(x)}{\Delta x},$$

where we use Δ to denote a change. For a linear function $f(x) = a + bx$, the slope of the function is equal to b, as described earlier (see Figure 19.2). This slope is a constant and does not depend on the value of x. For nonlinear functions, such as the quadratic function $f(x) = ax^2 + bx + c$, this is no longer the case. For example, if

$f(x) = x^2 + x$ then if we are at the point $x = 1$ and we increase x by 1 then we have

$$\frac{\text{Change in } f(x)}{\text{Change in } x} = \frac{(2^2 + 2) - (1^2 + 1)}{2 - 1} = 4,$$

while if we are at the point $x = 2$ and we increase x by 1 then we have

$$\frac{\text{Change in } f(x)}{\text{Change in } x} = \frac{(3^2 + 3) - (2^2 + 2)}{3 - 2} = 6.$$

Furthermore these values will change if we change the amount of the increase of x. Returning to the point $x = 1$, if we increase x by 0.1 then we have

$$\frac{\text{Change in } f(x)}{\text{Change in } x} = \frac{(1.1^2 + 1.1) - (1^2 + 1)}{1.1 - 1} = 3.1,$$

while if we are at the point $x = 2$ and we increase x by 0.1 then we have

$$\frac{\text{Change in } f(x)}{\text{Change in } x} = \frac{(2.1^2 + 2.1) - (2^2 + 2)}{2.1 - 2} = 5.1.$$

To define precisely the derivative of a function f at x we take the limit of the ratio $\frac{\text{Change in } f(x)}{\text{Change in } x}$ as the amount of change in x goes to zero. That is, we define the derivative of a function $f(x)$ as

$$\frac{df}{dx} = \lim_{k \to 0} \frac{f(x + k) - f(x)}{k}, \qquad (19.1)$$

and the derivative of f at x is well defined if the limit exists, in which case we say that f is **differentiable** at x. Another common notation for the derivative of a real-valued function $f(x)$ is $f'(x)$.

There are well-known formulas for the derivatives of commonly used functions. For example, the derivative of a quadratic function $f(x) = ax^2 + bx + c$ is $f'(x) = 2ax + b$. We can actually derive this using (19.1) as follows:

$$f'(x) = \lim_{k \to 0} \frac{[a(x + k)^2 + b(x + k) + c] - [ax^2 + bx + c]}{k}$$

$$= \lim_{k \to 0} \frac{[ax^2 + 2akx + k^2 + bx + bk + c] - [ax^2 + bx + c]}{k}$$

$$= \lim_{k \to 0} \frac{2akx + k^2 + bk}{k}$$

$$= \lim_{k \to 0} [2ax + k + b]$$

$$= 2ax + b.$$

Some other often-used derivatives (you can look up more in any standard calculus textbook) are

1. If $f(x) = a + \ln(x)$, where $\ln(x)$ is the natural logarithm, then $f'(x) = \frac{1}{x}$.
2. If $f(x) = x^k$ then $f'(x) = kx^{k-1}$.

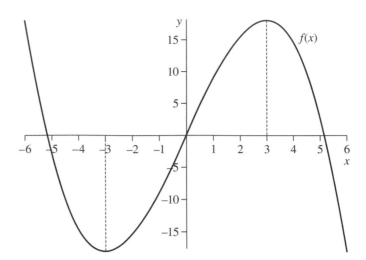

FIGURE 19.5 The function $f(x) = 9x - \frac{1}{3}x^3$.

Returning to the objective of maximizing a function, we can now state the following result:

Proposition 19.1 *If the function $f(x)$ is differentiable and if x^* is an optimum then $f'(x^*) = 0$.*

Therefore if x^* is an optimum then it is a necessary condition that the derivative of the function is equal to 0 at x^*, which is referred to as the **first-order (necessary) condition.** However, this is true for either a maximum or a minimum. Take for example the function $f(x) = 9x - \frac{1}{3}x^3$. The derivative of this function is $f'(x) = 9 - x^2$, which is equal to zero at two points: $x_1 = 3$ and $x_2 = -3$. By looking at Figure 19.5 we can see that $x_{max} = 3$ and $x_{min} = -3$. But how would we know this without seeing the graph?

To answer this question, it is useful to observe how the derivative changes around the optima. Just below x_{min} the derivative is negative; it reaches 0 at x_{min}; and then it becomes positive. That means that as x increases around x_{min} the derivative is increasing from negative to positive. This in turn implies that the *derivative of the derivative*, or the **second derivative,** denoted $f''(x)$ or $\frac{d^2 f}{dx^2}$, is positive. Similarly at x_{max} the second derivative is negative. Therefore

Proposition 19.2 *Let $f(x^*) = 0$. If $f''(x^*) < 0$ then x^* is a local maximum while if $f''(x^*) > 0$ then x^* is a local minimum.*

This is known as the **second-order (sufficient) condition** because it guarantees that the point in question is an optimum. We use the term "local" to mean that there may be more than one such point, in which case one local maximum may be better than another one (imagine a function with several humps). The derivative conditions will not identify which of several local maxima (or minima) is the *global* one—the actual highest (or lowest).

A differentiable function $f(x)$ is **concave** if its second derivative is negative for all values of x, and it is **convex** if its second derivative is positive for all values of x.

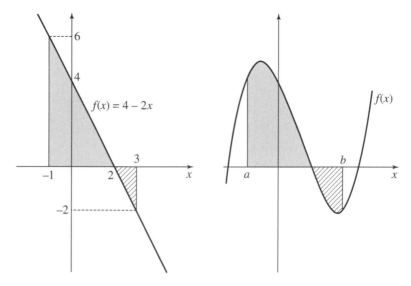

FIGURE 19.6 The definite integral: Two examples.

Looking back at Figure 19.4, the function in the left panel is convex while the function in the right panel is concave. Therefore

Proposition 19.3 *Let $f(x)$ be a concave (convex) differentiable function. If $f'(x^*) = 0$ then x^* is a global maximum (minimum).*

For practically all the analyses in this text, when functions over real variables represent payoffs we will resort to using concave payoff functions. As a consequence, when we are looking for a maximum it suffices to use the first-order condition.

19.3.3 Integration

The **indefinite integral** (sometimes called the **antiderivative**) of a real-valued function $f : \mathbb{R} \to \mathbb{R}$ is a function $F : \mathbb{R} \to \mathbb{R}$ whose derivative is equal to f, that is, $F'(x) = f(x)$. The indefinite integral of $f(x)$ is denoted as $F(x) = \int f(x)dx$, and the process of finding the indefinite integral (integration) is the opposite of the process of finding a derivative (differentiation). For example, if $f(x) = 2ax + b$ then $F(x) = ax^2 + bx + c$, and if $g(x) = 1$ then $G(x) = x$.

The **definite integral** of a function f over an interval $[a, b]$, denoted by $\int_a^b f(x)dx$, is defined informally as the total area "under" the function $f(x)$ when $f(x) > 0$, less the area "above" the function $f(x)$ when $f(x) < 0$, in between the points $x = a$ and $x = b$. For example, Figure 19.6 depicts two examples of definite integrals. The function $f(x) = 4 - 2x$ is depicted in the left panel. The definite integral $\int_a^b f(x)dx$ of this function, where $a = -1$ and $b = 3$, is equal to the gray shaded area (the total area under the function when $f(x) > 0$) minus the hatched area (the total area above the function when $f(x) < 0$). Because $f(x) = 4 - 2x$ is a linear function we can easily calculate these right-triangular areas. The gray shaded triangle has a height of length 6 and a base of length 3 so its area is 9, and the hatched triangle has a height of length 2 and a base of length 1 so its area is 1. Hence the definite integral is equal to 8. A nonlinear graphical example is depicted in the right panel of Figure 19.6.

A very well-known result called the **fundamental theorem of calculus** relates the definite integral of a function $f(x)$ over an interval $[a, b]$ to its indefinite integral evaluated at the points a and b. In particular if $f(x)$ is the derivative of $F(x)$ then

$$\int_a^b f(x)dx = F(b) - F(a).$$

It is common to use the notation $[F(x)|_a^b \equiv F(b) - F(a)$. To illustrate this with the example in Figure 19.6, in which $f(x) = 4 - 2x$, $a = -1$, and $b = 3$, note that $F(x) = 4x - x^2 + c$, where c can be any real number. We therefore have[3]

$$\int_a^b (4 - 2x)dx = \left[4x - x^2\right]\Big|_{-1}^3 = (4(3) - 3^2) - (4(-1) - (-1)^2) = 8.$$

This textbook makes very little use of integrals, but as some parts of the next section illustrate, computing the expected value of a continuous random variable (to be defined shortly) will require their use.

19.4 Probability and Random Variables

19.4.1 Basic Definitions

Some variables are not predetermined but instead depend on some form of chance. A **random variable** (or **stochastic variable**) describes such a chance-determined variable that can equal one of many different values (or outcomes), each with an associated **probability** (or likelihood). For example, a fair coin that is tossed in the air will lead to an outcome of either heads or tails, each with equal probability of $\frac{1}{2}$, which we denote as $\Pr\{\text{head}\} = \Pr\{\text{tail}\} = \frac{1}{2}$; a fair die that is rolled on a Monopoly board will result in an integer value between 1 and 6, each with equal probability of $\frac{1}{6}$; and the temperature in San Francisco tomorrow will be any one of the real numbers between 50 and 80, with some values more likely than others.

The coin or the die is an example of a **discrete random variable** because each can result in one of a discrete, in this case finite, set of values. In contrast the temperature in San Francisco tomorrow is an example of a **continuous random variable** because it can result in an outcome drawn from an interval or continuum of values, in this case [50, 80]. We define an **event** as an outcome or a set of outcomes that can occur. For example, if we roll a die, define the event A as "the die shows an odd number," and define the event B as "the die shows the numbers 1 or 2," then $\Pr\{A\} = \frac{1}{2}$ and $\Pr\{B\} = \frac{1}{3}$. The function that describes the relative frequency of the possible events that can be realized by a random variable is called a **probability distribution.** The set of possible outcomes is typically called the **sample space** of the distribution, and if the sample space consists of real numbers it is often called the **support** of the distribution.

For discrete and finite random variables such as the coin toss or the die roll, the probability distribution associates each outcome in the sample space with a

---

3. Notice that c will be added and subtracted, which implies that we can ignore it when we write down the definite integral.

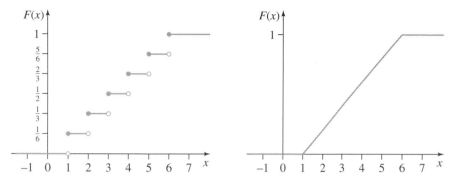

FIGURE 19.7 Discrete and continuous CDFs.

probability, which is a number between 0 and 1, and the sum of all the probabilities of the discrete outcomes must add up to 1. For example, if a fair die is rolled then the probability of each number between 1 and 6 being realized is $\frac{1}{6}$.

19.4.2 Cumulative Distribution and Density Functions

If, like the example of the roll of a die, all the outcomes of a random variable are numbers then the probability distribution can be defined by a **cumulative distribution function** (CDF), $F : \mathbb{R} \to [0, 1]$, which describes the probability that the realized value x will be less than or equal to a specified value, denoted by $F(x_0) = \Pr\{x \leq x_0\}$. For example, if the random variable is generated by the roll of a die then $F(4) = F(4.9) = \frac{2}{3}$ because the probability that the outcome is less than or equal to 4 is equal to the probability that it is less than 4.9, which is the probability that the roll of the die was either 1, 2, 3, or 4. Similarly $F(5) = F(5.43) = \frac{5}{6}$ and $F(6) = F(112.3) = 1$.

Any CDF F satisfies the following three properties:

1. If $x_2 > x_1$ then $F(x_2) \geq F(x_1)$.
2. $F(-\infty) = 0$ and $F(\infty) = 1$.
3. $1 - F(x_0) = \Pr\{x > x_0\}$.

The CDF of a discrete distribution, like that of the roll of a die, will be a "step" function consisting of jumps at each possible outcome of the sample space. This is illustrated by the CDF in the left panel of Figure 19.7. The CDF of a continuous distribution will rise gradually and continuously. An example for a random variable that is distributed over the interval [0, 6], with every number in the interval being equally likely to be the outcome, is depicted in the right panel of Figure 19.7. Such a continuous random variable that is equally likely to result in a number from a closed interval $[a, b]$ is called a **uniform distribution** over $[a, b]$. In the example $a = 0$ and $b = 6$.

Because continuous random variables such as the uniform distribution can take on one of an uncountable, infinite number of values, it follows that each outcome has a probability 0 of occurring. Such a distribution cannot be represented by a probability distribution and instead is defined by its CDF. For the uniform distribution over [0, 6] in the right panel of Figure 19.7 the CDF is given by

$$F(x) = \begin{cases} 0 & \text{for } x < 0 \\ \frac{1}{6}x & \text{for } 0 \leq x \leq 6 \\ 1 & \text{for } x > 6. \end{cases}$$

If the CDF F of a continuous distribution is differentiable then it has a derivative $f = F'$, which is called the **probability density function.** It therefore follows from the fundamental theorem of calculus that for $x_2 > x_1$

$$\int_{x_1}^{x_2} f(x)dx = F(x_2) - F(x_1),$$

implying that the probability that the random variable outcome is between x_1 and x_2 is equal to the definite integral of the probability density function in that range. For example, the density function for the uniform distribution over $[0, 6]$ is given by

$$f(x) = \begin{cases} 0 & \text{for } x < 0 \\ \frac{1}{6} & \text{for } 0 \leq x \leq 6 \\ 0 & \text{for } x > 6. \end{cases}$$

19.4.3 Independence, Conditional Probability, and Bayes' Rule

Given two events A and B, the notation $A \cap B$ (A intersection B) represents the event that both events A and B occurred. Intuitively two events are independent if the likelihood of one event occurring is not influenced by whether or not the other event occurred. Formally we say that two events A and B are **independent** if and only if the probability that they both occur is equal to the product of each event's probability, that is,

$$\Pr\{A \cap B\} = \Pr\{A\} \Pr\{B\}.$$

For example, imagine the roll of a die and the flip of a coin occurring simultaneously, with A being "die's number is even" and B being "coin shows head." First, we know that $\Pr\{A\} = \Pr\{B\} = \frac{1}{2}$, and letting $\neg A$ denote the event "A did not occur" we know $\Pr\{\neg A\} = \Pr\{\neg B\} = \frac{1}{2}$. Therefore the four possible combined events are $A \cap B$, $\neg A \cap B$, $A \cap \neg B$, and $\neg A \cap \neg B$. Because the roll of the die and the toss of the coin do not depend on each other (hence the term "independence"), it follows that the likelihood of each of these four combined events is $\frac{1}{4}$, so that $\Pr\{A \cap B\} = \Pr\{A\} \Pr\{B\} = \frac{1}{4}$.

In contrast consider the following situations. Two twins, Tweedledum and Tweedledee, always wear the same matching outfits. They have two sets of clothes, black and red, which are washed every night and prepared the next morning. Every morning the twins flip a coin to determine what they will wear: heads chooses black and tails chooses red. If A is the event "Tweedledum wears black" and B is the event "Tweedledee wears black" then $\Pr\{A\} = \Pr\{B\} = \frac{1}{2}$, but because they always wear the same clothes then $\Pr\{A \cap B\} = \frac{1}{2}$ as well, and the condition for independence is violated.

The **conditional probability of an event A given an event B** is the probability of A occurring if B is known to have occurred; it is commonly denoted by $\Pr\{A|B\}$. The notions of conditional probability and independence are tightly related. Consider the example of the die roll and coin toss. Because the events A (even die number) and B

(heads) are independent, the probability of A occurring does not depend on whether or not B occurred, and hence $\Pr\{A|B\} = \Pr\{A|\neg B\} = \Pr\{A\} = \frac{1}{2}$.

Things are different for Tweedledum and Tweedledee in choosing their clothes. As noted previously, $\Pr\{A\} = \frac{1}{2}$, but because they *always* choose to wear the same clothes it follows that $\Pr\{A|B\} = 1$ while $\Pr\{A|\neg B\} = 0$. We call the unconditional probability (also called the marginal probability), $\Pr\{A\}$, the **prior probability** of the event A because it is the prior belief that the event A occurs without knowing any additional information. We call the conditional probability $\Pr\{A|B\}$ the **posterior probability** of the event A after learning that B occurred.

Assuming that $\Pr\{B\} > 0$, the conditional probability of A given B is defined by

$$\Pr\{A|B\} = \frac{\Pr\{A \cap B\}}{\Pr\{B\}}. \tag{19.2}$$

This relationship is intuitive. If event B is known to have occurred then there are two possibilities: either A has occurred or it has not. We therefore have two possible combined states: $\{A \cap B\}$ and $\{\neg A \cap B\}$, with prior beliefs of $\Pr\{A \cap B\}$ and $\Pr\{\neg A \cap B\}$, respectively. Because either A or $\neg A$ must have occurred, and these events are mutually exclusive, it is easy to see that

$$\Pr\{B\} = \Pr\{A \cap B\} + \Pr\{\neg A \cap B). \tag{19.3}$$

As a consequence, if we learn that B has occurred, then our posterior conditional on this knowledge is the *relative likelihood* of $\{A \cap B\}$ occurring among all states in which B had occurred, which is given by (19.2).

If we multiply both sides of equation (19.2) by $\Pr\{B\}$ we obtain

$$\Pr\{A \cap B\} = \Pr\{A|B\} \Pr\{B\},$$

and by symmetry it follows that

$$\Pr\{A \cap B\} = \Pr\{B|A\} \Pr\{A\}. \tag{19.4}$$

Now we can replace the numerator of (19.2) with the right side of (19.4) to obtain the formula for **Bayes' rule** as follows:

$$\Pr\{A|B\} = \frac{\Pr\{B|A\} \Pr\{A\}}{\Pr\{R\}}. \tag{19.5}$$

Equation (19.5) can be generalized as follows. Let Z be the set of all possible outcomes from a random variable and let $\{A_1, A_2, \ldots, A_n\}$ be a partition of the set Z. It follows that

$$\Pr\{B\} = \sum_{i=1}^{n} \Pr\{A_i \cap B\}, \tag{19.6}$$

because the events $\{A_1, A_2, \ldots, A_n\}$ are mutually exclusive and they cover the whole state space. Now using (19.4) we can rewrite (19.6) as follows:

$$\Pr\{B\} = \sum_{i=1}^{n} \Pr\{B|A_i\} \Pr\{A_i\},$$

which implies a more general version of Bayes' rule as follows:

$$\Pr\{A_i|B\} = \frac{\Pr\{B|A_i\}\Pr\{A_i\}}{\sum_{i=1}^{n}\Pr\{B|A_i\}\Pr\{A_i\}}. \tag{19.7}$$

The version of Bayes' rule in (19.7) is heavily used in Section 15.2 to define beliefs of players in their information sets. To demonstrate this, imagine two events, A_1 and A_2, that are known to player 1 and not to player 2, although player 2 knows the probabilities $\Pr\{A_1\}$ and $\Pr\{A_2\}$. After learning which of these two events occurred, player 1 can choose B_1 or B_2. Player 1 can condition his choice on the event that he observes, and his choice can also be stochastic. Let $\Pr\{B_j|A_i\}$ be player 2's belief that player 1 chooses B_j conditional on observing A_i. It follows from Bayes' rule that

$$\Pr\{A_i|B_j\} = \frac{\Pr\{B_j|A_i\}\Pr\{A_i\}}{\sum_{i=1}^{2}\Pr\{B_j|A_i\}\Pr\{A_i\}}.$$

Note that Bayes' rule is well defined only if the denominator is positive, which occurs if $\Pr\{B_j|A_i\}\Pr\{A_i\} > 0$ for at least one i.

19.4.4   Expected Values

The **expected value** (also called the **mean** or the **expectation**) of a discrete real-valued random variable is equal to the probability-weighted average of all the possible values that can be realized. Informally we can interpret the expected value as the average of many independent repetitions of an experiment that is represented by the random variable. For example, imagine that a coin is tossed so that when heads occurs you get $2, while if tails occurs you lose $1. By definition the expected value is equal to $\frac{1}{2}(2) + \frac{1}{2}(-1) = 0.5$. Now imagine that the coin is tossed 100 times, so that given the probability of heads being $\frac{1}{2}$ you would expect to win $2 about 50 times and lose $1 about 50 times, resulting in a total income of about $50(2) + 50(-1) = \$50$. Hence on average you would win $0.50 per toss. This shows that the expected value may not be "expected" in the sense that it may be impossible (receiving $0.50).

More formally imagine that a discrete real-valued random variable x can take on the values $x_i \in \{x_1, x_2, \dots, x_n\}$, each x_i with probability p_i. The expected value of the random variable is given by

$$E[x] = \sum_{i=1}^{n} p_i x_i.$$

For example, imagine that a die is rolled and you receive an amount of money v equal to the number shown by the die. In this case your expected value is equal to $E[v] = \sum_{i=1}^{6} \frac{1}{6} i = 3.5$.

For a continuous real-valued random variable we cannot compute a weighted average in the same way. Let x be a continuous real-valued random variable. If the CDF $F(x)$ is differentiable so that the density function $f(x)$ exists then the expected value of x is defined as

$$E[x] = \int_{-\infty}^{\infty} x f(x) dx.$$

Intuitively this is the continuous version of the weighted average of a discrete real-valued random variable. The density function plays the role of the weights, and the realization x is the value. For example, if x is a random variable that is uniform over the interval $[a, b]$ then $f(x) = \frac{1}{b-a}$ for all $x \in [a, b]$ and zero elsewhere, so that

$$E[x] = \int_a^b x \frac{1}{b-a} dx = \left[\frac{x^2}{2(b-a)} \right]_a^b = \frac{b^2 - a^2}{2(b-a)} = \frac{a+b}{2}.$$

It follows that the expected value of a uniform random variable is the "midpoint" of the support of the distribution.

References

Akerlof, George (1970). "The Market for Lemons: Quality Uncertainty and the Market Mechanism," *Quarterly Journal of Economics* **89(2):** 488–500.

Austen-Smith, David, and Jeffrey S. Banks (1996). "Information Aggregation, Rationality, and the Condorcet Jury Theorem," *American Political Science Review* **90(1):** 34–45.

Bain, Joe S. (1949). "A Note on Pricing in Monopoly and Oligopoly," *American Economic Review* **39(2):** 448–464.

Bar-Isaac, Heski, and Steven Tadelis (2008). "Seller Reputation," *Foundations and Trends in Microeconomics* **4(4):** 273–351.

Baron, David, and John Ferejohn (1989). "Bargaining in Legislatures," *American Political Science Review* **83(4):** 1181–1206.

Bernheim, B. Douglas (1984). "Rationalizable Strategic Behavior," *Econometrica* **52(4):** 1007–1028.

Bertrand, Joseph (1883). "Théorie Mathématique de la Richesse Sociale," *Journal des Savants,* pp. 499–508.

Black, Duncan (1948). "On the Rationale of Group Decision-making," *Journal of Political Economy* **56(1):** 23–34.

Blackwell, David (1965). "Discounted Dynamic Programming," *Annals of Mathematical Statistics* **36(1):** 226–235.

Brandenburger, Adam, and Eddie Dekel (1993). "Hierarchies of Beliefs and Common Knowledge," *Journal of Economic Theory* **59(1):** 189–198.

Camerer, Colin F. (2003). *Behavioral Game Theory: Experiments on Strategic Interaction.* Princeton, NJ: Princeton University Press.

Cassidy, Ralph, Jr. (1967). *Auctions and Auctioneering.* Berkeley: University of California Press.

Cho, In-Koo, and David M. Kreps (1987). "Signaling Games and Stable Equilibria," *Quarterly Journal of Economics* **102(2):** 179–221.

Clarke, Edward H. (1971). "Multipart Pricing of Public Goods," *Public Choice* **11(1):** 17–33.

Coase, Ronald (1960). "The Problem of Social Cost," *Journal of Law and Economics* **3(1):** 1–44.

Cournot, Augustine (1838). *Recherches sur les Principes Mathématiques de la Théorie des Richesses.* Paris. [English translation: *Researches into the Mathematical Principles of the Theory of Wealth* (N. Bacon, trans.). New York: Macmillan, 1897.]

Crawford, V., and J. Sobel (1982). "Strategic Information Transmission." *Econometrica* **50(6):** 1431–1451.

Cremer, Jacques (1986). "Cooperation in Ongoing Organizations," *Quarterly Journal of Economics* **101(1):** 33–50.

Downs, Anthony (1957). *An Economic Theory of Democracy.* New York: Harper and Row.

Epstein, David, and Peter Zemsky (1995). "Money Talks: Deterring Quality Challengers in Congressional Elections," *American Political Science Review* **89(2):** 295–308.

Feddersen, Timothy J., and Wolfgang Pesendorfer (1996). "The Swing Voter's Curse," *American Economic Review* **86(3):** 408–424.

———— (1998). "Convicting the Innocent: The Inferiority of Unanimous Jury Verdicts under Strategic Voting," *American Political Science Review* **92(1):** 23–35.

Friedman, James W. (1971). "A Non-cooperative Equilibrium for Supergames," *Review of Economic Studies* **38(1):** 1–12.

Fudenberg, Drew, and Eric Maskin (1986). "The Folk Theorem for Repeated Games with Discounting and Incomplete Information," *Econometrica* **54(2):** 533–554.

Fudenberg, Drew, and Jean Tirole (1984). "The Fat-Cat Effect, the Puppy-Dog Ploy, and the Lean and Hungry Look," *American Economic Review* **74(2):** 361–366.

———— (1991). *Game Theory.* Cambridge, MA: MIT Press.

Gardner, Roy (2003). *Games for Business and Economics.* New York: John Wiley & Sons.

Gilligan, Thomas, and Keith Krehbiel (1987). "Collective Decision-Making and Standing Committees: An Informational Rationale for Restrictive Amendment Procedures," *Journal of Law, Economics, and Organization* **3(2):** 287–335.

Gintis, Herbert (2000). *Game Theory Evolving.* Princeton, NJ: Princeton University Press.

Govindan, Srihari (2003). "A Short Proof of Harsanyi's Purification Theorem," *Games and Economic Behavior* **45(2):** 369–374.

Green, Edward J., and Robert H. Porter (1984). "Noncooperative Collusion under Imperfect Price Information," *Econometrica* **52(1):** 87–100.

Greif, Avner (2006). *Institutions and the Path to the Modern Economy: Lessons from Medieval Trade.* Cambridge: Cambridge University Press.

Groves, Theodore (1973). "Incentives in Teams," *Econometrica* **41(3):** 617–631.

Hardin, Garrett (1968). "The Tragedy of the Commons," *Science* **162(3859):** 1243–1248.

Harsanyi, John (1967–68). "Games of Incomplete Information Played by 'Bayesian' Players, I–III," *Management Science* **14(3):** 159–182, 320–334, 486–502.

———— (1973). "Games with Randomly Disturbed Payoffs: A New Rationale for Mixed-Strategy Equilibrium Points," *International Journal of Game Theory* **2(1):** 1–23.

Herrero, Maria (1985). "A Strategic Theory of Market Institutions," unpublished doctoral dissertation, London School of Economics.

Hotelling, Harold (1929). "Stability and Competition," *Economic Journal* **39(1):** 41–57.

Hurwicz, Leonid (1972). "On Informationally Decentralized Systems," in Roy Radner and C. B. McGuire, eds., *Decision and Organization,* 425–459. Amsterdam: North-Holland.

Jehle, Geoffrey A., and Philip J. Reny (2011). *Advanced Microeconomic Theory* (Third Edition). New York: Prentice Hall.

Keynes, John Maynard (1936). *The General Theory of Employment, Interest and Money.* London: Macmillan.

Klein, Benjamin, and Keith B. Leffler (1981). "The Role of Market Forces in Assuring Contractual Performance," *Journal of Political Economy* **89(4):** 615–641.

Kreps, David M. (1988). *Notes on the Theory of Choice.* Boulder, CO: Westview Press.

———— (1990a). *A Course in Microeconomic Theory.* Princeton, NJ: Princeton University Press.

———— (1990b). "Corporate Culture and Economic Theory," in J. E. Alt and K. A. Shepsle, eds., *Perspectives on Positive Political Economy,* 90–143. Cambridge: Cambridge University Press.

Kreps, David M., and Robert Wilson (1982). "Sequential Equilibria," *Econometrica* **50(4):** 863–894.

Kreps, David M., Paul R. Milgrom, John D. Roberts, and Robert Wilson (1982). "Cooperation in the Finitely Repeated Prisoners' Dilemma," *Journal of Economic Theory* **27(2):** 280–312.

Krishna, Vijay (2002). *Auction Theory.* San Diego: Academic Press.

Kuhn, H. W. (1953). "Extensive Games and the Problem of Information," in H. W. Kuhn and A. W. Tucker, eds., *Contributions to the Theory of Games II,* 193–216. Princeton, NJ: Princeton University Press.

Laibson, David (1997). "Golden Eggs and Hyperbolic Discounting," *Quarterly Journal of Economics* **112(2):** 443–477.

Luce, R. Duncan, and Howard Raiffa (1957). *Games and Decisions.* New York: John Wiley & Sons.

Mailath, George J., and Larry Samuelson (2006). *Repeated Games and Reputations: Long-Run Relationships.* Oxford: Oxford University Press.

Mas-Colell, Andreu, Michael D. Whinston, and Jerry R. Green (1995). *Microeconomic Theory.* Oxford: Oxford University Press.

McKelvey, Richard D., and Thomas R. Palfrey (1992). "An Experimental Study of the Centipede Game," *Econometrica* **60(4):** 803–836.

Mertens, Jean-Francoise, and Shmuel Zamir (1985). "Formulation of Bayesian Analysis for Games with Incomplete Information," *International Journal of Game Theory* **14(1):** 1–29.

Milgrom, Paul R. (2004). *Putting Auction Theory to Work.* Cambridge: Cambridge University Press.

Milgrom, Paul, and John Roberts (1982). "Limit Pricing and Entry under Incomplete Information: An Equilibrium Analysis," *Econometrica* **50(2):** 443–459.

Milgrom, Paul R., and Robert J. Weber (1982). "A Theory of Auctions and Competitive Bidding," *Econometrica* **50(5):** 1089–1122.

Milgrom, Paul R., Douglass C. North, and Barry R. Weingast (1990). "The Role of Institutions in the Revival of Trade: The Law Merchant, Private Judges, and the Champagne Fairs," *Economics and Politics* **2(1):** 1–23.

Murphy, Kevin M., Andrei Shleifer, and Robert W. Vishny (1989). "Industrialization and the Big Push," *Journal of Political Economy* **97(5):** 1003–1026.

Muthoo, Abhinay (1999). *Bargaining Theory with Applications.* Cambridge: Cambridge University Press.

Myerson, Roger (1981). "Optimal Auction Design," *Mathematics of Operations Research* **6(1):** 58–73.

——— (1991). *Game Theory: Analysis of Conflict.* Cambridge, MA: Harvard University Press.

Nasar, Sylvia (1998). *A Beautiful Mind.* New York: Simon and Schuster.

Nash, John (1950a). "Non-Cooperative Games," *Annals of Mathematics* **54(2):** 286–295.

——— (1950b). "The Bargaining Problem," *Econometrica* **18(2):** 155–162.

O'Donoghue, Ted, and Matthew Rabin (1999). "Doing It Now or Later," *American Economic Review* **89(1):** 103–124.

Osborne, Martin J., and Ariel Rubinstein (1990). *Bargaining and Markets.* San Diego: Academic Press.

——— (1994). *A Course in Game Theory.* Cambridge, MA: MIT Press.

Palacios-Huerta, Ignacio, and Oscar Volij (2009). "Field Centipedes," *American Economic Review* **99(4):** 1619–1635.

Pearce, David G. (1984). "Rationalizable Strategic Behavior, and the Problem of Perfection," *Econometrica* **52(4):** 1029–1050.

Piccione, M., and A. Rubinstein (1997). "On the Interpretation of Decision Problems with Imperfect Recall," *Games and Economic Behavior* **20(1):** 3–24.

Riley, John G., and William F. Samuelson (1981). "Optimal Auctions," *American Economic Review* **71(3):** 381–392.

Rosenthal, Robert W. (1981). "Games of Perfect Information, Predatory Pricing and the Chain Store Paradox," *Journal of Economic Theory* **25(1):** 92–100.

Rubinstein, Ariel (1982). "Perfect Equilibria in a Bargaining Model," *Econometrica* **50(1):** 97–109.

Savage, Leonard J. (1951). *The Foundations of Statistics.* New York: John Wiley & Sons.

Schwalbe, Ulrich, and Paul Walker (2001). "Zermelo and the Early History of Game Theory," *Games and Economic Behavior* **34(1):** 123–137.

Selten, Reinhard (1975). "Reexamination of the Perfectness Concept for Equilibrium Points in Extensive Games," *International Journal of Game Theory* **4(1):** 25–55.

Shaked, Avner, and John Sutton (1984). "Involuntary Unemployment as a Perfect Equilibrium in a Bargaining Model," *Econometrica* **52(6):** 1351–1364.

Spence, Michael A. (1973). "Job Market Signaling," *Quarterly Journal of Economics* **87(3):** 355–374.

Ståhl, Ingolf (1972). *Bargaining Theory.* Stockholm: Stockholm Research Institute.

——— (1977). "An N-Person Bargaining Game in the Extensive Form," in R. Henn and O. Moeschlin, eds., *Mathematical Economics and Game Theory,* Lecture Notes in Economics and Mathematical Systems, Vol. 141, 156–172. Berlin: Springer-Verlag.

Strotz, Robert H. (1956). "Myopia and Inconsistency in Dynamic Utility Maximization," *Review of Economic Studies* **23(3):** 165–180.

Tadelis, Steven (1999). "What's in a Name? Reputation as a Tradeable Asset," *American Economic Review* **89(3):** 548–563.

Tversky, Amos, and Daniel Kahneman (1981). "The Framing of Decisions and the Psychology of Choice," *Science* **211(4481):** 453–458.

Vickrey, William (1961). "Counterspeculation, Auctions and Competitive Sealed Tenders," *Journal of Finance* **16(1):** 8–37.

von Neumann, John, and Oskar Morgenstern (1944). *Theory of Games and Economic Behavior,* Princeton, NJ: Princeton University Press.

von Stackelberg, Heinrich (1934). *Marketform und Gleichgewicht* [Market Structure and Equilibrium]. Vienna: Springer-Verlag.

Index

Page numbers for entries occurring in figures are followed by an *f* and those for entries in notes, by an *n*.